THE OSPREY

THE OSPREY

TIM MACKRILL

T & AD POYSER
LONDON · OXFORD · NEW YORK · NEW DELHI · SYDNEY

Dedicated to my sons, Harry and Laurie,
who were both born while I was researching and writing this book.

T & AD POYSER
Bloomsbury Publishing Plc
50 Bedford Square, London, WC1B 3DP, UK
29 Earlsfort Terrace, Dublin 2, Ireland

BLOOMSBURY, T & AD POYSER and the T & AD Poyser logo are trademarks of
Bloomsbury Publishing Plc

First published in the United Kingdom 2024

Copyright © Tim Mackrill, 2024
Cover © John Davis, 2024
Diagrams © Julian Baker, 2024, pages 10, 18, 37, 57, 64, 89, 111, 112, 118,
119, 126, 128, 130, 131, 137, 145, 149, 150, 154, 159, 169, 181, 209, 224.

Tim Mackrill has asserted his right under the Copyright,
Designs and Patents Act, 1988, to be identified as Author of this work.

For legal purposes the Acknowledgements on p. 294
constitute an extension of this copyright page.

All rights reserved. No part of this publication may be reproduced or transmitted in any form or
by any means, electronic or mechanical, including photocopying, recording, or any information
storage or retrieval system, without prior permission in writing from the publishers.

Bloomsbury Publishing Plc does not have any control over, or responsibility for, any third-party
websites referred to or in this book. All internet addresses given in this book were correct at the
time of going to press. The author and publisher regret any inconvenience caused if addresses have changed
or sites have ceased to exist, but can accept no responsibility for any such changes.

A catalogue record for this book is available from the British Library.

Library of Congress Cataloguing-in-Publication data has been applied for.

ISBN: HB: 978-1-4729-1990-8
TPB: 978-1-4729-9261-1
ePub: 978-1-4729-1991-5
ePDF: 978-1-4729-9262-8

2 4 6 8 10 9 7 5 3 1

Design by Mark Heslington Ltd, Scarborough, North Yorkshire
Printed and bound in Turkey by Elma Basim

To find out more about our authors and books visit
www.bloomsbury.com and sign up for our newsletters.

Contents

Foreword by Roy Dennis 6

Preface 7

1 A citizen of the world 9

2 Hunting 26

3 Dispersal and settlement patterns 48

4 Nesting and courtship 71

5 Incubation and chick-rearing 97

6 Migration 124

7 Wintering behaviour 163

8 Our relationship with the Osprey 187

9 Longevity and survival 241

Appendix 1 Osprey breeding attempts in the Rutland Water area 2001–2022 271

Appendix 2 Selected Osprey viewing sites in the UK 273

References 275

Acknowledgements 292

Index 293

Foreword
By Roy Dennis

The Osprey has become a well-known and well-loved species in the British Isles and throughout the world, so to open a book that gives new insights into this iconic raptor is refreshing. More refreshing still, though, is the hope it brings us at this time of great concern for the natural world. With human damage to the climate and to biodiversity evident across the planet, the Osprey is bucking the trend, most of its populations having increased in the last 50 years.

Tim Mackrill brings us the science and the stories behind the bird's success. Sharing with us the skills both of the Ospreys as they migrate across the planet and of the researchers and Osprey workers who chart their progress, he demonstrates the power of the Osprey to unite us in a common goal.

So much has changed since I saw my first Osprey at Loch Garten in the Scottish Highlands in April 1960. It's not just that we now have well over 300 pairs in the UK rather than just one; it's that there has been an extraordinary growth in our knowledge of and relationship with this species. I remember watching the male Osprey that spring so long ago and wondering – just as he was – when or even if the female would arrive. Compare that with the present day, when tiny satellite transmitters provide us with precise information hour by hour, minute by minute, about the individuals that we are studying. We now know the very moment at which an Osprey has left its wintering quarters in West Africa, the exact time at which it is gliding towards its usual nest to start another breeding season. Nest cameras, migration mapping and social media tell the stories of many pairs of Ospreys, and for some they are like part of the family. The contact isn't just virtual: long-lived and individually recognisable birds such as Ospreys allow us to form special bonds with pairs nesting close to our homes. Across the planet, birders check them in, count the number of young and record their departure each autumn, while more and more of us have the chance to marvel at an Osprey's dramatic plunge into water to catch a fish.

I first met Tim Mackrill when the first young Scottish Ospreys were translocated to Rutland Water (where this year, he tells me, 11 pairs have raised chicks). Some years later he came north to do research on Osprey fishing ability for his degree. Now, having gained a PhD on Osprey migration, he is my highly valued colleague in the Roy Dennis Wildlife Foundation, carrying out nature conservation projects such as the reintroduction of the White-tailed Eagle to the Isle of Wight and the translocation of Ospreys to the south coast of England. As a true champion of the Osprey, he is the perfect birder to update us all the species' life and times. I am delighted to recommend this masterclass of a book to anyone interested in this very special bird and the story it has to tell.

Preface

The Osprey *Pandion haliaetus* has a global distribution, but my early encounters with this bird started very close to home. In 1996 a pioneering project to restore the Osprey to central England began at Rutland Water, a large reservoir situated between Leicester and Peterborough. The following year, as an enthusiastic local schoolboy, I signed up as a volunteer. In early August I found myself sitting in an old caravan that doubled as the project's hide, to monitor eight juvenile Ospreys that had been translocated from northern Scotland the previous month by Roy Dennis. It was a thrilling experience to watch the young birds making short flights around the release site at Lax Hill. I was joined that morning by Paul Stammers, who has been a great friend ever since, and together we identified each newly-released bird by its colour ring.

One of the Ospreys that Paul and I logged that day was 03(97). We did not know it at the time, but that particular individual would go on to play a very important part in the Osprey story in Rutland. Four years later I was sitting in another hide, this time overlooking the first Osprey nest in central England for 150 years: 03(97) had built the nest the previous summer and paired with an unringed female. I watched enthralled as she tore up pieces of a Roach *Rutilus rutilus* that had been delivered by her mate, and delicately offered them down into the nest-cup. I could see movement between the sticks of the large nest and then suddenly I could make out the downy head of the chick. It turned out that 2001 would be the first year of 15 that 03(97) bred at the nest we called Site B. In that time he reared a total of 32 chicks with three different females, and many of his offspring went on to breed in the local area, and further afield, helping to re-establish Ospreys in central England and Wales.

Looking back, I realise how influential were my early years at Rutland Water. Not only did it begin an interest that is strong as ever today, but I was fortunate to meet people like Roy Dennis and Tim Appleton. Both are advocates of a proactive approach to conservation and had instigated the Rutland project despite considerable opposition, including from some within the conservation sector. Thanks to their perseverance, the Rutland project not only re-established Ospreys in central England and Wales, but also became a model for other Osprey translocations across Europe, which have helped to restore the Osprey to parts of its former range where it was eradicated by the actions of humans, particularly historical persecution.

I went on to manage the Rutland Osprey Project for more than 10 years, monitoring newly established breeding pairs, building artificial nests, working with landowners, farmers, gamekeepers, foresters and many others to ensure each breeding pair had the best chance of success. My time at Rutland Water also demonstrated the great interest that the general public have in Ospreys. Weekly boat trips on the *Rutland Belle* in search of fishing Ospreys were oversubscribed, and thousands of people came to the Lyndon Visitor Centre to view the nest in nearby Manton Bay. In addition, a livestream of the Manton Bay Ospreys was viewed online by thousands of people all over the world. On one memorable occasion I was in the hide overlooking the nest when an Australian woman walked in. I asked her why she was in the UK and she said, 'Well, my family think I've come to see them, but actually it's the Ospreys I really wanted to see!' There can be few other species that capture the public's imagination in the way that Ospreys do. This was demonstrated by the enthusiasm with which schools participated in an education programme linking students in different countries along Osprey migration flyways.

I have been privileged to enjoy watching Ospreys in many places around the world, including

Senegal and The Gambia where many of the birds from the UK spend the winter. I made almost annual visits with colleagues from Rutland Water for 10 years before the Covid-19 pandemic put a temporary stop to international travel. A particular highlight of the trips was in 2015 when we chanced upon one of the Rutland Water birds at its wintering site at the Sine-Saloum Delta. Such was my excitement that I nearly capsized the boat that myself and my fellow travellers were using to explore the mangrove-fringed waters of the delta. For me, seeing that bird, a grandson of 03(97), was as significant as my early experiences at Rutland Water.

Bird migration has always fascinated me and through my work at Rutland Water I was able to complete a PhD at the University of Leicester in which I analysed the migratory journeys of satellite-tagged Ospreys from England and northern Scotland. This was thanks in no small part to the generosity of Roy Dennis, who allowed me to use data he had collected over many years. I have been lucky to have Roy as a friend and mentor since first meeting him aged 15, and I continue to work with him today on projects to restore species like the Osprey and White-tailed Eagle *Haliaeetus albicilla*.

This book is the result, in part, of careful observation and conservation of Ospreys at Rutland Water for more than 20 years, during which time a self-sustaining population has become established. I never tire of seeing a newly arrived Osprey back at its nest in the spring, or watching a juvenile take to the air for the first time later in the summer. John Wright, whose photos and artwork illustrate this book throughout, deserves special mention for the many hours he spent observing the Rutland Ospreys during our time working together. John was also a companion on visits to West Africa, France and Spain in search of migrant Ospreys. He is the most outstanding field ornithologist I know, with an exceptional eye for detail that few can match. We were supported at Rutland Water by a dedicated team of volunteers who helped protect nests, collected valuable data, and shared their interest in Ospreys with members of the public who came to the Lyndon Visitor Centre and hides overlooking Manton Bay.

There are many other committed Osprey researchers and conservationists around the world whose work has contributed greatly to this book. In fact, one of the great joys of researching and writing it has been to read so many fascinating studies, which cover every aspect of the Osprey's ecology and conservation. I am fortunate to call friends many of the people whose work I have referenced. The Osprey is a bird that knows no borders, and this is reflected in the friends I have been fortunate to make over the years in many parts of the world. I am greatly indebted to everyone who responded to my queries about their research and who shared photographs.

I hope that my great enthusiasm for Ospreys comes through on these pages and that I have done justice to the remarkable array of studies that have been undertaken. I have included many of our own observations from Rutland Water, encounters with specific birds, and interesting movements of Ospreys we satellite-tagged. There are still elements of the Osprey's life that we do not yet fully understand, but the immense amount of research carried out means there can be few species that are as well studied. The recovery of this bird during the latter part of the twentieth century and in the first decades of the twenty-first century has been heartening at a time of growing concerns about the fast-developing climate and biodiversity emergencies. The Osprey should be seen as a beacon of hope, demonstrating that with a proactive approach to the restoration of nature we can make positive changes for generations to come.

CHAPTER 1
A citizen of the world

It is late March and a gentle breeze is blowing across the blue-grey water. A lone Sand Martin *Riparia riparia* zips low across the gentle ripples, to a soundtrack of the staccato song of a Chiffchaff *Phylloscopus collybita*. A hunting Osprey appears overhead, its striking white underside illuminated by the spring sun. It circles, then turns into the wind, hovering characteristically as it stares intently at the water below. After a few seconds it moves a couple of metres further along, then hovers again. Suddenly, it folds its wings and descends, arrow-like, thrusting its talons forward a split second before it crashes into the water – almost completely disappearing from view. The Osprey struggles, splayed on the surface and then, with a few flaps of its powerful wings, lifts clear of the water, grasping a silvery fish. The Osprey almost drops its catch, but in a single skilful motion, adjusts its grip so that the fish is facing forwards, and then performs a mid-air shake, sending a shower of water droplets spiralling downward. There are few spectacles in the natural world as impressive or dramatic as a fishing Osprey.

My own experience with the Osprey began at Rutland Water in central England, but this bird is truly a citizen of the world. It is one of only six landbird species to occur on every continent except Antarctica during the course of the year, with most populations migrating south for the winter (BirdLife International 2022). The northernmost-occurring Ospreys breed within the Arctic Circle in Finnish Lapland, while some Scandinavian individuals migrate as far south as the southern tip of South Africa. Hunting Ospreys can be seen from Argentina to Australia, from Siberia to Senegal. The scientific name for Osprey, *Pandion haliaetus*, is derived from the mythical Greek king of Athens, Pandion I, and the ancient Greek word *haliaietos*, *hals* meaning 'salt' or 'the sea' and *aetos* meaning 'eagle'. This is a reference to the fact that the Osprey, which is exclusively piscivorous, is often a bird of coastal habitats, but in fact it is an opportunistic forager that readily catches fish in both freshwater and marine environments.

Most northern populations of Ospreys are migratory, with individual birds making long and perilous journeys across oceans and deserts twice a year. Adult Ospreys, which may live to 20 years or more, often remain faithful to the same breeding and wintering sites throughout their life and, after spending the winter apart, are reunited with the same mate upon their return. Whether atop a pine tree in Siberia or a telegraph pole in North America, a newly returned Osprey sitting resplendent beside its nest is a sure sign that spring has arrived.

This is a bird whose history is inexorably linked to our own. Humans have been responsible for catastrophic declines of Osprey populations across the world; but in more recent times, we have helped the Osprey return to places where it was eradicated by our ancestors.

Distribution and taxonomy

As mentioned above, the Osprey is one of only six landbird species that occurs on every continent, except Antarctica. This group also includes Great Egret *Ardea alba*, Cattle Egret *Bubulcus ibis*, Glossy Ibis *Plegadis falcinellus*, Barn Owl *Tyto alba* and Peregrine Falcon *Falco peregrinus* (Del Hoyo *et al.* 1992). Its cosmopolitan distribution reflects its ability to exploit variable breeding habitats across its range, and to adapt its migratory strategy according to local conditions. While most northern populations of Ospreys are migratory, those from lower latitudes often make only limited post-breeding movements, or are sedentary throughout the year. The distribution and main migratory flyways of the Osprey is shown in Figure 1.1.

The Osprey is considered sufficiently distinct from other raptor species within the order Accipitriformes, most notably family Accipitridae – which includes eagles, hawks, kites, buzzards, harriers and Old World vultures – in terms of morphology and plumage characteristics to be classified as a family of its own, Pandionidae (Sclater and Salvin 1873). The Osprey can be recognised as a species in the fossil record as far back as the Miocene, 12 to 13 million years ago, when it was thought to be widespread in North America, which is considered to be the centre of origin for the species (Poole 2019). Genetic analysis by Monti *et al.* (2015) suggests that the species spread from America into the Indo-Australasian region, via the Bering Strait and Pacific coast of Asia, including Japan. From there, rapid range expansion resulted in the colonisation of eastern Asia, Europe and Africa. This pattern of dispersal is reflected in the degree of genetic divergence evident in different Osprey populations around the world today, with European birds more closely related to Ospreys from the Indo-Pacific region than they are to birds from North America. Monti *et al.* (2015) identified four different clades of Ospreys globally, indicating that the diversification into four lineages began around 1.16 million years ago in the Early Pleistocene when the American group became distinct. This was followed by the Indo-Australasian group 730,000 years ago, and further divergence into Asian and European clades approximately 640,000 years ago. Further analysis by Monti *et al.* (2015) indicated that European Ospreys likely experienced a further notable expansion that began about 10,000 years ago. Such a scenario was also evident in a review of Holocene fossils of Ospreys found in Eastern Europe (Zachos and Schmölke 2006).

Most taxonomists regard the Osprey *Pandion haliaetus* (Linnaeus, 1758) as the only extant species within this family, with four distinct subspecies recognised. *P. h. haliaetus* occurs across Europe, northwest Africa and Asia north of the Himalayas. Within Europe, the Osprey's range, which extends from Finnish Lapland south to the Mediterranean islands, is indicative of the species' ability to exploit a variety of different habitats for both breeding and foraging. Northern populations in Scandinavia and Russia breed in forested areas, and often hunt exclusively in

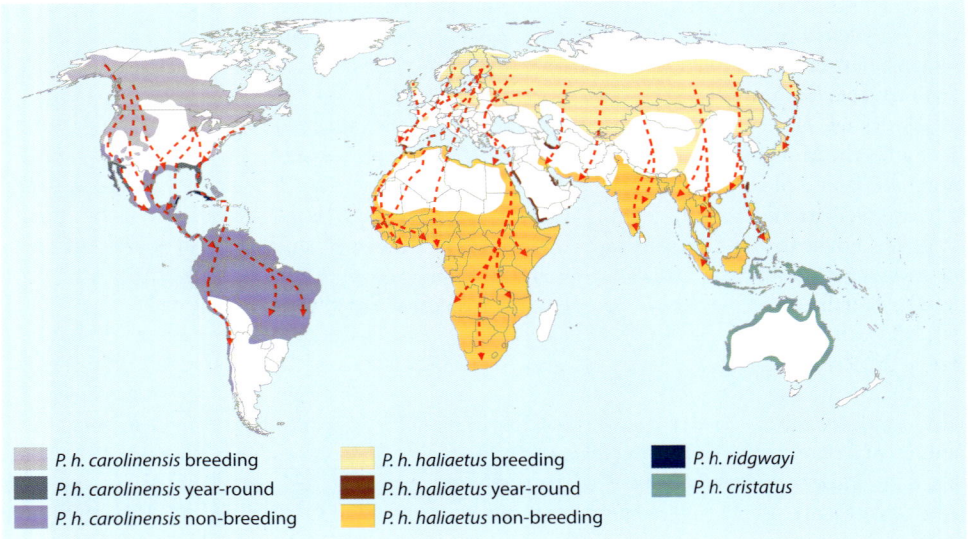

Figure 1.1. Map showing Osprey distribution and its main migratory flyways. Northern populations are migratory, whereas southern populations are sedentary or make limited post-breeding movements (Original map by Alan Gililand, redrawn and updated by Julian Baker).

inland lakes and rivers, whereas birds that breed in the Mediterranean region build nests on sea cliffs and forage along coasts. Despite this extensive geographical distribution, the European population is disjointed and *P. h. haliaetus* remains absent from large parts of its former range due to historical persecution and the effects of harmful insecticides, such as Dichlorodiphenyltrichloroethane (DDT) (Schmidt-Rothmund *et al.* 2014). Encouragingly, it is now beginning to recover and is expanding across much of Europe, aided by translocation projects, which have been undertaken in six different countries (see chapter 8). The majority of Ospreys in Europe are migratory with birds from Western Europe (UK, Sweden, Norway, Germany, France) migrating south through Iberia and across the Sahara before wintering in West Africa, from Mauritania, south to Ivory Coast and Ghana (Hake *et al.* 2001, Dennis 2008, Mackrill 2017). Those from populations further east in Europe, including Finland, Russia, Latvia and Estonia, often use a more easterly migration route that takes them south through Eastern Europe and the Middle East and then across the Sahara to eastern and southern Africa (Väli and Sellis 2016, Babushkin *et al.* 2019, Østnes *et al.* 2019). However, satellite tracking and colour-ringing studies have revealed that some birds from these eastern populations favour the western migration route to West Africa (Saurola 2021). An increasing number of northern European birds have also begun wintering in Iberia and the Mediterranean, rather than sub-Saharan Africa (Martín *et al.* 2019). During the winter, these birds inevitably encounter the Ospreys that inhabit lower latitudes in Europe. Most of those breeding in the Mediterranean region, in Corsica and the Balearics, as well as parts of the north coast of Africa in Algeria and Morocco tend to be sedentary, or make only limited post-breeding movements, although some individuals do migrate to countries south of the Sahara (Monti *et al.* 2018a). Isolated resident populations also occur on the Canary Islands (Siverio *et al.* 2011) and further south on Cape Verde (Palma *et al.* 2020).

Likewise, further to the east, resident populations of *P. h. haliaetus* occur along the coastlines of the Arabian Peninsula. They are widespread along the Red Sea, with breeding pairs broadly distributed along the coasts and islands of Saudi Arabia and Yemen as well as Egypt, Sudan and Eritrea. Smaller numbers occur along the coasts of both Yemen and Djibouti on the Gulf of Aden, and also on the Arabian Sea coast of Oman. Further sedentary populations can be found around the Persian Gulf in the United Arab Emirates, Qatar, Bahrain and northeast Saudi Arabia (Fisher *et al.* 2001).

The range of *P. h. haliaetus* at more northern latitudes spreads east in a wide arc across Kazakhstan, Mongolia, Siberia, China and east as far as Japan, with breeding pairs on the northern island of Hokkaido as well as other areas further south. These populations are generally less well studied than others further west, often because of the remote areas they inhabit, but are known to migrate south to winter in the Indian subcontinent and South-East Asia, including Myanmar, Thailand, Vietnam and Malaysia (BirdLife International 2022).

The North American race of the Osprey, *P. h. carolinensis*, occupies a broad range of habitats, including both fresh water and salt water, across the United States and Canada (Poole 1989, Poole 2019, Bierregaard *et al.* 2020). As in Europe, North American Ospreys suffered extensively from the effects of harmful insecticides during the mid-twentieth century, but have since recovered strongly, again aided by translocations, and the breeding Ospreys in the United States and Canada now account for approximately two-thirds of the world's population (Poole 2019). There are now breeding records or nesting attempts for all of the lower 48 states of the United States, with core populations along the Gulf and Atlantic coasts (Bierregaard *et al.* 2020). The highest densities occur on the Atlantic coast, with over 9,000 breeding pairs around Chesapeake Bay, a vast estuarine system in the states of Maryland and Virginia. In Canada, the range has expanded northward in recent decades, with pairs breeding as far as 68.3°N at Inuvik in the Northwest Territories (Poole 2019).

The northern populations of Ospreys in North America are all migratory. Ospreys that breed in eastern parts of the United States and Canada tend to migrate south along the East Coast before heading across the Atlantic from Florida to Cuba, and then continuing through Haiti and the Dominican Republic, before crossing the Caribbean Sea to Colombia or Venezuela. Some birds winter in these northern parts of South America, while others keep travelling south to winter in the Amazon Basin in Brazil. In contrast, Ospreys that breed in western parts of North America use an alternative route that takes them south through Mexico and Central America towards Peru, with some individuals wintering as far south as south-central Chile. Meanwhile, Ospreys from Midwestern states are known to use three distinct routes to wintering sites in Mexico, Central and South America and the Caribbean (Bierregaard *et al.* 2020).

As is the case in Europe, Ospreys that breed in southern parts of North America tend to be sedentary. Florida supports extremely high concentrations of breeding Ospreys – with some inland lakes home to the densest colonies in the world. Here individual flat-topped cypress trees (*Taxodium* spp.) can sometimes hold multiple pairs (Poole 2019). In Mexico, significant breeding populations occur in the northwest of the country on Baja's Pacific coast and along the Gulf of California (Henny and Anderson 2004).

Subspecies *P. h. ridgwayi* occurs in the Caribbean region, with the population distributed from the east coast of Yucatán in Mexico, south into Belize and east to Cuba, the Bahamas, and the Turks and Caicos Islands. All of the birds in these populations are sedentary, and present in all areas throughout the year. The stronghold for *P. h. ridgwayi* is thought to be Cuba, although this population is little studied (Poole 2019).

Pandion h. cristatus is found in Oceania and the Indo-Pacific region, with breeding populations in Australia, Indonesia, Philippines, the Palau Islands, New Guinea, the Solomon Islands and New Caledonia (Poole 1989, Bretagnolle *et al.* 2001, Mees 2006, River *et al.* 2007, Mittermeier *et al.* 2013). Significantly, these are the only Ospreys that breed south of the Equator. In Australia, nesting Ospreys are found in most coastal areas, with the only exceptions being the southern coasts of Western Australia and Victoria, where cold ocean currents limit fish availability (Dennis and Clancy 2014). As in other parts of the world, the distribution of Ospreys in Australia emphasises their ability to adapt to local conditions and breeding opportunities. For example, Ospreys are widely distributed along the remote rugged western coastline of Western Australia, where birds breeding on coastal cliffs or offshore rocks often do not come into contact with people. Yet they also thrive on the east coast, where the species has learnt to live in a very different landscape alongside humans in Queensland and New South Wales, often nesting on artificial structures such as electricity pylons.

There has been much recent debate about the status of *P. h. cristatus* following a genetic analysis of three of the four subspecies, *P. h. haliaetus*, *P. h. carolinensis* and *P. h. cristatus*, that was undertaken by Wink *et al.* (2004). In their study, the authors amplified the mitochondria Cytochrome b gene by Polymerase Chain Reaction (PCR – an acronym familiar to many from the pandemic) and then sequenced it. They identified genetic differences of between 2 per cent and 4 per cent, which likely represents divergence 1–2 million years ago. They argue that because genetic differences of some closely related eagle species, such as Greater Spotted Eagle *Clanga clanga* and Lesser Spotted Eagle *Clanga pomarina*, and Spanish Imperial Eagle *Aquila adalberti* and Imperial Eagle *Aquila heliaca*, are in the range 1.7–2.1 per cent, and thus less divergent than the Osprey subspecies, that the four *Pandion haliaetus* subspecies should actually be regarded as separate species, even though morphological and plumage differences (upon which the existing classification is based) are small. Christidis and Boles (2008) accepted this and elevated the subspecies *cristatus* to a separate species, the Eastern Osprey *Pandion cristatus*. However, this proved to be controversial and the most recent BirdLife Australia Working List v3 assigns subspecies status for the 'Eastern Osprey' *Pandion haliaetus cristatus* (BirdLife Australia 2019).

In a more extensive genetic study of Ospreys, Monti *et al.* (2015) used samples from 31 countries across five continents, encompassing the whole of the species' distributional range, and extracted and amplified DNA by PCR for the mitochondria Cytochrome b gene and NADH dehydrogenase subunit 2 (ND2). They concluded that the Osprey is structured in four main genetic groups, representing quasi non-overlapping geographical regions, but that these did not entirely correspond to the four subspecies. They found an Indo-Australasia group thus corresponding with *P. h. cristatus*, and a Europe-Africa group in line with *P. h. haliaetus*. However, they also identified a new lineage for *P. h. haliaetus* in northeast Asia (Siberia and Japan), and only identified a single lineage for the Americas, encompassing both *P. h. carolinensis* and *P. h. ridgwayi*. As with the study undertaken by Wink *et al.* (2004), genetic differences between these clades (1.5–2.6 per cent) were in the range that has been used by taxonomists for designating distinct raptor species. An additional finding of Monti *et al.* (2015) was that nucleotide diversity was low – at 1.0 per cent – compared to other raptors, such as the Bearded Vulture *Gypaetus barbatus* (2.9 per cent). Thus, they conclude that the decision to split Ospreys into different species should consider factors other than morphological differences and Mitochondrial DNA (mtDNA), including nuclear genes (i.e. genes located in the cell nucleus, rather than the mitochondria) and specific behavioural aspects such as migration that could act as reproductive barriers between geographically distant populations. Subsequent analysis by Monti *et al.* (2018d) by means of DNA microsatellite markers identified three main groups, Australasia, North America and Palearctic, between which gene flow is almost non-existent, but not the fourth phylogenetic clade in eastern Asia that was evident in the mtDNA analysis (Monti *et al.* 2015). They also identified what they considered two geographical sub-entities in the Palearctic, with a first subgroup from Japan to northeast Europe, and a second in the Mediterranean. The two subgroups were interconnected, but estimated gene flow was considered limited and asymmetric. This is likely influenced by the highly philopatric behaviour of male Ospreys, in particular, who usually return to their natal area to breed, whereas females may disperse between populations, as discussed in chapter 3. It is currently unclear at what timeframe the observed differences occurred, but geographical isolation of southern populations due to the highly fragmented nature of the European Osprey population compared to historical times, covered in chapter 8, may also be a factor.

Characteristics and identification

Regardless of the specific subspecies, the Osprey is a distinctive bird. A mid-sized raptor, it is larger than the Common Buzzard *Buteo buteo* but smaller than most eagles, with a wingspan of approximately 1.5m. Ospreys have relatively long, narrow wings, and fly with strong, steady wingbeats interspersed with long glides. They hold their wings in a shallow 'M' shape, giving them a distinctive flight profile.

Females are slightly larger than males, with longer, broader wings, a heavier bill and thicker tarsus. They may weigh up to 2kg, and tend to be around 20–30 per cent heavier than males, but this varies between subspecies (Muriel *et al.* 2010a). Adult birds of both sexes have a prominent dark mask around a yellow eye and are uniform brown above and white below, with varying amounts of brown streaking on the head, breast and underwings. The beautiful underwing markings of Ospreys, which are unique to each individual and change very little year-to-year, are best studied in photographs. The greater underwing-coverts are dark brown-black and form a wing-band that merges with a dark carpal patch and contrasts with the white, variably brown spotted forearm. The inner hand is relatively pale, but the wing-tips are dark. The underside of the primaries and secondaries are barred, often with a dark trailing edge, which is most apparent in the secondaries. The extent of barring and streaking in the underwings varies between the sexes and subspecies (Forsman 1999, Strandberg 2013). Females tend to be more heavily marked on the underwings than males, and often have a pronounced breast-band. There is considerable variation between individuals, however. Sometimes a breeding pair can be virtually indistinguishable from each other, even when perched together, while on other occasions the differences are clear.

It is also important to consider the plumage and morphological differences between the four subspecies. Adult *P. h. haliaetus* and *P. h. carolinensis* are the most similar, but there are some key, if somewhat subtle, differences, as examined by Strandberg (2013) in a detailed analysis of the ageing, sexing and subspecies identification of Ospreys. *P. h. carolinensis* tend to be more blackish-brown on the upperparts than *P. h. haliaetus*, and the greater underwing-coverts and primary underwing-coverts are also darker, and the carpal patch more extensive than *P. h. haliaetus* – the latter because it includes the median under primary coverts, which are spotted in *P. h. haliaetus*. The breast-band is usually more prominent in *P. h. haliaetus* and brown spotting in the lesser and median underwing-coverts also more pronounced. Interestingly, *P. h. haliaetus* breeding around the Red Sea and Arabian Peninsula are distinct from northern populations, with pale brown upperparts, clean underwings and only limited breast streaking. Strandberg (2013) suggested that further investigation is required to determine whether birds breeding in this area should be considered a separate subspecies.

Whereas to the casual observer, *P. h. haliaetus* and *P. h. carolinensis* are difficult to separate in the field, *P. h. cristatus* is more distinctive. Perhaps the most characteristic feature is its smaller size, with a wing-length 12–14 per cent shorter than the other subspecies (Poole 2019). *P. h. cristatus* also has fewer dark feathers on the head, leaving a characteristic 'bandit mask', especially in males, and a white crown in both sexes. The upperparts are also paler brown than in the other subspecies. Both sexes of *P. h. cristatus* have very clean underwings, with unmarked median and lesser underwing-coverts and white areas at the base of the primaries, with no prominent banding (Strandberg 2013). *P. h. ridgwayi* is larger than *P. h. cristatus* but has similarly clean underwings, with a dark carpal patch, and a white crown with a less pronounced stripe through the eye, giving it a very white-headed appearance. Unlike *P. h. cristatus* there is only a limited breast-band in females and this feature is absent in males. Photographs of adult birds from the four subspecies are shown in Figure 1.2.

Sexual dimorphism in Ospreys is somewhat variable, and usually a combination of jizz – the overall shape, posture and flying style – and plumage features are required to confidently separate individuals in the field. The larger size of females is most evident in flight when their broader wings, longer tail and more pronounced head gives them a bulkier appearance than slimmer males which, as Strandberg (2013) points out, appear 'more athletic' in appearance. However, the overall size of males and females does overlap, and so it is usually necessary to consider differences in plumage as well.

The extent of the brown breast-band varies between the subspecies, but in all cases it tends to be broader and darker in females, and so is a good place to start. It is important to consider, though, that there is overlap between the sexes. Strandberg (2013) concluded that about one in ten males of *P. h. haliaetus* show a female-like breast-band, and Poole (1989) estimated that he could sex 50–70 per cent of pairs in North America when considering breast-band alone.

When Ospreys are viewed in flight, the extent of spotting on the underwing-coverts can be a reliable means by which to separate the sexes, particularly in *P. h. haliaetus*. Males appear to have much cleaner underwings when seen at relatively close range with binoculars, because

Figure 1.2. Comparison of the four Osprey subspecies. Adult *Pandion h. haliaetus* (top left) and *P. h. carolinensis* (top right) are very similar, but *P. h. carolinensis* tends to be more blackish brown on the upperparts, with a paler breast. *P. h. ridgwayi* (bottom left) has a distinctive white crown, while *P. h. cristatus* (bottom right) is smaller and has fewer dark feathers on the head, giving it a distinctive 'bandit mask' (Clockwise from top left: © MikeLane45/iStock, © FLPA/Alamy, © Stephanie Jackson/Alamy, © Iain Brownlee/Alamy).

they lack the extensive brown spotting of females. A typical *P. h. haliaetus* male will have just a few small brown spots on the median underwing-coverts and completely clean axillaries, whereas females usually have two rows of larger more extensive brown spots on the median underwing-coverts, which continue through the axillaries and often onto the flanks (Figure 1.3). *P. h. carolinensis* males tend to have completely white median underwing-coverts or very limited spotting, whereas females usually have a cluster of more pronounced brown spots – sometimes in two rows as in female *P. h. haliaetus*. Both *P. h. cristatus* and *P. h. ridgwayi* are less well marked on the underwings, with spotting in the median underwing-coverts more defuse but, again, usually more prominent in females.

The final plumage characteristic worth considering is the carpal patch, which is generally more extensive in females. Figure 1.3 shows a comparison of the plumage and overall structure of adult and juvenile (early second calendar year) *P. h. haliaetus*.

Juvenile Ospreys can be easily identified from adults through numerous differences. Perhaps the most obvious difference is the beautiful pale fringing to the upperparts and pale tips to the flight feathers, as evident in Figure 1.4. When fresh, these feathers give juvenile Ospreys a distinctly mottled appearance, which acts as excellent camouflage in the nest. There are also clear differences in the underwing markings of juveniles. The greater underwing-coverts are

Figure 1.3. Juvenile (early second calendar year (2cy)) and adult *P. h. haliaetus*, showing variations in plumage and overall structure. Clockwise from top left: 2cy male, 2cy female, 2cy female, adult female, adult female, adult male, adult male, adult male (© John Wright).

barred and, as a result, form a less distinct band than in adults. The barring on the underside of the secondaries is distinct and regular, and bordered by a narrow dark trailing edge. This gives the inner wing a pale appearance, particularly as juveniles lack the increasingly dark inner secondaries of adults. Juveniles also tend to show a more evenly spotted carpal patch compared to adults. Further, juveniles have a yellow-buffish wash to the underwing-coverts and a characteristic orange to reddish eye, which varies between subspecies.

As in the adults, there are some subtle differences between juveniles of subspecies. Strandberg (2013) notes that juvenile *P. h. carolinensis* look very pied compared to *P. h. haliaetus*, and lack the whitish patch that juvenile *P. h. haliaetus* often show on the upperwing-coverts. Juvenile *P. h. haliaetus* are paler around the eye and have an orange iris, whereas the iris colour of *P. h. carolinensis* is distinctly red-orange. *P. h. haliaetus* usually has black gorget streaks which are either lacking or very fine in *P. h. carolinensis*, and a blue-grey upper gape which is black in *P. h. carolinensis* (Strandberg 2013). Like adults of the same subspecies, *P. h. cristatus* and *P. h. ridgwayi* have whiter crowns than either *P. h. carolinensis* or *P. h. haliaetus*.

Sexing juvenile Ospreys by plumage is sometimes possible, depending on the presence and extent of spotting and barring on the underwing-coverts and axillaries, which is more extensive in females, although considerable overlap occurs. With experience, males and females can also be separated by size and jizz. As with adults, juvenile female Ospreys look heavier and broader-winged in flight, while their bulkier bill is evident when perched. A considerable size difference is often evident between males and females if you see them perched together, but it is more difficult to separate the sexes when observing single birds.

Many juvenile Ospreys are ringed as nestlings, and this presents a unique opportunity to study sexual dimorphism in detail, particularly with regard to specific morphometric differences. The larger size of females is usually apparent if nestlings are ringed at 35–45 days of age, at which point skeletal growth is complete and maximum body weight attained, but feathers are still growing. Females tend to be 200–300 grams heavier than males, with some females (particularly *P. h. haliaetus* and *P. h. carolinensis*) weighing in excess of 1,800g, although this varies with food supply. This criterion was used by Schaadt and Bird (1993) to determine the sex of nestlings in a study of growth rates of North American Ospreys. Poole (1989) also concluded that body weight is a reliable criterion for sexing nestlings older than 30–35 days.

Size differences are often pronounced, but other morphological features can also be used to sex nestlings. Females of all subspecies have longer wings, tails, claws and bills (Poole 1989). Muriel *et al.* (2010a) examined this subject by analysing sex-based variation in four morphometric measurements: body mass, length of the flattened wing chord, tarsus length and forearm length (from the front of the folded wrist to the proximal extremity of the ulna) of 114 juvenile Ospreys from Germany, Finland and Scotland, 91 of which were translocated to Andalucía in southern Spain as part of a reintroduction programme. Measurements were taken when young were 40–73 days old, which is close to the average fledging date for Ospreys (50–60 days, as detailed in chapter 5). They found that, as expected, males were smaller in all morphometric measurements, with forearm and tarsus the most dimorphic characteristics (Figure 1.5). The mean body mass of males was 1,356.8g, compared to 1,538.50g among females; forearm length was a mean 185.6g in males and 198.5g in females; tarsus length was 53.6mm in males and 57.9mm in females; and wing chord was 426.6mm in males and 447.4mm in females. Forearm and tarsus length were considered the most reliable predictor of sex in juvenile Ospreys, because unlike other aspects, including body mass, they are not dependent on other factors such as food supply or the number of days since hatching (assuming the bird is 40 days of age or older).

Figure 1.4. Adult male (left) and a newly fledged juvenile female *P. h. haliaetus*. The characteristic pale fringing on the upperparts of the juvenile are diagnostic (© John Wright).

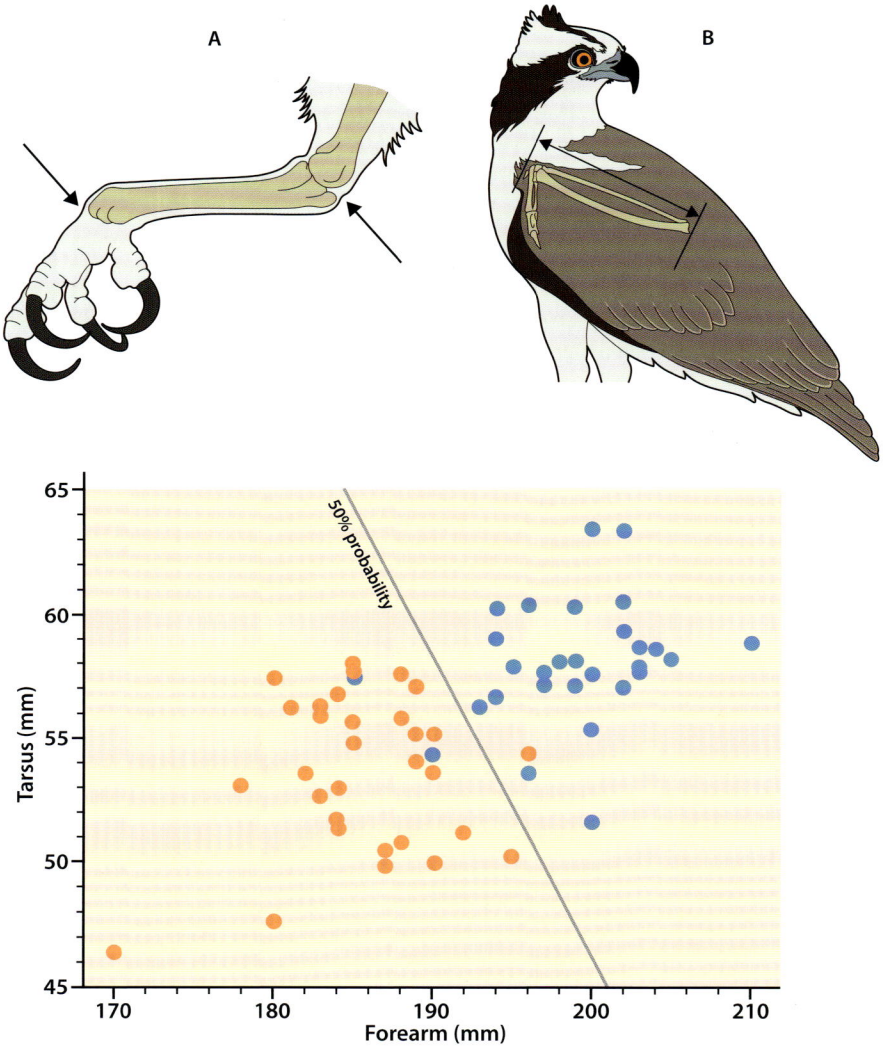

Figure 1.5. Scatterplot showing the tarsus length (A) and forearm length (B) of 61 young Ospreys. These morphometric measurements were found to be reliable predictors of sex. Females are indicated by blue circles and males by orange circles; straight line shows the discriminant threshold between males and females, with a 50 per cent probability of classification; two females were misclassified as males and one male as a female (reproduction of figures from Muriel et al. 2010).

Moult

By November or December of their first year most juvenile Ospreys from migratory northern populations will have reached suitable wintering grounds, and they begin their post-juvenile moult. This begins in the primaries, with the innermost primary (P1) replaced first – usually sometime between November and January of the first winter (Figure 1.6). Moult then proceeds in a wave through the 10 primaries to the wing-tip (Prevost 1983, Forsman 1999), with approximately one primary replaced every month. Primary moult takes approximately 12 months to complete, but a second wave – again beginning from the innermost primary – will normally begin before the first has been completed. This means that by the latter part of their second calendar year, immature Ospreys may have two simultaneous moult waves underway (Forsman 1999). Secondary moult begins between one and three months after the first primary has been moulted. In his extensive study of the moult of overwintering Ospreys in Senegal, Prevost (1983) found that this began at secondaries one and five, progressing in towards the bird's body, and also from the outermost secondary, progressing outwards. Moult tends to be more variable in the secondaries, meaning that any pattern tends to be lost within a few months. Juvenile Ospreys will usually replace all of their secondaries by the early part of their third calendar year (i.e. January/February of their second winter), but secondary moult continues until their first migration north. Moult of the tail feathers begins one to three months after primary moult has been initiated, beginning with the central tail feathers. The moult then continues outwards in each half of the tail, although any symmetry evident in the early stages tends to be lost fairly quickly. All of the tail feathers are usually replaced before the first wave of primary moult is completed. Prevost (1983) also found that growth rate of primaries was 5.7 mm per day, compared to 3.1 mm per day for secondaries and 3.2 mm per day for tail feathers.

By spring of their second calendar year, the upperparts of juvenile Ospreys appear more uniformly brown due to wear, with the distinctive pale tips of the upperparts far less prominent. However, juvenile greater coverts and secondaries are still visible, meaning that juvenile birds retain prominent barring on the underwings. Some juvenile secondaries are retained until early in the third calendar year, at which point they are clearly distinct from adult-type secondaries with the dark terminal band. By the time Ospreys return to the breeding grounds, usually in the late spring of their third calendar year, they are not separable from older adult birds, having completed the full post-juvenile moult.

Immature birds (i.e. those in their second or early third calendar year) only have one or two waves of primary moult evident in each wing at any one time, but adults can have up to four (Prevost 1983, Forsman 1999). This ensures that all of the primaries are renewed annually, while also allowing the bird to suspend moult during key periods. Moult in Ospreys is an excellent example of the trade-off between the need to replace worn feathers but not in a way that significantly impair flying ability, particularly during key periods such as migration. It is unsurprising, therefore, that Prevost (1983) found that moult was most intensive between October and December, once migrant Ospreys had arrived on their wintering grounds. During this period adult Ospreys tend to be highly

Figure 1.6. Juvenile Ospreys initiate their first moult during their first winter. As indicated, this young female had already replaced three inner primaries in each of her wings as well as two central tail feathers by late January of her second calendar year (© John Wright).

sedentary, favouring the same perches for extensive periods each day and fishing once early in the morning, and again in late afternoon or early evening. This presents an ideal opportunity to moult primary feathers and several waves of moult are often present in each wing (Figure 1.7). These waves are evenly spaced in the primaries, with each moulted or growing feather separated by two or three complete feathers. The next feather in a wave is only dropped once the one preceding it has reached 80 per cent or more of its final length (Prevost 1983). The growth of secondaries, on the other hand, frequently overlaps and close to half of the secondaries can be missing or growing at peak periods of moult on the wintering grounds.

Figure 1.7. Winter is a good opportunity for adult Ospreys to moult. John Wright found these feathers at an adult male's regular roosting site at Langue de Barbarie, Senegal in January 2011 (© John Wright).

The position of dropped or growing feathers is often identical in each wing, but sometimes adult Ospreys will have one fewer wave of moult in one wing than the other. Moult intensity declines in January and February and is usually completely suspended in March prior to departure on northward migration. Moult in tail feathers is closely correlated with the primaries, and tends to follow a similar pattern, peaking in the early winter and then ceasing before the onset of the spring migration (Prevost 1983). Nevertheless, there is usually no clear left–right symmetry in the tail moult of adult birds.

Once they have arrived on the breeding grounds, females are able to resume moult again because they are highly sedentary during incubation and chick-rearing periods, when they are reliant on their partner for food. Obvious gaps in both wings are often apparent if you see a breeding female take to the air during this time, with two or more waves of moult occurring simultaneously. It is rarer for breeding males to moult primary feathers during the breeding season when they are required to provision their mate and young. When moult resumes after it has been interrupted, it always continues at the point where it left off.

Ospreys as individuals

Over time and with a great deal of practice, it is possible to identify individual adult Ospreys by distinctive plumage characteristics. Ospreys are usually faithful to the same nest and wintering site every year, and one of the great joys of monitoring them is getting to know individual birds very well – particularly those that live into their teens or early twenties. As described earlier, young Ospreys undertake a complete post-juvenile moult, which begins in their first winter and is usually completed early in their third calendar year. At this point they have attained full adult plumage; this features the unique head, breast and underwing markings they retain for the rest of their life. This means that even if an Osprey is not colour-ringed, it is possible to identify a returning bird in the spring by its plumage. This has been greatly aided in recent years by the installation of high-definition cameras on Osprey nests and advances in digital photography, which make it possible to catalogue individual birds within a population.

One of the most distinctive features of the Osprey, and specifically *P. h. haliaetus* and *P. h. carolinensis*, is the dark markings on the head. It has been known for some time that these dark brown streaks and spots vary between individuals, and generally do not change significantly from year to year. Bretagnolle *et al.* (1994) analysed the pattern of black marks that appear on the dorsal, ventral and lateral aspects of the head of 33 individuals breeding on the coastal cliffs of Corsica and found the dorsal markings, in particular, varied noticeably between birds. The head patterns of each bird generally consisted of a single large dark mark on the nape, another on the forehead and a variable number of small spots on the crown; but the number, size and shape of these markings were variable and unique to each individual. Furthermore, the head patterns of known ringed individuals remained unchanged between two successive seasons, although it appeared that some of the black markings were larger in the second season, indicating that the size of the marks may increase with age (Bretagnolle *et al.* 1994). This method was also employed successfully by Englund and Greene (2008) in a study of breeding Ospreys in Minnesota, but Dennis (2007) found it too unreliable in windy conditions. The unringed female that first bred at the Manton Bay nest on Rutland Water was readily identified by her distinctive head marking when she returned each spring (Figure 1.8).

Another characteristic that can vary between individuals is the dark flecking pattern that appears in the yellow iris of adult Ospreys. Dennis (2007) produced detailed sketches of the 'clock-face' location of these dark flecks from a hide situated 20–40 metres from active nests on Kangaroo Island in South Australia. Comparison with sketches made in successive years revealed a level of consistency that enabled some individuals to be identified using this characteristic alone. Similarly, an unringed female Osprey that bred at the Loch of the Lowes in Scotland for 24 years was identified on her return each spring by distinctive iris markings, a task made easier in the bird's latter years by the installation of a high-definition camera on the

Figure 1.8. The unringed female, known as Maya, that first bred at the Manton Bay nest at Rutland Water in 2010 was readily identified each spring by her characteristic crown pattern, which formed an obvious cross and was better defined than that of her first mate, 5R(05) (© John Wright).

nest. Cameras such as this have enabled even greater scrutiny of iris markings. Dennis (2007) noted that iris darkening makes identification of some older birds impossible using this method but, as the bird at Loch of the Lowes demonstrates, this is not always the case.

As previously described, the underwing markings of Ospreys varies between the sexes and individuals. This is particularly the case in *P. h. haliaetus*, which generally exhibits clear sexual dimorphism in the extent of brown-black streaks and spots in the underwing-coverts. Males usually have just a few small brown spots on the median underwing-coverts and completely clean axillaries, whereas females tend to have two rows of larger more extensive brown spots on the median underwing-coverts, which continue through the axillaries and often onto the flanks. As with head markings, this spotting can be very distinctive, and once an Osprey has attained its full adult plumage, the size and shape of the streaks vary only slightly between moult cycles. Both male and female *P. h. haliaetus* have a dark carpal patch, which is generally more extensive in females. Males, however, tend to show more extensive areas of white in the greater under primary coverts, with limited brown spots and streaks which form a narrow central band on some birds (Figure 1.9). Females tend to have a more barred appearance, with two broad dark bands across the greater under primary coverts. By photographing known individuals on an annual basis, John Wright was able to determine that these underwing markings changed very little from year to year (Figure 1.10). John built up an extensive library of

Figure 1.9. Variations in the carpal patch and greater underwing-coverts of Ospreys at Rutland Water. Individuals can be identified by these markings, which do not change year-to-year (© John Wright).

A citizen of the world

Figure 1.10. Photographs showing the consistency of carpal patch and greater underwing-coverts of an adult male at Rutland Water, 03(97) (left–right taken in 2009, 2013 and 2015) (© John Wright).

photographs of all adult birds in the Rutland Water population and was able to identify birds exclusively by the shape and extent of markings in the underwing-coverts and carpal patch. This proved an extremely reliable method to identify returning unringed individuals and birds in flight whose colour ring was obscured.

Vocalisations

The Osprey has a range of calls, and it is possible to identify at least four different types of vocalisation. Perhaps the most evocative sound that can be heard in an Osprey colony during the breeding season is the high-pitched call of a displaying male. A distinctive *eep eep eep* betrays the presence of a male high in the sky. He rises and falls as he utters the cry, often hanging in the air with feet dangling before diving down and rising up again, like a rollercoaster (Figure 1.11). Sometimes the displaying bird is so high that the call is only just audible, and it can take several minutes to locate the bird – a mere speck in the aether. Unpaired males give this call to advertise themselves to passing females, but they also display to other males in an effort to establish territory. As a result, younger males tend to display far more than older birds with an established mate. Sometimes females will give a similar call if a rival female passes too close to their nest, but any associated display flight is rarely as spectacular as the one performed by males.

A short, sharp *tioop* is uttered by breeding birds in response to the sight of an intruding Osprey approaching the nest, and is referred to as a 'guard call' by Poole (1989) and Bretagnolle and Thibault (1993). If the intruding bird is some distance away then this call acts as a warning not to come any closer. If this is not heeded, the call increases in both frequency and intensity as the intruder approaches. Bretagnolle and Thibault (1993) refer to this as the 'excited call', and the difference between this and the initial, more relaxed, guard call is evident in sonograms produced by both Poole (1989) and Bretagnolle and Thibault (1993). Bretagnolle and Thibault also identified a 'screaming call', which often follows the guard and excited calls and is usually given in flight as a breeding bird chases an intruder away. The frequency of intrusions at breeding nests generally increases as the summer progresses, when later-arriving non-breeding birds are present in larger numbers. Sometimes small groups of non-breeding birds will fly from nest to nest and the air is filled with a cacophony of *tioops* and *chips*. At times like this, breeding males need to be particularly vigilant, and this explains why Bretagnolle and Thibault (1993) found that guard, excited and screaming calls (all relating to nest defence) accounted for 79 per cent of the vocalisations uttered by males.

Adult birds can be equally territorial on the wintering grounds, and may be particularly aggressive towards newly arrived juveniles searching for somewhere to spend their first winter. If the initial warning *tioop* is ignored then, just as in the breeding season, calls increase in frequency and intensity and may eventually lead to a chase. A similar call is sometimes given when two birds are hunting close together. In fact, some Ospreys become highly possessive over specific hunting grounds, and will chase others away, using their full repertoire of guard, excited and screaming calls to warn off potential competitors.

During the breeding season, male and female Ospreys have clearly defined roles. The male's key responsibilities are to provide a regular food supply and to defend the nest, while the female undertakes the majority of incubation and then broods and protects the young. A key call in the female's repertoire, therefore, is the food solicitation call. This is a series of repetitive notes, which Bretagnolle and Thibault (1993) describe as being on a continuum of intensity ranging from low to very high. In their study of Corsican Ospreys they found that the presence of the male at the nest prompted different levels of food solicitation from the female, according to her mate's motivation and time since the last feed. The longer the gap since the last fish, the more intensive the food-begging calls. Unsurprisingly, the sight of the male flying in with prey or perched close to the nest with a fish prompts the most intensive calls. The response of the male to the food-begging calls of his mate varies massively. This solicitation call often prompts the male to go fishing, but on other occasions he will remain unresponsive, leading to louder and more persistent calls from the female. Overall, Bretagnolle and Thibault (1993) found that food solicitation accounted for 54.8% of calls given by female Ospreys during the breeding season.

A final call heard at Osprey nests during the breeding season is a repeated *de de de* alarm, given when people or a potential predator approaches the nest. This is similar to the

Figure 1.11. An adult male displaying with a Rainbow Trout at Rutland Water, while giving the distinctive *eep eep eep* display call (© John Wright).

food-begging call, but shriller. It is given by both sexes but there is sexual dimorphism, with the female's call audibly stronger and deeper than the male (Poole 1989, Bretagnolle and Thibault 1993). Like the intruder call, the alarm call is first uttered when a threat is identified by the bird. It may be quiet at first, but rapidly increases in volume, intensity and pitch as the threat draws nearer, culminating in squeals and shrieks very similar to a display call.

Juveniles are very quiet when they first hatch, although Bretagnolle and Thibault (1993) identified a call similar to the adult's guard call given by very small chicks being fed. This is sometimes audible from the many nest cameras that stream live from Osprey nests all around the world. As they become larger, Osprey chicks get more vocal, and begin to food-beg like their mother. Chicks begin giving the food solicitation call before they fledge, and then continue throughout the post-fledging period when they remain dependent on their parents for food. As such, the noisy food-begging of juveniles in late summer is a characteristic sound of Osprey nests worldwide. Sometimes a hungry juvenile will continue calling even once the male has delivered a fish to the nest and it is eating, so strong is the instinct to give the call. As it is essential that juvenile Ospreys feed well before they depart on migration, the food-begging call is vital for young birds in migratory populations. Juveniles also give a version of the 'excited call' described by Bretagnolle and Thibault (1993) when exercising their wings in the last few days prior to fledging. At this point the strength of the young Ospreys is palpable and the call seems to reflect their desire to leave the nest and take to the air for the first time.

Conclusions

- The Osprey is exclusively piscivorous and readily catches marine and freshwater fish.
- It has a cosmopolitan distribution and is one of only six landbird species that occur on every continent except Antarctica.
- Four subspecies are recognised, based on morphology and plumage variations. The status of *P. h. cristatus*, which occurs in Oceania and the Indo-Pacific region, has caused some recent debate and is regarded by some as a separate species, the Eastern Osprey *Pandion cristatus*.
- Northern populations are migratory, while those at lower latitudes are sedentary or undertake limited post-breeding movements.
- Female Ospreys are 20–30 per cent heavier than males and tend to have a more prominent breast-band and more well-marked underwings.
- Juvenile Ospreys take approximately 18 months to moult into full adult plumage, and are usually not separable from adults by spring of their third calendar year, when they typically return to potential breeding areas for the first time.
- Some adults can be identified by specific head markings, the dark flecking patterns in the yellow iris, and the shape and extent of markings on the underwing-coverts and carpal.
- The Osprey has a range of calls. It is possible to identify at least four different types including a high-pitched display call, a guard call, a food-solicitation call and an alarm call.
- The Osprey's distribution and annual movements are driven by its specialist piscivorous diet, which is considered in more detail in chapter 2.

CHAPTER 2
Hunting

The Osprey's prowess as a hunter has been well known for centuries. Indeed, the species' penchant for taking fish from medieval trout ponds was one of the reasons it was widely persecuted in the United Kingdom and across Europe. Fortunately, we now live in more enlightened times and, although the threat of persecution lingers in some places, the sight of an Osprey plunging into the water at great speed, and then lifting off a few seconds later with a fish grasped in its talons, is much more likely to excite and inspire people than to create animosity. As a specialist piscivore, the Osprey is supremely well adapted for catching fish in a variety of habitats, from coasts to inland lakes and rivers. Indeed, the migratory lifestyle of many of the world's Ospreys means that this ability to forage successfully in different habitats is an essential life skill. An Osprey that breeds in the boreal forests of Scandinavia will catch freshwater species in inland lakes during the breeding season, but may then switch to a diet consisting entirely of saltwater fish on its wintering grounds on the coast of Africa.

The hunt

A hunting Osprey may use a variety of techniques which vary according to local conditions. In areas with wide expanses of open water it will typically fly at a height of between 20 and 40 metres. The Osprey's long, narrow wings give hunting birds a graceful appearance. As mentioned in chapter 1, slow, steady wingbeats are interspersed with prolonged glides, during which the wings are held in a shallow M shape. This gives Ospreys a characteristic flight profile and makes them relatively easy to pick out, even at a considerable distance.

One of the key adaptations Ospreys possess is extraordinary vision, and in particular, high visual acuity, which enables a hunting bird to locate fish from a considerable height above the water. This is attained through two key anatomical factors: eye size and a high density of photoreceptors within the retina, as described by Mitkus *et al.* (2018). Ospreys, like other predatory raptors, have relatively large eyes with a long anterior focal length, and hence, a large retinal image which aids visual acuity. An Osprey's eye functions in much the same way as ours. Light enters through the transparent cornea, lens and vitreous body before reaching the retina. Here a very high density of photoreceptors is located in the foveae, tiny pits in the macula of the retina. Most birds of prey, such as Ospreys, have a deep central fovea and a shallow temporal fovea. The deep central fovea allows for the highest visual acuity in the lateral visual field – which is likely used for detecting prey from considerable distance – while the temporal fovea is thought to facilitate sharp vision in the frontal field of view.

If the bird sees fish moving in the water below, it will often turn into the wind and hover with its legs dangling, a series of powerful wingbeats enabling it to maintain its position for a few seconds while it stares intently at the water (Figure 2.1). If the fish are out of range it will move on, but may pause soon afterwards to hover again. If it is windy, the bird will always turn to face into the wind, holding its position with remarkable poise even in gusty conditions. Wind can actually help a hunting Osprey because hovering is energetically costly for such a large bird, particularly when there is little or no wind, and this is all related to physics. The power requirements for a bird in normal horizontal flight are related to its airspeed (the speed of the bird in relation to air, rather than the ground). Flying at either very high or very low airspeeds necessitates high power in order to counteract the differing effects of gravity and drag, but intermediate flight speeds are less energetically demanding (Pennycuick 1969, Alerstam 2000). This is directly applicable to a hovering Osprey, because maintaining a stationary airborne position in different wind speeds is equivalent to flying forwards in still air, at the corresponding airspeeds. For example, if a bird is hovering into a wind of 8 metres per second (m/s) it will require the same power as a bird flying forwards in still air at 8 m/s. As such, hovering in still air conditions requires a high amount of power, but the power requirement decreases with increasing wind speed, down to a minimum at the wind speed equivalent to the optimum speed for an Osprey in flapping flight, which was estimated to be 10.3 m/s for an average female (Strandberg et al. 2006). Furthermore, in strong winds Ospreys are sometimes able to hold their position above the water with barely any flapping at all.

The ability of Ospreys to adapt their hunting technique according to the wind conditions was demonstrated by Strandberg et al. (2006) at Lake Hammarsjön, Scania, southern Sweden. They described four different techniques which varied in use according to the strength of the wind. The mean hovering time was 5.77 seconds for males, and 6.65 seconds for females, but this varied according to wind speed. At the lowest wind speeds (<1 m/s) Ospreys did not hover at all, and hovering times were distinctly shorter at low wind speeds (0–5 m/s) compared to either moderate (5–10 m/s) or high (10–15 m/s) wind speeds. Hovering with flapping flight was used at wind speeds of 1–10 m/s, but when the wind speed exceeded 8 m/s the birds were also able to incorporate 'glides' into the hovers, whereby the hunting bird could maintain its position above the water for a few seconds with wings open rather than flapping, thereby minimising energy expenditure. Hovering by pure gliding was also recorded in the strongest winds, although the birds did not hunt once the wind speed exceeded 14.5 m/s. There were also sex-based differences, which relate to the larger size of females and the associated higher power

Figure 2.1. Hunting Ospreys will frequently stop and hover, staring intently into the water below (© John Wright).

required for hovering. Males were able to hover for longer in weak winds than females, and hovering times increased more steeply for males with increasing wind speed than for females. Intriguingly, hovering time peaked at a wind speed of 10.5 m/s for males, but there was no maximum for females. This may be because the larger size of females enables them to better utilise the gliding technique and thus increase the overall duration of hovers in strong winds (Strandberg *et al.* 2006).

Sometimes a hovering bird will stoop, dropping in altitude a few metres before hovering again. It may repeat this manoeuvre two or three times, before finally folding its wings and descending towards the water, its eyes fixed on its target throughout. Birds typically dive from a height of 20–30 metres (Snow and Perrins 1998), but it can range from five metres to 75 metres or even more (Prevost 1982, Ferguson-Lees and Christie 2001). Ospreys will also sometimes dive from perches overlooking water (Prevost 1982, Ferguson-Lees and Christie 2001). There seems to be some degree of individual variation in the use of perch-diving. For instance, at Horn Mill Trout Farm in Rutland, central England, where a photography hide has been set up next to a small pond in order to afford spectacular views of diving Ospreys, some individuals (recognisable by colour rings) always dive from a tree overlooking the water, whereas others prefer to hunt from the wing (Figure 2.2). This does not appear to have any effect on strike rates of the individual birds.

The pale underside of Ospreys helps them to remain inconspicuous while they are circling or hovering over water, but hunting birds also adapt the angle of the dive according to local conditions and the fish species they are targeting, in order to avoid detection. On the coast of West Africa, hunting Ospreys often avoid hovering and tend to favour a shallow dive – which may be close to horizontal – when they are targeting fast-moving surface fish such as mullet and garfish (Prevost 1982). This appears to reduce the likelihood of the fish detecting their approach, and means that on occasions birds hunting in this way barely get their feet wet as they pluck the fish from the water. It also probably also reduces the risk of predation by crocodiles, which are capable of catching Ospreys if they are struggling with a large fish on the surface of the water. Bartosik (2009), meanwhile, noted that Ospreys fishing in Florida and Texas in the United States usually dived at an angle of 45°. The author hypothesised that this may be necessary to enable the diving Osprey to avoid being speared by sharp spines in the dorsal fins of species such as Pinfish *Lagodon rhomboides*. If the Osprey dived from directly above it would be difficult to dodge these sharp spines, but by approaching from a shallower angle, diving birds can strike the unprotected lateral side of the fish, with their grip aided by the flexibility of the outer toe, which is reversible (see below). On other occasions Ospreys dive from a much steeper angle, which can be almost vertical (Carpenteri 1997). Prevost (1982) found that Ospreys favoured a steeper dive and hovered more frequently when targeting slow-moving fish

Figure 2.2. Some Ospreys will hunt from perches overlooking the water, a tactic often used by birds at Horn Mill Trout Farm in Rutland, such as this individual (© Geoff Harries).

such as bass, grunts and jackfish, which may be slightly below the surface of the water, but still within range of the Osprey's long legs. Prevost (1982) also watched some Ospreys flying 1–5km out to sea off the coast of West Africa where they targeted flying fish. Here they would rise to an estimated height of approximately 300m in order to locate their prey, before slowly losing altitude and then diving from approximately 100m. Recent advances in satellite tracking have cast further light on this behaviour. A juvenile female that Roy Dennis satellite-tagged in northern Scotland in 2019, for example, regularly made flights up to 12km out to sea while it was wintering on the coast of Senegal, potentially targeting flying fish (Dennis 2019).

Even as the Osprey descends towards the water, it is still able to abort a dive at the very last moment if necessary. An Osprey only stands a chance of making a catch if the fish is on or just below the surface, and so if the intended target senses impending danger and darts away, the Osprey will pull out of the dive, sometimes just a few centimetres above the water, regardless of the angle of descent, and swoop away. However, if the fish remains within range, the bird throws its talons forward, so that, as it prepares to strike, its feet are just millimetres in front of its eyes with the toes splayed (Figure 2.3).

As it prepares to strike, the diving Osprey has to contend with the effects of refraction: the process whereby light waves bend as they enter the water, making the fish appear closer than it really is. It is not known exactly how Ospreys manage this but their exceptional visual acuity is likely a key factor. A nictitating membrane (a transparent third eyelid) helps protect each eye as the Osprey hits the water, and its long legs enable it to reach down to a depth of one metre. As it makes contact with its target, the razor-sharp talons are clamped firmly into the body of the fish, with the bird's grip aided further by spicules (small spines) on the underside of the toe pad (Snow and Perrins 1998). A further key attribute that enhances the ability of a diving Osprey to grasp hold of a fish is that it can rotate the fourth toe antero-posteriorly: in other words, reverse it. Ospreys usually hold their toes in an anisodactyl arrangement, whereby digits II–IV face anteriorly (forward) and digit I posteriorly (backwards). Unlike other raptors, however, they are also able to switch to a zygodactyl arrangement (where digits II and III face forward and digits I and IV backwards), described by Raikow (1985) as semi-zygodactyl. The process of reversing the fourth digit is aided by various key anatomical adaptations: the fact that all toes are of equal length; the lack of membrane between it and digit III (Tsang 2012); the shape of the outer trochlea (constituting part of the joint between the lower leg and the toes) that permits wide lateral movement in the fourth toe (Jollie 1976); and well-developed foot muscles (Hudson 1948). The ability to reverse the fourth toe in this way provides crucial manoeuvrability in the foot, which is important when the bird first grasps a fish. In an extensive analysis of 1,138 photographs of Ospreys in various positions, Sustaita *et al.* (2019) demonstrated that they were 5.7 times more likely to have their toes positioned in a zygodactyl arrangement when grasping an object in flight than when perched.

Figure 2.3. A diving Osprey throws its talons forward as it prepares to strike the fish below (© John Wright).

A sure sign that an Osprey has captured a fish is the sight of it resting on the water with wings outstretched as it ensures it has a firm grasp, before mustering sufficient energy to lift its catch out of the water. Ospreys have oily feathers to avoid getting waterlogged while grappling with a fish (Ferguson-Lees and Christie 2001). This is vital as it can sometimes take several attempts for the bird to get airborne again (Figure 2.4). Fish caught typically weigh 150–300g (Prevost 1982, Poole 1989) but much heavier fish, weighing up to 2kg (equal in weight to an adult female Osprey), have been recorded in some exceptional cases (Ferguson-Lees and Christie 2001). In a lengthy study of foraging Ospreys in West Africa, Prevost (1982) found that there was considerable variation in the size of fish caught, with a range of 42–1,017g. Getting back into the sky with a large fish is no easy task, but a series of powerful, almost horizontal wingbeats usually enables the bird to lift off (Figure 2.5). Conversely, if the fish is simply too heavy then the bird will release its grip and try again. Occasionally, hunting Ospreys have been observed using their wings like oars in order to swim to shore if a fish is too heavy to lift out of the water (Mackrill *et al.* 2013). Remarkably, when diving into a shoal of small fish, a hunting Osprey will sometimes catch two fish in a single dive, with a fish grasped in each foot (Figure 2.6).

Although Ospreys will preferentially grasp most fish with both sets of talons, it is sometimes necessary for the bird to take off while holding the fish in only one foot (Figure 2.7). It is here that the ability to switch to a zygodactyl arrangement is particularly important because it greatly improves the ability of the bird to keep hold of its catch. Sustaita *et al.* (2019) found that the odds of observing zygodactyly in a single-foot grasp were 4.1 times greater with fish

Figure 2.4. Ospreys will often rest on the water surface with wings outstretched before mustering the strength to lift the fish out of the water (© Geoff Harries).

Hunting

Figure 2.5. A series of powerful wingbeats usually helps the hunting Osprey get airborne again after a successful dive, like this one at Rutland Water (© John Wright).

Figure 2.6. If they target a shoal, Ospreys sometimes catch two fish in a single successful dive (© John Wright).

Figure 2.7. Ospreys sometimes take off with the fish held in only one foot, then manoeuvre it into a better position in the air (© John Wright).

compared to other objects. Ospreys share this semi-zygodactyl arrangement of the toes with owls, who as Sustaita *et al.* (2019) point out, also feed on prey that is difficult to capture: usually fast-moving nocturnal small mammals. While stealth plays a key role in both cases too, the morphological adaptations of Ospreys and owls appear to be indicative of the ecological, adaptive origins of semi-zygodactyly – especially, the ability to catch slippery fish or fast-moving small mammals. Nevertheless, there is still much to understand about the semi-zygodactyl feet of Ospreys, including how the rotation of the outer toe is controlled. Sustaita *et al.* (2019) suggest it may be driven entirely by the action of the musculus abductor muscle controlling the fourth digit, or by the morphology of the tarsometatarso-phalangeal joint and tendons in the foot.

Once the Osprey is airborne, a characteristic head and body shake in mid-air removes any excess water, and the bird repositions its catch so that the head of the fish is facing forward. This is not an easy task given that the fish is nearly always alive and often writhing strongly. One foot is positioned near the head of the fish, while the second grasps close to the dorsal fin. This

Figure 2.8. Red Kites on occasion attempt to steal fish from Ospreys in the Rutland Water area, but they are rarely successful. Here an adult Red Kite is chasing established breeding female 5N(04) (© John Wright).

means that as it carries its catch away the Osprey has one foot directly in front of the other. Allen *et al.* (2018) suggest that the front foot is used to pith (i.e. pierce or sever the spinal cord) and kill the fish, while the rear foot helps to stabilise it. Holding the fish in this manner improves aerodynamics, which is particularly important when carrying a heavy load into a headwind. It also helps the Osprey to evade the advances of kleptoparasites. White-tailed Eagles *Haliaeetus albicilla* and Bald Eagles *Haliaeetus leucocephalus* are particularly skilled at stealing fish from Ospreys, with some chases lasting five minutes or more (Allen *et al.* 2018). There is some evidence that Ospreys actively avoid nesting close to White-tailed Eagle territories as a result (Shoji *et al.* 2011). Herons, and large gulls may also attempt to steal fish from Ospreys (Ferguson-Lees and Christie 2001) and a Great Cormorant *Phalacrocorax carbo* was observed stealing a fish from an Osprey at Rutland Water while it was grappling with its catch on the surface of the water (Mackrill *et al.* 2013). We also see Red Kites *Milvus milvus* attempting to steal fish from Ospreys in the Rutland Water area, but they are rarely successful (Figure 2.8). Allen *et al.* (2018) found that Ospreys appear to preferentially fly with the left foot in front of the right. They suggest that these findings imply a lateralised brain function likely underpins Osprey flight stance, and is necessary due to the complexity of movements required to catch and then keep hold of a live fish.

Factors influencing dive success

An Osprey is not successful with every dive, and the outcome of each plunge into the water is influenced by a variety factors. Dive success can range from as little as 20% to up to 90%, but more typically in the range of 50–70% (Prevost 1982, Flemming and Smith 1990).

AGE

Age is a key factor determining Osprey dive success. Juvenile Ospreys begin making dives into the water as early as five days after they have fledged and have been observed catching fish 11 days after making their first flight (Edwards 1989), but this is exceptional and it takes some time for young birds to become as adept as adults. There is some evidence that social learning may play a limited role in the development of foraging mechanics. For example, Edwards (1989) found that capture success of young with siblings at Newnan's Lake in Florida was initially greater than singleton young during the post-fledging period. However, given that many nests are located away from water, it seems that the development of foraging skills is largely instinctive and refined with practice in the weeks and months after fledging. Many early dives have no intended target, and are simply a way for young birds to practise this essential life skill. Recently fledged juveniles also quickly learn to hover in the same way as adults, but as Strandberg *et al.* (2006) demonstrated they are not as adept at utilising this technique, with hovers limited to periods when the wind speed ranged from 2m/s to 12m/s. Young Ospreys usually remain dependent on their parents for food until they depart on migration, meaning that many birds do not catch their own fish until they have begun travelling south (Figure 2.9). Even then it can take six months or more before they begin catching with the same degree of success of adults.

Prevost (1982) found that only 19% of the dives made by immature Ospreys of six months of age in the mouth of the Senegal River in northern Senegal were successful, compared to 40% of dives by adults. Interestingly, time per prey capture did not vary between the two age classes because adults took longer between dives (9.2 minutes) than juveniles (6.5 minutes), implying that adult birds were better at judging when to dive. I have been fortunate to witness this myself in Senegal and The Gambia during regular winter trips since 2011. In the fish-rich waters of the Senegal and Gambian coast, adult Ospreys often catch species such as mullet with ease, but I have watched juveniles making repeated dives and failing each time, before retreating, exhausted, back to the shore to recuperate.

Figure 2.9. Juveniles rarely begin catching their own fish until they disperse away from their natal nest-site. It takes several months for them to attain the foraging skills of adults (© John Wright).

WEATHER CONDITIONS

The ability of adults to judge when best to dive is exemplified by the response of birds to differing weather conditions. Weather, and wind in particular, has also been shown to have an effect on Osprey foraging, although this varies according to local factors.

As explained above, some wind can actually aid Osprey foraging because it makes hovering less energetically costly. This is significant because Grubb (1977) found that the success rate of Ospreys diving from a hover was 50 per cent greater than from inter-hover flight. However, as wind speed increases other factors come into play which hamper the ability of hunting Osprey to locate their prey. In a study of Ospreys foraging at Corn Creek Marsh in Creston, British Colombia, Machmer and Ydenberg (1990) found that as wind speed increased, and water surface conditions deteriorated, foraging flights consisted of more gliding and less flapping (as per Strandberg *et al.* 2006), and the probability that a hunt ended with a capture decreased markedly. Furthermore, the detrimental effects of the wind increased at wind speeds exceeding 6.2m/s, and the cost of capturing a fish when the wind speed was more than 7m/s actually exceeded energetic gain. Similarly, Grubb (1977) found that increased rippling on the water surface decreased capture rates at Lake George in Florida, as did increases in cloud cover. He hypothesised that, in each case, this was because fish became more difficult to spot. Indeed, the success rate of individual dives did not vary; Ospreys simply dived less when there were more ripples on the water surface, presumably because fish remained out of sight for longer. This again emphasises the ability of adult birds to judge when best to plunge into the water.

In windy conditions, Ospreys adapt their hunting behaviour to maximise their chances of a successful capture. At Rutland Water, where I have enjoyed watching Ospreys for many years, strong winds can be problematic for hunting Ospreys due to large areas of exposed open water. To counteract this, hunting birds always favour bays and inlets on the leeward shore on windy days. Here the water is less disturbed and the birds use the wind to hover, often for prolonged periods, rather than search more widely around the reservoir as they would on calmer days. They may also abandon fishing at the reservoir altogether, in favour of an easier catch at the nearby trout farm where wind is unlikely to be a factor. This is especially important for males at the height of the breeding season when they are required to catch four or five fish every day. Hunting Ospreys must continually assess such trade-offs.

Size of prey

The size of the intended prey species can also have an influence on dive success. Ospreys find it easier to catch larger fish which are easier to grasp than smaller ones of a similar shape, as these can more easily evade capture. In his study of Ospreys foraging on the coast of West Africa, Prevost (1982) was able to demonstrate that fishing Ospreys specifically targeted large individuals of Fimbriated Herring *Ethmalosa fimbriata* (240–260g) even though netting indicated that they were scarcer than smaller fish of the same species. Moreover, dive success was significantly correlated with fish size, with the success rate of hunting Ospreys catching fish 350–500g up to twice as great as when prey was smaller (150–250g). Nevertheless, time per capture was also significantly correlated with fish weight, due to the fact that smaller fish were more abundant. Thus, it was necessary for Ospreys to hunt for longer in order to catch fish of the preferred size – another clear example of the trade-offs a fishing Osprey is required to make, and the ability of experienced adult birds to be selective in when they make a dive.

Tidal stage

Estuaries are one of the best places to watch fishing Ospreys, and here hunting birds use the tide to their advantage. Flemming and Smith (1990) found that Ospreys fishing in Antigonish Harbour in Nova Scotia favoured the mid-tide, with the number of foraging birds and dive success peaking at this tidal amplitude. Prevost (1977), meanwhile, who studied Osprey foraging behaviour at the same location, found that dive success varied among locations within tidal periods, due to differences in water depth. At high tide, when water is too deep, Winter

Flounder *Pseudopleuronectes americanus,* a bottom dweller and favourite prey item of North American Ospreys, are out of range. Conversely, at low tide the flounder are forced into deeper channels, where they are usually safe from hunting Ospreys. Flounder have been shown to move extensively at mid-tide (Tyler 1971) in search of marine worms, and it is during this period that they are particularly susceptible to predation from Ospreys. At Antigonish Harbour, 70 per cent of dives made by Ospreys at this time were successful regardless of whether it was an ebbing or incoming tide, compared to 56 per cent at low tide and 53 per cent at high tide (Flemming and Smith 1990).

PREY SPECIES

The intended prey species and its ecology is another factor that influences Osprey dive success. Benthic-feeding fish, such as flounders, are generally the easiest group for Ospreys to catch, assuming they are within range (Swenson 1979). It seems that species like flounders, which are behaviourally adapted to foraging in the muddy bottoms of estuaries, have a limited ability to sense attack from above, and instead rely on camouflage to go undetected. However, the high visual acuity of the Osprey's eyesight means that the slightest puff of mud – created as a flounder feeds – can alert a hunting bird to a potential catch. Faster-moving piscivorous fish, such as Northern Pike *Esox lucius* and Brown Trout *Salmo trutta* are more difficult to catch because their feeding habits make it necessary for them to be fast-moving and agile and thus better adapted to evading capture by a hunting Osprey. Swenson (1979) quantified this relationship by creating a 'prey species foraging index' for 13 studies of Osprey foraging success. Fish were broadly grouped into three categories according to foraging behaviour: benthic feeders, fish that feed on limnetic organisms (but not other fish), and piscivorous fish (obligate and facultative). The index varied from 0.0 if the diet consisted exclusively of benthic fish, to 2.0 if it was limited to piscivorous fish only. This index was then plotted against the relevant Osprey dive success rate. There was a statistically significant correlation between dive success and the prey species foraging index. Ospreys feeding exclusively on benthic fish had a dive success rate of 70 per cent, but this figure dropped to 20–40 per cent with a higher proportion of piscivorous fish (1.5–2.0 on the prey species index).

Duration of foraging trips

In optimal habitat and under prime foraging conditions adult Ospreys are capable of catching fish very quickly, but on other occasions it can take an hour or more for a bird to make a catch. Prevost (1982) found that time per capture of Ospreys foraging in the coastal waters of Senegal and The Gambia was typically 20–30 minutes, varying from a minimum of 12.9 minutes in shallow water at the mouth of the Senegal River during October to December, to a maximum of 61.3 minutes along the sandy shore near north of Dakar from January to March. Meanwhile, in South Humboldt Bay in California in the United States, Ospreys took a mean 11.8 minutes to make a successful catch, with 56 per cent of fishing efforts requiring only a single dive, 18 per cent needing two dives and 6 per cent requiring three (Ueoka and Koplin 1973). It is notable that 63 per cent of fish taken were surfperch (*Embiotocidae* spp.) which, as benthic feeders, are easier to catch. Conversely, in areas where Ospreys are feeding on piscivorous fish, foraging times tend to be longer. In my own study area at Rutland Water where Rainbow Trout *Oncorhynchus mykiss* often forms a significant proportion of the diet, foraging flights regularly exceed an hour in duration and it is common for breeding males to carry out several unsuccessful dives before eventually making a catch. On one particular occasion a breeding male,

03(97), was seen to make 17 unsuccessful dives during two hours of continuous fishing (Mackrill et al. 2013). Although this was exceptional, monitoring by volunteers revealed that the average duration of foraging trips by this particular male was 99 minutes in 2008, when Rainbow Trout constituted 46 per cent of the catch (Mackrill et al. 2013). The proportion of Rainbow Trout was considerably higher than previous years, when Roach *Rutilus rutilus* was the most commonly caught species. For example, in 2004 Rainbow Trout constituted just 13 per cent of the diet, compared to 45 per cent Roach, and the average duration of foraging trips was 60 minutes. This emphasises the difficulty of catching piscivorous species rather than benthic feeders or species such as Roach that congregate in shoals close to the surface. The duration of foraging trips is increased further on windy days at Rutland Water, and a similar scenario has been recorded in both estuarine and freshwater environments. Machmer and Ydenberg (1990) reported a 41 per cent increase in foraging time at Corn Creek Marsh in British Colombia, Canada in a wind speed of 8m/s compared to 0m/s, while Strandberg et al. 2006 found that Ospreys did not attempt to hunt at all when wind speed exceeded 14.5m/s at a freshwater lake in southern Sweden. Intriguingly, Grubb (1977) noted that Ospreys tended to catch more mullet on days when the surface of the water at Lake George in Florida was rippled. He hypothesised that this may be because they are more conspicuous than other fish caught at the lake, such as crappies, due to their silvery colouration and habit of occurring in shoals close to the surface.

Rain, like wind, makes fishing more difficult for hunting Ospreys, and they often cease hunting if rain becomes too heavy, as noted at Antigonish Harbour in Nova Scotia by Flemming and Smith (1990).

Timing of foraging flights

There is no evidence that time of day has any direct effect on Osprey fishing success, but there are often clear patterns in the timing of foraging flights. Ospreys are usually active soon after dawn, and most studies have identified a peak in Osprey foraging during the morning and another later in the afternoon/evening. Prevost (1982), for example, noted a morning peak in numbers of Ospreys fishing at two sites in Senegal during January–March, with a second less-pronounced increase later in the day. Ospreys have greater foraging requirements when provisioning young, but the same broad patterns are still evident. Flemming and Smith (1990), for instance, noted a peak in activity between 05:00 and 08:00 in the morning, and again between 17:00 and 20:00 in the evening in Nova Scotia. A similar trend is evident at Rutland Water, with breeding males regularly returning to nests with fish before 06:00, particularly when they have chicks to feed. Once young have hatched there is a clear upturn in the male's fishing effort, and he will make regular forays from the nest throughout the day, usually every two to three hours. Monitoring at several nests in the Rutland Water area by volunteers has enabled the time of fish deliveries to be studied in detail. For example, volunteers recorded a total of 257 fish deliveries to a nest referred to as Site B in Rutland in 2009. Observers were present throughout the season from 6am until 8pm, and on a 24-hour basis while the birds were incubating. There was a clear morning peak in fish deliveries between 06:00–08:00 (total 51 fish) and a second, more drawn-out and slightly lower afternoon/evening peak between 15:00 and 20:00 (119 fish). Monitoring revealed that fish deliveries to a second Rutland nest, known as Site N, followed a similar pattern but timings were slightly different. The morning peak did not occur until 11:00–12:00, and the evening peak happened between 18:00 and 20:00 (see Figure 2.10).

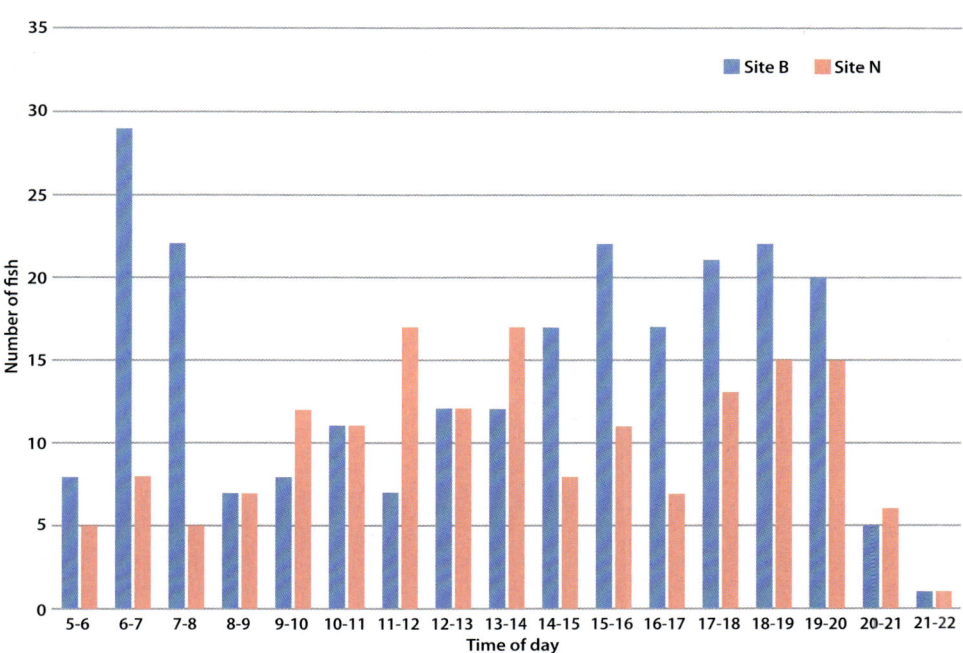

Figure 2.10. Time of fish deliveries to two nests at Rutland Water during 2009 (data courtesy of Leicestershire and Rutland Wildlife Trust).

Diet

Ospreys are opportunistic hunters, and will take a variety of live fish from a range of different saltwater and freshwater habitats, usually favouring the most abundant and readily caught species. There are occasional examples of Ospreys taking alternative prey: an Osprey was seen with a small American Alligator *Alligator mississippiensis* in North America (Poole 1989), John Wright watched an Osprey catch (and then drop) a Signal Crayfish *Pacifastacus leniusculus* at Rutland Water (Mackrill *et al.* 2013), and a live Little Grebe *Tachybaptus ruficollis* was brought back to the Loch Garten nest in northern Scotland in 2022 (I. Perks pers. comm. 2022). However, such instances are very rare and perhaps a case of mistaken identity by the Osprey.

The chief requirement for any hunting Osprey is that fish are within easy reach of the surface, and as such they often favour shallower water. Any fish, of suitable size, swimming up to a depth of one metre is within range. The migratory habits of many of the world's Ospreys means that the diet of individual birds can be highly diverse during the course of a single calendar year, reflecting the changing availability of prey as birds move between northern breeding grounds and wintering sites further south. The ability of experienced Ospreys to adjust their hunting technique, including varying the angle of dive, according to the particular traits of their intended prey species and the type of habitat in which they are hunting, is thus an essential life skill that enables them to prosper throughout the year.

The diet of Ospreys has been studied around the world. Most older studies relied upon identifying fish remains recovered from nests after the breeding season, but in recent years the installation of high-definition cameras on nests has greatly facilitated studies of breeding pairs.

Most of the Ospreys that breed at northern latitudes tend to fish in freshwater lakes (Marquiss *et al.* 2007). Here they catch a range of piscivorous and non-piscivorous species that inhabit the limnetic zone. Brown Trout and introduced Rainbow Trout are favoured prey of Ospreys fishing inland lakes in northern Europe, but a range of other species are taken, including Roach (Figure 2.11), Perch *Perca fluviatilis* and Northern Pike (Carss and Brockie 1994, Ivanovski 2012, Mackrill *et al.* 2013). Roach are a favoured species in many areas because, as described above, they tend to congregate in shoals near the water's surface, leaving them vulnerable to predation by hunting Ospreys. I have enjoyed many sunny evenings watching Ospreys targeting shoals of Roach at Rutland Water (Mackrill *et al.* 2013). Ospreys are also known to take fish that have become infested with *Ligula intestinalis*. This tapeworm, which is found in several species, including Roach, has a severe effect on its host's behaviour and vision, rendering it vulnerable to predation by Ospreys and other piscivorous species (Brown *et al.* 2002, Babuskin and Kuznicov 2012). Northern Pike are most readily caught in the spring when they come into shallows to spawn (Mackrill *et al.* 2013).

Artificially stocking reservoirs for angling has helped Ospreys in many areas. At Rutland Water, stocked Rainbow Trout constitute close to 50 per cent of the diet of some pairs during the course of the breeding season, and certain birds also visit the local trout farm. A photographic hide was installed at the trout farm following an increase in the number of visits by hunting Ospreys and this has proved highly successful, offering excellent photographic opportunities and valuable additional income for the owner (Mackrill *et al.* 2013, Mackrill 2019) (Figure 2.12). Stocked trout are often particularly important early in the breeding season when some birds return as early as mid-March. At this time in the early spring, cold snaps can make fishing a challenge, but the availability of recently stocked fish provides a readily available food source. Similarly, in Strathspey in northern Scotland, Pike and Brown Trout were the main species brought to the famous Loch Garten nest during the 1970s and 1980s, but by 1996 Rainbow Trout had exceeded both species (Dennis 2008).

Swenson (1979) demonstrated that piscivorous fish such as Northern Pike are more difficult for Ospreys to catch, and so it is perhaps no surprise that in some parts of Europe, Ospreys favour benthic feeders. The ecology of these species mean that they are usually out of range for a fishing Osprey in freshwater lakes, but the increasing prevalence of aquaculture, particularly in Eastern Europe has offered an alternative food supply in many countries. In Latvia, for example, benthic feeders such as Common Carp *Cyprinus carpio*, Tench *Tinca tinca* and Prussian Carp *Carassius gibelio* constitute the bulk of Osprey diet (Kalvans and Bajinskis 2016). The favourable growing conditions provided for these species in fishponds means

Figure 2.11. Roach are a favoured prey species of many Ospreys in Northern Europe, including at Rutland Water (© John Wright).

they quickly reach suitable prey size, and the fact that they are kept in high densities alters their behaviour and makes them more accessible to Ospreys than in natural waters (Kalvans and Bajinskis 2016). A similar situation occurs in Estonia, where it is estimated that 14–18 per cent of the Osprey population rely on carp farms (Tuvi and Väli 2006), as well as Germany (Müller *et al.* 2005) and Poland (Woźniak *et al.* 2022). Benthic feeders are also susceptible to predation when they are spawning in shallow water. This is the case at Rutland Water, where Bronze Bream *Abramis brama* and Silver Bream *Blicca bjoerkna* are frequently caught during this stage of their breeding cycle, but less commonly at other times of the year. There have also been instances of Tench being caught while they are spawning, including on one occasion when a breeding male caught seven in a single day (Mackrill *et al.* 2013) (Figure 2.13). The diet of Ospreys in different parts of Europe is shown in Table 2.1.

Figure 2.12. Introduced Rainbow Trout are frequently caught by Ospreys in Northern Europe (© Geoff Harries).

The species caught by Ospreys in freshwater habitats in North America differ from those in Europe, but similar trends are evident. Ospreys nesting near deeper, colder waters tend to rely on trout (Poole 1989). For example, Swenson (1978) found that Cutthroat Trout *Salmo clarki* constituted 93 per cent of the diet of Ospreys on Yellowstone Lake in Wyoming, with Ospreys tending to favour immature fish (25–35cm). Here Ospreys foraged over deep water in response to the habitat preference of the immature trout. In contrast, Ospreys breeding around warmer, shallow lakes, such as those in the western United States, tend to catch carp, White Sucker *Catostomus commersonii* and Brown Bullhead *Ameiurus nebulosus* (Van Daele and Van Daele 1982, Poole 1989).

Figure 2.13. Benthic feeders, such as Tench, are caught when they enter shallow water, such as when they are spawning (© John Wright).

Many breeding populations of Ospreys occur close to the coast, and here estuaries in particular offer rich foraging grounds. Marquiss *et al.* (2007) noted that Ospreys breeding at lower latitudes tend to be coastal breeders, taking advantage of warm sea surface temperatures and the related availability of surface-dwelling fish. In an extensive analysis of the literature, they demonstrated that coastal Ospreys were exclusively marine foragers where the mean Sea Surface Temperatures (SSTs) in June were between 18°C (Corsica) and 31°C (Farasan, Red Sea), and primarily marine foragers where SSTs were 17°C (Cape Province) and 13°C (southwest Scotland). Conversely, Ospreys were primarily freshwater foragers where June SST was 11°C.

Coastal Ospreys have learnt to be opportunistic, taking advantage of seasonal variations in fish abundance, and changing their hunting habits accordingly. In southern Scotland, for instance, Flounder, a favoured species of coastal Ospreys, are more difficult to catch in the spring because they spawn in deep water. As a result, Garfish *Belone belone* and Grey Mullet tend to be favoured early in the breeding season, with Flounder caught more frequently from June and Sea Bass from July (Marquiss *et al.* 2007). A similar pattern is evident in the colder waters of northern Scotland, where Sea Trout *Salmo trutta* morpha *trutta* are caught throughout in river estuaries and along the coast of Moray from spring onwards, with Flounder increasingly prevalent later in the breeding season (Dennis 2008). In West Wales, where SSTs are warmer, Grey Mullet are a favoured catch throughout the breeding season and constituted 47.4 per cent of the diet of a breeding pair at the Dyfi estuary between 2011 and 2014, with Sea Trout (22.6 per cent) and Flounder (19.8 per cent) the next most commonly caught species (Evans 2014). Grey Mullet and Sea Trout are particularly important when Ospreys first return in the spring, with the proportion of Flounder increasing through the season, as observed in other localities. The Flounder catch peaks in August when it may constitute close to 50 per cent of the

Table 2.1. Key prey species of Ospreys in European countries

Location	Key prey species and percentage of diet	Study
Finland (C)	Bream (63.7%), Roach (17.6%)	Häkkinen (1978)
Finland (SW)	Bream (39.6%), Rudd *Scardinius erythrophthalmus* (22.7%)	Häkkinen (1978)
Latvia	Common Carp (47.2%), Tench (14.3%), Prussian Carp (11.3%)	Kalvans and Bajinskis (2016)
England (C)	Rainbow Trout (35%), Roach (29%)	Mackrill *et al.* (2013)
Wales (C)	Grey Mullet (47.4%), Sea Trout *Salmo trutta* (22.6%), Flounder *Platichthys flesus* (19.8%)	Evans (2014)
Scotland (NE)	Northern Pike (64%) and Perch (36%)	Carss and Godfrey (1996)
Scotland (NE)	Rainbow Trout (37%), Roach (21%), Perch (18%), Brown Trout (13%), Pike (11%)	Carss and Brockie (1994)
Scotland (SW)	Flatfish Pleuronectidae spp. (36.8%), Grey Mullet *Chelon* spp./*Liza* spp. (14.2%), Brown Trout (10%)	Marquiss *et al.* (2007)
Germany (NE)	Bream (64.9%), Tench (20.5%)	Müller *et al.* (2005)
Belarus	Pike (29.2%), Bream (25.6%), Perch (12.4%)	Ivanovski 2012
Russia (NW)	Bream (28%), Blue Bream *Abramis ballerus* (24%), Roach (23%), Orfe *Leuciscus idus* (14%)	Babuskin and Kuznicov (2012)
Corsica	Grey Mullet Mugilidae (72.9%), Bream (12.8%), Saupe *Sarpa salpa* (10.8%)	Francour and Thibault (1996)

diet. In southern England, Grey Mullet is also the preferred catch of Ospreys on the South Coast, where the old English name for the Osprey was 'mullet hawk' (Dennis 2008).

As you travel further south through Europe, mullet (Mugilidae spp.) are an increasingly important food source for coastal Osprey populations, and they are a staple in the diet of breeding and wintering Ospreys in many regions of the world (Figure 2.14). Mugilidae are a family of ray-finned fish found worldwide in coastal temperate, subtropical and tropical waters (Nelson 2006) and consist of 20 genera and 79 species (Fricke *et al.* 2020). While mullet may reach a size of 120cm, they are typically around 30cm in length (Harrison 2002), and thus an optimum size for foraging Ospreys (Prevost 1982, Poole 1989). Most species of mullet are euryhaline and this allows them to inhabit a range of coastal waters, brackish lagoons, estuaries and fresh water (Harrison 2002). Mullet are opportunistic feeders: adults have frequently been shown to consume primary producers, including fresh or detrital plant material and microphytobenthos, but they may also feed on meiofauna (Lebreton *et al.* 2011). Many mullet migrate to littoral areas in the spring, where they use the tide to forage across mudflats, travelling along creeks and channels accessible at high tide (Laffaille *et al.* 2000). This explains why they are often a favoured prey item of European Ospreys when they first return to the breeding grounds each year. Furthermore, mullet frequently congregate in large shoals close to the surface (Harrison 2002). In some coastal areas the water seems to bubble and shimmer with the silvery, darting shapes of mullet as they break the surface. This makes them an easy target for hunting Ospreys.

The value of mullet is exemplified by the diet of Ospreys in Corsica, where the local breeding birds build their nests on spectacular sea cliffs. The fish assemblages around the coast of this Mediterranean island are rich and varied, but mullet are the only resident species that gather in shoals within range of hunting Ospreys. As such, it is no surprise that they were found to constitute 73 per cent of the Osprey's diet (Francour and Thibault 1996). The global nature of the importance of mullet to Ospreys is reflected in the fact that studies based at sites along the Gulf of California in Mexico (Cartron and Molles 2002) and the north coast of New South Wales in Australia (Clancy 2005) also found that mullet were a key prey species.

It is not only on the breeding grounds that Ospreys hunt mullet. Prevost (1982) studied the diet of Ospreys wintering and, in the case of second-calendar-year birds, summering along the coast of Senegal and The Gambia in West Africa. In these fish-rich waters, where Prevost identified a total of 54 fish species being taken by hunting Ospreys, mullet, along with Fimbriated Herring or Bonga Shad *Ethmalosa fimbriata* (previously referred to as *Ethmalosa dorsalis*) dominated the diet. Six different mullet species were taken – Flat-head Grey Mullet *Mugil cephalus*, Banana Mullet *Mugil bananensis*, White Mullet *Mugil curema*, Sicklefin Mullet *Liza falcipinnis*, Grooved Mullet *Chelon dumerili* and Largescaled Mullet *Parachelon grandisquamis* – with one or two of these species tending to dominate the diet at different sites. In shallow coastal tidewaters, mullet species made up 60 per cent of the diet between January and May, while at the mouth of the Senegal River the percentage rose to 85 per cent during the same period. Mullet also dominated in mangrove habitats, constituting 89 per cent of fish taken in open mangroves between January and May and 83 per cent between June and December. The only location where Prevost (1982) found that mullet was not the most important prey item was along areas of rocky shore, when Ospreys switched to other species. Ospreys also take advantage of seasonal changes in fish populations in West Africa. For example, Mediterranean Flying Fish *Cheilopogon heterurus* were taken while they were present between December and March on days when they were pushed closer to the shore by strong westerly winds (Prevost 1982).

Colleagues and I have been fortunate to study fishing Ospreys in both The Gambia and Senegal regularly since 2011 and our fieldwork indicates that mullet remain the favoured prey of

Figure 2.14. Mullet are a staple in the Osprey diet in many parts of the world, including along the coast of West Africa (© John Wright).

Ospreys in the region. Many of the Ospreys that winter in places such as the vast Sine-Saloum delta in Senegal roost in mangroves, and then fly out to sand islands at the mouth of the delta during the day. The water around these islands is often alive with mullet. Adult birds, who are well practised in the art of catching them, frequently make a successful catch within minutes of leaving their favoured perch. John Wright analysed 150 of his photographs of Ospreys carrying or eating fish to compare with Prevost's findings (Figure 2.15, Table 2.2). As expected, mullet comprised more than three-quarters of the fish in these photographs.

One notable change, however, was that John's photographs showed only a single Bonga Shad. This is a species exploited by artisanal fisheries along the coast of West Africa, but recent increases in industrial-scale fisheries by European and Chinese interests may be impacting stocks (chapter 9). Senegal Needlefish *Strongylura senegalensis* and other needlefish species now seem to be caught in greater numbers than during Prevost's study. These piscivorous species occur in coastal areas and swim near the surface, making them vulnerable to predation by Ospreys even though they are capable of moving very rapidly (Edwards *et al.* 2001). They were the second-most frequently caught group of species, after mullet.

The ability of Ospreys to adapt their diet to local availability of fish species is evident in Cape Verde, an isolated archipelago off the West African coast. The remains of 32 fish species were found in a study of the resident population of Ospreys that fish in the tropical coastal waters (Martins *et al.* 2011) – but diet was dominated by only six species, which have been rarely or never found in the diets of Ospreys elsewhere: the Pompano *Trachinotus ovatus* (24.1 per cent), Two-winged Flying Fish *Exocoetus volitans* (16.6 per cent), Atlantic Trumpet Fish *Aulostomus strigosus* (14.6 per cent), Mediterranean Parrotfish *Sparisoma cretense* (11.2 per cent), Madeiran Sardinella *Sardinella maderensis* (10.2 per cent) and Keel-jawed Needlefish *Tylosurus acus* (6.8 per cent). This reflects the speciality of the fish communities that occur here, with species composition varying spatially around the islands. For example, a study carried out by Rosiane Fortes, a student at Cabo Verde University, demonstrated that there are significant differences in the species caught on the wind-exposed northern and northeastern coasts of São Nicolau

Figure 2.15. Ospreys catch a range of fish off the coast of Gambia and Senegal (left to right, top: flying fish sp., White Mullet; second row: mullet sp., Largehead Hairtail; third row: Sompat Grunt, needlefish sp.; bottom row: needlefish sp., Tilapia.) (© John Wright).

Island, compared to the more sheltered southern and southwestern shores. Furthermore, higher prey abundance and diversity on the exposed coasts supports a larger number of breeding pairs in these areas. This general trend is replicated on other islands in the archipelago (L. Palma pers. comm. 2022).

While it is true that the fish species caught by Cape Verde Ospreys are local specialities, there are similarities in terms of ecology and morphology with species favoured by Ospreys fishing in other tropical or subtropical waters. The Blue-spotted Cornetfish *Fistularia commersonii*, for instance, a favoured prey item of Ospreys at Tiran Island (Safriel *et al.* 1985), situated

Table 2.2. Fish caught by Ospreys in The Gambia and Senegal, identified from photographs taken by John Wright

Fish (family, genus or species)	Percentage of diet
Mullet spp.	76.7%
Needlefish spp.	11.3%
Mediterranean Flying Fish	2.0%
Largehead Hairtail *Trichiurus lepturus*	1.3%
Sompat Grunt *Pomadasys jubelini*	1.3%
Grouper spp.	1.3%
Tilapia spp.	1.3%
Catfish spp.	1.3%
Bonga Shad	0.7%
unidentified	2.8%

at the mouth of the Gulf of Aqaba separating Egypt and Saudi Arabia, is a ray-finned member of the Syngnathiformes broadly similar to the Atlantic Trumpetfish. Likewise, needlefish are also consumed in both localities as well as in the Arabian Gulf (Beech 2003), and on the coast of West Africa as described above.

As in Europe, North American Ospreys exploit certain species depending on their seasonal availability. Along the south coast of New England, for example, Winter Flounder constitute approximately half of the fish caught by Ospreys, with herring (*Alosa* spp.) and Menhaden *Brevoortia tyrannus* accounting for 20 per cent each (Poole 1984). Ospreys appear to preferentially target the latter two species when they occur in large shoals close to the surface and, in the case of herring, as they migrate along shallow streams to freshwater spawning grounds. However, during periods when these fish are not available Ospreys switch back to Flounder (Poole 1984). Similarly, in eastern Nova Scotia in Canada, Ospreys catch Pollock *Pollachius virens* and herring when they move into estuaries each spring, but rely on Flounder later in the season when these species are not present (Greene *et al.* 1983).

Osprey diet also varies spatially in some locations. Chesapeake Bay, a vast estuary on the East Coast of the United States, with a shoreline of 13,000km, is thought to support the largest concentrations of breeding Ospreys anywhere in the world (Poole 2019). Glass and Watts (2009) found that the diet of local breeding Ospreys varied according to the salinity of the water. Catfish Ictaluridae spp. and Gizzard Shad *Dorosoma cepedianum* constituted 80 per cent of the diet of Ospreys inhabiting upper-estuarine sites where the salinity is lower, whereas Sea Trout *Cynoscion* spp., Atlantic Menhaden, Spot *Leiostomus xanthurus* and Atlantic Croaker *Micropogonias undulatus* (total 76 per cent) were the favoured species in lower-estuarine sites where salinity is higher. Glass and Watts (2009) suggest that the higher energy content of fish caught by Ospreys at the upper-estuary sites may be responsible for their faster population growth rates recorded in these areas.

Spatiotemporal differences in foraging behaviour

In recent years, our understanding of Osprey foraging behaviour has been greatly enhanced by developments in satellite-tracking telemetry. This has given new insights into the behaviour of breeding males while they are provisioning young and of migrant Ospreys at stop-over and wintering sites. Male Ospreys are normally the sole providers for their family during the breeding season, and so build up a detailed knowledge of all potential fishing sites in their usual home range. Most Osprey nests tend to be within a few kilometres of water, but telemetry studies have demonstrated that males will sometimes fly much further in search of a meal.

Ospreys breeding near the Moray coast in northern Scotland are known to take a range of freshwater and saltwater fish, and satellite tracking by Roy Dennis has enabled the behaviour of individual males to be studied in detail. Analysis of these data demonstrates that males typically have a range of favoured fishing localities, and that these sites are not always the closest waterbodies to their nests. For example, two males, Talisman and Nimrod, which reared chicks at neighbouring nests situated just 1.4km apart in 2010, had a broadly similar pattern of freshwater and marine foraging but their choice of foraging localities differed significantly. The satellite data showed that both birds tended to favour freshwater fish during April and May. Talisman fished in trout lochs within 2km of his nest, whereas Nimrod made daily visits to a series of small lochs 10km to the south-west. Nimrod only visited the lochs favoured by Talisman once during this period, despite the fact they were just 2km from his own nest. During May, Talisman began making frequent trips to lochs 10.5km to the east, and regularly visited Findhorn Bay on the Moray coast, a flight of 10km from his nest. Nimrod began to make much more frequent visits to the coast from late May onwards. Like Talisman, he also visited Findhorn Bay, but he usually fished a 15km stretch of coastline between Findhorn and Nairn Bar, sometimes as frequently as three times per day. The satellite data reveal that when fishing at the coast Nimrod usually remained within 125 metres of the shore, flying at an average altitude of 52 metres (range 22–162 m). These foraging flights were a minimum 23km round-trip for Nimrod. In contrast, Talisman rarely fished along the coast in this way, instead concentrating on the shallow tidal waters of Findhorn Bay. Such differences seem to be due to the preferences of the individual birds. Nevertheless, it was clear that marine fish constituted a key part of the diet while they were provisioning young in June until they departed in September. Both birds made daily visits to the coast, while also catching freshwater fish in their favoured haunts on a frequent basis. During September, the two birds appeared to feed exclusively on marine fish until their departure (on 12th and 23rd respectively). Such seasonal changes in diet have also been recorded in North America. For example, Sandford, an adult male from Massachusetts, fished inland fresh waters in May and saltwater coasts and estuaries in June (Poole 2019).

Over the course of the summer Nimrod covered a total of 344km^2, while the corresponding figure for Talisman was 322km^2. Their core home ranges, encompassing all favoured fishing locations were considerably smaller, covering 100km^2 and 80km^2 respectively. However, these core home ranges were significantly larger than another bird satellite-tagged by Roy Dennis. Red 8T, a male breeding 50km south-west in landlocked Strathspey, had a core home range of just 27km^2 during the same summer. This male's favoured fishing sites were a loch situated 1.25km from the nest and Rothiemurchus Fishery, a flight of 3km. This latter location became particularly important from mid-July onwards, while he was provisioning two chicks.

Satellite tracking has also shown that, on occasion, breeding males make surprisingly long journeys for fish. For instance, AW(06), a five-year-old male that we satellite-tagged at a nest close to Rutland Water in 2011, made 50km round-trips to reservoirs in Northamptonshire at least once every week while he was provisioning three young (Mackrill et al. 2013). This was despite his nest being just over 2km from his favoured fishing site. It seems likely that in this case AW was visiting sites he had been familiar with before settling to breed. The energetic demands of flying that distance back to the nest with a fish would have been considerable.

Breeding males have high energy requirements during the breeding season, and will sometimes eat a fish away from the nest, before catching another to take back to their waiting family. Studies have shown that, where possible, Ospreys favour fish that are high in lipids and have few bones because they offer considerably more calorific value (Prevost 1982). This is one of the reasons mullet is a favoured catch of Ospreys around the globe. Fish that are high in lipids also help females to get into the best possible breeding condition in the spring. Poole (2019) calculated that hunting time required by breeding male Ospreys in southern New England in order to meet the energetic requirements of themselves and their brood of three young and a mate were six to seven times greater than the wintering males studied by Prevost (1982) in Senegal. Other factors come into play too. As described earlier, Machmer and Ydenberg (1990) found that when wind speed exceeded 7m/s the energetic costs of male Ospreys catching fish in British Colombia actually exceeded energetic gain.

Social interactions at foraging sites

At relatively large sites, such as Findhorn Bay, it is common to see several Ospreys fishing simultaneously, particularly in late summer when large numbers of flounder are caught on a rising or falling tide. At smaller sites, however, individual birds can become highly protective and territorial. The Osprey population in the Rutland Water area increased from a single pair in 2001 to 11 pairs in 2022, following a successful translocation project. Over the intervening years a series of different male Ospreys have become dominant at the local trout farm described earlier. The first bird, 03(97), held fort each year until he failed to return in 2016. That gave a rival male, 28(10) – who had begun breeding at a site nearby – the opportunity to visit more frequently. Sensing the value of the site, 28(10) would often return to the fish farm immediately after delivering a fish to the nest, in order to perch in a prominent location and chase off rival birds that attempted to fish there. Similar behaviour has also been recorded at Rothiemurchus Fishery in Scotland. There was one particular unpaired Osprey that spent most of the summer around the fishery when I was undertaking the fieldwork for my undergraduate dissertation on Osprey foraging. It would often chase rival birds every time they attempted to fish.

The presence of other hunting Ospreys can sometimes be an advantage, however. Flemming et al. (1992) studied fishing Ospreys in Antigonish estuary in Nova Scotia, Canada where Winter Flounder constituted 90 per cent of the diet. Males of 29 breeding pairs located within 6km of the study site regularly foraged in the estuary. Ospreys foraging in the shallow estuary typically hovered for between one and three minutes before making a dive. Hunting in this manner is a highly visible behaviour, and hovering birds were frequently joined by conspecifics, to form 'flocks' within which two or more birds remained within 200 metres of each other. Individuals often hovered within 25 metres of two or three others for up to 10 minutes in this way (Flemming et al. 1992). This behaviour had clear advantages, because the mean search time per dive of birds foraging in flocks of 2–9 individuals (114.1s) was significantly shorter than birds hunting on their own (163.6s) and may have been particularly beneficial when environment conditions made fish more difficult to locate.

One of the benefits of breeding close to other Ospreys is that individual birds can take cues from neighbours. Greene (1987), who also studied colonially nesting Ospreys on the coast of Nova Scotia, found that the sight of neighbouring birds returning to a colony with certain species of shoaling fish, such as Herring and Pollock, prompted breeding males to fly off in the direction from which the neighbour had arrived. Flounder did not elicit the same response, indicating that the prospect of locating shoals of fish may have been the main motivation. Furthermore, Flemming *et al.* (1991) found that Ospreys breeding in colonies on utility poles in coastal areas of Nova Scotia raised more young than solitary nesters. The colonially nesting birds bred a mean 0.7km from their neighbours, usually on utility poles, and had unobstructed views of 4–9 other nests. These nests were a mean 1.7km from the nearest estuary. Solitary nesters, meanwhile, were located an average of 6.7km from other breeding pairs, and those near the coast were situated a mean 2.4km from the closest estuary. Coastal nesters were more successful in terms of fledged young per occupied nest than birds breeding inland, but it was notable that the most successful birds of all were those nesting in colonies at the coast (1.53 per occupied nest). Flemming *et al.* (1991) suggest that this may have been due to enhanced opportunities for social foraging which, as already described, may increase capture rates.

Conclusions

- The Osprey is an exclusively piscivorous species.
- Fish are caught by plunge diving, usually from a height of approximately 10–40 metres, after hovering or gliding above the water. Hunting Ospreys may also dive from a perch.
- The angle of dive may be adjusted according to the type of fish the Osprey is targeting, or the depth of the water.
- Ospreys have long legs and highly adapted talons to grasp fish swimming on or just below the surface of the water. They can switch to a zygodactyl arrangement of their toes – where digits II and III face forward and digits I and IV backwards – to provide greater manoeuvrability in the foot. Fish are always carried away headfirst to improve aerodynamics.
- Dive success is influenced by a range of factors including age, weather conditions, size of prey, tidal stage and the intended prey species.
- Foraging flights can vary from a few minutes to over an hour. There is usually a peak of foraging activity soon after dawn and another during late afternoon/early evening.
- Ospreys are opportunistic hunters, and will take a variety of live fish from a range of different saltwater and freshwater habitats, usually favouring the most abundant and readily caught species. They catch benthic feeders when they are close to the surface of the water, and also faster-moving piscivorous species. Ospreys breeding at northern latitudes may take freshwater fish in lakes during the summer and then switch to exclusively saltwater fish in winter.
- Male Osprey provision their mate and offspring during the breeding season and regularly travel more than 10km to favoured foraging sites.
- Some birds can become highly territorial over specific foraging sites, but in other situations the presence of other foraging Ospreys can help individuals to locate fish.

CHAPTER 3
Dispersal and settlement patterns

One of the most evocative sounds to be heard anywhere that has breeding Ospreys is the *eep eep eep* call of a displaying male. Sometimes the call is barely audible as the male sky dances high above. The displaying bird advertises its presence to passing females or rival males by dangling his legs, or grasping a fish, and then rising and falling like a rollercoaster, while uttering the distinctive high-pitched call. Trying to locate a displaying male way up in the sky can be a surprisingly difficult task. Often they are just a mere speck, such is the altitude that they are flying, but it is well worth persisting. Watching a male Osprey performing this spectacular aerial display against a cloudless blue backdrop is a part of the Osprey summer of which I never tire (Figure 3.1).

Establishing a territory and finding a mate is not an easy task for young Ospreys attempting to enter the breeding population, but it has profound implications for lifetime fitness and reproductive success. These individual choices also have an impact at a population level, helping to shape settlement patterns and, consequently, the geographic distribution of the species.

In ecology, there are two key types of dispersal. Natal dispersal is the movement of an individual from the place it hatched to the site of its first breeding; breeding dispersal is the

Figure 3.1. Males often display while carrying a fish. Here, a Rutland Water male is carrying a Roach while giving the distinctive *eep eep eep* call (© John Wright).

movement from the first to any subsequent breeding site (Greenwood and Harvey 1982). Dispersal serves to minimise competition for resources and to avoid inbreeding (Gandon and Michalakis 2001), and influences both the distribution and abundance of a species (Johnson and Gaines 1990). There are selective pressures against doing so, however. Breeding in new areas involves a loss of familiarity with resources and conspecifics and, in some cases, predators (Clobert et al. 2001).

Natal dispersal is usually most pronounced in migrant species (Paradis et al. 1998, Newton 2010) because migratory individuals are able to prospect for breeding sites during migration and pre-breeding movements (Weatherhead and Forbes 1994). Nevertheless, some migrants display surprisingly high levels of natal philopatry (i.e. where individuals return to breed in the same area they were born). In a study of colonially nesting Lesser Kestrels *Falco naumanni* in southern Spain, 57 per cent of returning birds bred in their natal colony (Negro et al. 1997). However, the actions of a few individuals who move long distances have since been shown to be responsible for the colonisation of unoccupied patches and the founding of new metapopulations (Serrano and Tella 2012).

The Osprey is a species that lends itself to the study of dispersal and settlement patterns because it is migratory across much of its breeding range and often chooses to nest in prominent locations. Studies have provided fascinating insights into the movements of colour-ringed or satellite-tagged birds, and the associated impact on metapopulation dynamics and the geographical distribution of the species.

The simplest and most cost-effective way of monitoring dispersal is by marking individual birds. Ospreys have been colour-ringed in the UK since the 1960s in a long-running scheme organised by Roy Dennis. Young Ospreys are fitted with darvic (a type of PVC) rings with a unique alpha-numeric code as nestlings, and this enables individuals to be identified in future years. A total of 4,980 Ospreys were ringed in the UK between 1967 and 2020 (Robinson et al. 2021), and a large proportion of these birds were fitted with colour rings in addition to the standard British Trust for Ornithology (BTO) metal ring. Similar research has been undertaken elsewhere in the Osprey's range, aided by recent advances in digital photography and high-definition nest cameras which have greatly increased the number of sightings of known birds. As such, both the natal and breeding dispersal of Ospreys is well understood.

Natal dispersal

The long-term colour-ringing study in the UK, and research across much of the Osprey's range, has demonstrated that males generally show a high degree of natal philopatry, whereas females frequently disperse to join other populations away from their natal area. 59 per cent of 29 colour-ringed Scottish males returned to breed within 25km of their natal nest, compared to just 6 per cent of 34 females. Furthermore, 71 per cent of females bred 50km or more from their natal nest, compared to just 14 per cent of the males (Dennis 2008). In addition, at least one ringed female from Sweden and two females from Norway have joined the Scottish breeding population over the years. It is rarer for males to join other breeding populations in this way, but a breeding male at a nest Strathspey in 1963 was ringed as a chick in a nest near Stockholm (Dennis 2008).

Similarly, Saurola (2005) reported that the natal dispersal of males (median = 27km, maximum = 433km, n = 38) ringed in Finland and subsequently recaptured as breeders was considerably less than females (median = 133km, maximum = 534km, n = 34). Three females and one male ringed as nestlings in Sweden have been recaptured as breeders in Finland, 380–480km away from their natal sites (Saurola 2005). Likewise, immigration of German females has been key in the expansion of the Osprey population in central France, with some birds dispersing >900km from their natal site (Schmidt and Wahl 2001, Wahl and Barbraud 2014).

Sex-based differences in natal dispersal are also evident in United States Ospreys. Martell et al. (2002) found that the mean natal dispersal of male Ospreys in Minnesota, where a population was re-established by translocation, was 27km (range 1–65km) for males and 384km for females (range 8–1,075km). Meanwhile in New England, more than twice as many males nested within 10km of their natal site than females, and all males settled within 50km. In contrast, more than 20% of females dispersed 50km or more from their natal site (Poole 1989).

While at a population level it is clear that males show greater natal philopatry than females, we are beginning to understand more about the factors underlying the choices that individual birds make. It is likely that a range of things influence where Ospreys settle to breed for the first time. The availability of nest-sites, for one, plays a central role. In New England in the United States for example, Spitzer et al. (1983) found that only two out of every 10 males dispersed more than 10km from their natal site, and none dispersed further than 50km. At the time of this study there was an abundance of nesting opportunities in New England, boosted by the provision of artificial nests. Thus, it was a relatively simple task for young birds to establish a nest-site of their own (Poole 1989). Young birds that fledge from nests in established colonies may be forced to disperse further, due to a lack of local breeding opportunities (Dennis 2008).

A recognition of landscape patterns may also be a key contributory factor. The German Osprey population has increased significantly in recent decades and now stands at over 750 breeding pairs (D. Schmidt pers. comm. 2022). In a study of nest-site selection between 1995 and 2006, Bai et al. (2009) noted a gradual shift from forest-dominated to agricultural-dominated landscapes. The traditional explanation would be that the preferred patches – within forests – were mostly occupied, forcing later settlers to the edge. However, it is possible that the population was undergoing a reshaping of habitat selection with new preferences being formed, particularly as nests surrounded by more agricultural land had higher breeding success; this was probably due to more efficient foraging in eutrophic lakes. If nest-site selection of Ospreys was influenced by the recognition of landscape patterns in the nestling stage, the preference for open landscapes would spread within the population.

Another key driver in determining where young birds settle to breed is the presence of other Ospreys. In some parts of the world Ospreys are highly communal, to such an extent that in Florida there can be more than one nest on the same tree (Poole 2019). Even in Europe, where the population density is usually lower than in North America, conspecific attraction appears to be important in shaping settlement patterns. This is particularly evident in new and expanding populations where it is sometimes possible to analyse the decisions of almost every individual within the population (Morandini et al. 2019).

In recent decades Ospreys have been successfully restored to Spain and England through a combination of translocations and artificial nest building. In Spain, breeding Ospreys have been re-established in Andalucía through a successful reintroduction project. A total of 191 birds were translocated from Germany, Scotland and Finland over the course of 10 summers, beginning in 2003, and released at two sites, 125km apart: Barbate Reservoir in the south, and Odiel Marshes, close to the city of Huelva (Muriel et al. 2010b, Morandini et al. 2019). By 2016 the breeding population had reached 23 territorial pairs, with natal dispersal distances calculated for 23 ringed breeding birds. All 13 ringed males returned to breed in their natal colony, but there was greater dispersal of females. Only three of 10 ringed females bred in their natal area, with four females from Cádiz moving to Huelva, one female from Huelva found breeding in Cádiz and two females each from Morocco and the Balearic Islands recorded breeding in

Cádiz and Huelva. Overall, females showed significantly higher natal dispersal distances than males, with median values for natal dispersal distance 9.6km in males, and 183.6km in females (Morandini *et al.* 2019).

LESSONS FROM ENGLAND AND WALES – UNDERSTANDING THE PROCESS OF NATAL DISPERSAL

In the UK, the Osprey population is synonymous with the Scottish Highlands, but in recent decades Ospreys have expanded south into England and Wales through a combination of natural recolonisation and translocations. The vast majority of birds in these small but expanding populations are colour-ringed and all known breeding sites are monitored closely. As in Spain, natal dispersal of all colour-ringed breeders is known, and recent advances in digital photography and high-definition nest cameras, coupled with a significant interest in Ospreys in the UK, means that many individuals are identified when they return north for the first time, usually during their third calendar year. This has made it possible to monitor the movements of known individuals before they settle to breed, and helps shed new light on the factors underlying natal dispersal.

Young Ospreys in migratory populations usually fly north to potential breeding areas in the spring of their third calendar year and typically begin arriving during May (Dennis 2008). It is very unusual for Ospreys to breed at two years of age, and so this first journey north is more a reconnaissance mission, helping young birds to learn the migration and to scope potential breeding sites. Most Ospreys do not breed until they are at least three years of age, and some birds have to wait considerably longer (Dennis 2008, Mackrill *et al.* 2013), but it is now clear that the process of entering the breeding population usually begins at two years of age when young birds return to potential breeding areas for the first time.

Sightings of colour-ringed birds, and satellite-tracking data, have shown that young Ospreys wander widely when they first return as two-year-olds. It seems that one of the key purposes of these explorations is to search for potential breeding sites, usually in areas that already hold breeding Ospreys. One of the most striking sounds that you might hear at an Osprey nest is the repeated *tioop* guard call uttered by breeding birds when an intruding Osprey appears. As described in chapter 1, this call acts as a warning, growing in volume and intensity as the intruder approaches. Returning two- and three-year-old Ospreys can be highly inquisitive and often very bold, sometimes attempting to land on or close to active nests, particularly if the breeding male is away fishing. This provides an excellent opportunity to identify returning young birds, particularly if the nest is equipped with a camera. These intrusions are usually no more than a nuisance to the breeding pair, but they act as a valuable learning experience for young birds. The long-term colour-ringing of Ospreys in the UK has revealed that, given the opportunity, first-time breeders will preferentially take over an established nest, rather than build their own from scratch (Dennis 2008, Mackrill *et al.* 2013), and so these early explorations help young birds to map out existing breeding sites. Sightings of colour-ringed individuals have demonstrated that young birds often range over a very large area as two-year-olds and that eventual settlement patterns are greatly influenced by these explorations.

The extent to which young birds wander, and the subsequent influence this has on natal dispersal, is exemplified by the movements of young birds from the Rutland Water population, which was established by the translocation of 64 Scottish Ospreys to the reservoir in the East Midlands of England between 1996 and 2001, and a further 11 birds in 2005 (Mackrill *et al.* 2013). All juveniles – both translocated and wild-fledged – have been colour-ringed and subsequent re-sightings of these birds have facilitated detailed analysis of the explorations and subsequent natal dispersal of different individuals. Of particular interest has been the movement of young birds between central England and North and West Wales. Dispersal and metapopulation connectivity are key elements of population dynamics, and this has been exemplified by the movements of young birds between these two areas, which lie approximately 200km from each other.

Ospreys had been absent from England for over 150 years prior to the Rutland Water translocation project (Mackrill *et al.* 2013), and from Wales for more than four centuries (Evans 2014). The first pair bred successfully at Rutland Water in 2001, and then, unexpectedly, Ospreys were discovered breeding at two sites in Wales in 2004. Surprisingly, the males at both Welsh sites were identified as birds that had been translocated to Rutland Water. First, 07(97) – 07 being the inscription in the colour ring, and 1997 the year of release – was breeding at a site near Welshpool in Mid Wales with a colour-ringed female from the Black Isle in northern Scotland; while a second male, 11(98), had paired with an unringed female in the beautiful Glaslyn Valley in North Wales (Figure 3.2). Neither of the two Rutland males had been seen in the East Midlands since their first migration, and the fact they had settled to breed 167km and 232km respectively from their natal site (in this case, the release site) was surprising. It was notable,

Figure 3.2. Translocated male 11(98) was found breeding in North Wales with an unringed female in 2004; the two birds are shown here in John Wright's illustrations (© John Wright).

however, that they had returned to the same latitude as Rutland Water, rather than continuing north to Scotland.

While most young male Ospreys settle to breed close to their natal site, those that disperse further often have a significant influence on the distribution of the species. Studies in Scotland have shown that these pioneering birds can result in new colonies becoming established if they are joined by other wandering Ospreys (Dennis 2008), and this proved to be the case in Wales. The Welshpool pair failed to return after raising a single chick in 2004, but 11(98) and his unringed mate raised a total of 26 chicks to fledging in a further 10 years of continuous breeding at the site. By 2014, the last year that 11(98) bred at Glaslyn, the Welsh population had increased to four pairs, and that summer 10 young fledged successfully.

Following the re-establishment of Ospreys in Wales, further birds from Rutland Water have joined the population, but unlike in 2004, all have been female. While the pre-breeding movements of the two translocated males were not known, multiple sightings of several females in the years and months before they bred for the first time has demonstrated the range and extent of exploratory flights and their significance in terms of eventual natal dispersal. The behaviour of these individuals corroborates the notion that the probability of an individual settling to breed in its natal population is greatly influenced by the availability of potential mates with suitable territories (Dale 2001; Lenz et al. 2007). A good example is 12(10), later named Glesni, a female that fledged from a nest in Rutland in 2010. The first sighting of the young female on her return to the UK was on 21st May 2012, when she was identified intruding at a nest on the Montgomeryshire Wildlife Trust's Cors Dyfi nature reserve in Mid Wales (Figure 3.3). The breeding female at the site on the Dyfi estuary was another Rutland female, 03(08), who had paired with an unringed male, known as Monty, in 2011. Glesni was quickly seen off by the breeding birds, and there were no further sightings for two months. Eventually, she was photographed intruding at the Manton Bay nest at Rutland Water, close to her own natal nest, on 23rd July. It was not clear how long she had been present in Rutland at that stage, but six days later she was seen at Cors Dyfi again. She became a regular visitor thereafter, even perching on the nest with the newly fledged juveniles on one occasion during August (Figure 3.3) (Evans 2014).

The dispersal capabilities of young Ospreys are exemplified by the fact that both adults and juveniles regularly travel in excess of 250km per day during migration (Mackrill 2017). When viewed in this context, it is clear that exploratory flights between nests in geographically isolated metapopulations are easily achievable, and provide vital knowledge of potential breeding opportunities. This was illustrated well the following spring – 2013 – when Glesni initially returned to Rutland but, finding an absence of unpaired males, moved west to Wales. The breeding female at Cors Dyfi, who coincidentally was Glesni's aunt, had failed to return, and Glesni ousted a rival unringed female after a ferocious three-day battle and paired with the established male (Evans 2014). She subsequently went on to rear two chicks that year, and a total of 12 chicks at the site between 2013 and 2017.

Glesni herself failed to return to Cors Dyfi in 2018 and, remarkably, her place was then taken by a third Rutland female, her cousin, 3J(13). 3J(13) had been seen in both Rutland and Wales when she first returned, during summer 2015, and again the following spring. The ease with which Ospreys travel between the two areas was exemplified well on 19th April that year, when John Wright observed 3J(13) leaving a nest on Lagoon 4 at Rutland Water where she had been courted by a five-year-old male. At 3pm she took off from the nest, circled to gain height and then headed purposefully west. At 8:50pm that evening she landed on the Glaslyn nest in North Wales, having flown 231km in less than six hours. Finding that nest occupied, she returned to the East Midlands the next day and then laid eggs in the nest on Lagoon 4 on 5th May. However, she and her equally inexperienced mate were ousted by a pair of Egyptian Geese *Alopochen*

Figure 3.3.a. Female 12(10), or Glesni as she later became known, perched on the Cors Dyfi nest with the single juvenile that fledged that year (© Emyr Evans).

Figure 3.3.b. Glesni was subsequently chased off by the resident female 03(08), Nora (front), herself also a Rutland Water bird. Glesni replaced Nora as the breeding female at the site in 2013 (© Emyr Evans).

aegyptiaca soon afterwards. Following the failure, 3J(13) headed west once again and was seen intruding at the Cors Dyfi nest on 14th May. She became a regular sight in Wales thereafter and, in April 2018, when Glesni failed to return, 3J(13) became the new breeding female at Cors Dyfi. Her persistence, coupled with extensive knowledge of established nests, had enabled her to become the occupier of an exceptional breeding site.

Two other birds, 24(10) and 5F(12), completed a quintet of Rutland-fledged females that bred successfully in Wales between 2011 and 2020. This represented 42 per cent of Rutland wild-fledged females that were known to have returned to breed in the UK at the time. Seven other females had settled in the Rutland Water area, and bred a mean 11km from their natal nest. This neatly demonstrates how the presence of other breeding Ospreys and, in particular, the availability of mates are key drivers in determining the dispersal and settlement patterns of females (Dale 2001; Lenz et al. 2007). In subsequent years, CJ7(15) became the first female to breed on the South Coast of England for two centuries when she paired with a translocated male, as described in chapter 8, while two other females settled in northwest England and the Scottish Borders.

As the examples of the two translocated males that bred in Wales indicate, males have the potential to have a greater impact on the settlement pattern of Ospreys because they build nests, or take over vacant old ones. Artificial nests have been used with good effect to encourage Ospreys to spread to new areas, but the two males that bred in Wales in 2004 constructed nests for themselves. Such pioneering individuals are relatively rare, however, and males who breed away from their natal site usually do so in another Osprey colony, as exemplified by another colour-ringed bird from Rutland Water. S2(15) (Figure 3.4) fledged from the famous Manton Bay nest at Rutland Water in 2015. Two years later he was photographed just south of Zwolle in the Netherlands on 26th May. Subsequent sightings in late August in northeast Belgium indicated that he had spent the rest of the summer in the Low Countries. That winter, S2(15) was photographed at his wintering site at the Sine-Saloum delta in Senegal by visiting Osprey enthusiast Alison Copland. We wondered if he would return to the UK following Alison's exciting discovery, but instead the young male went back to the Netherlands again, where he was seen for the first time on 1st May 2018 at the Biesbosch, a vast freshwater delta of 8,000 hectares where Ospreys began breeding in 2016. He was subsequently sighted at various locations around the Netherlands and, in 2019, built a nest on an electricity pylon in the Biesbosch, where he was joined by a ringed female from Germany, but did not breed. Although that nest was taken over by Peregrines prior to the Ospreys' return in 2020, they moved to a dead tree that S2(15) had first shown interest in the previous summer, and went on to breed successfully, raising three chicks (Figure 3.4). The nest was 387km from S2's natal site at Rutland Water and 550km from the nest where the female has been reared in Mecklenburg Vorpommern in northern Germany in 2016. S2(15) was the first Rutland male to breed elsewhere. The average natal dispersal of 19 other wild-fledged males that fledged from local nests between 2004 and 2016 was just 12km (maximum 40km). The fact that Ospreys were already present and breeding in the Biesbosch was likely a key determining factor in S2(15) settling there. He would have encountered these birds during his explorations since summer 2017, and just as the females in Wales, used this knowledge to good effect. S2's sibling from 2015, S1, attempted to breed for the first time in 2020, but he nested much closer to home – just 7km from his natal nest. This male had spent time at Poole Harbour in Dorset during his first summer back in the UK, but was eventually drawn home to Rutland Water, as is the case with most male Ospreys.

A similar scenario has unfolded in Kielder Forest in Northumberland, where Ospreys returned to breed in 2009. The original pair were unringed and were probably birds that had moved across the border from southern Scotland. However, the breeding males at two newly

Figure 3.4. Males tend to show greater natal philopatry than females, but S2(15) from Rutland Water bred successfully at De Biesbosch National Park in the Netherlands for the first time in 2020 (© Rob Braat).

occupied sites in 2011 were identified as brothers from the Glaslyn nest in North Wales. Yellow 37 fledged in 2005 – the first year that Rutland-translocated male 11(98) and his unringed mate bred successfully – and White YA hatched in the same nest two years later. It is 268km from Glaslyn to Kielder, a very similar to distance to the 232km their father had dispersed from Rutland Water to settle in North Wales. Another of 11(98)'s male offspring, Black 80, who fledged in 2006, bred for the first time at Threave Castle in Dumfries and Galloway, southern Scotland in 2009, 220km north of his natal nest, and two further males from the same nest have bred in Cumbria. A 2008 male, White YC, bred successfully for the first time at Roudsea Wood in the South Lakes in Cumbria in 2014, 160km north-east, and Blue 9C, a male that hatched in 2014 – the last year that 11(98) bred at Glaslyn – bred at another site in Cumbria, 162km from his natal nest, in 2018. A sixth of 11(98)'s offspring, White 91, a female that fledged in 2009, bred in Perthshire, Scotland for the first time in 2014. The fact that all of these birds dispersed away, particularly as five were male, is highly unusual, especially considering that none of 11(98)'s offspring returned to breed in Wales. One wonders whether the fact their father (and, presumably, mother) behaved in a similar fashion is merely coincidental or betrays some underlying genetic urge to disperse greater distances than normal.

Satellite tracking – adding detail to our knowledge of natal dispersal

The colour-ringing studies in England and Wales have revealed much about the extent to which young birds wander before they settle to breed, but satellite tracking provides even greater detail on day-to-day movements. A young male satellite-tagged by Roy Dennis in northern Scotland exemplified this well, and supplied some of the earliest evidence of how far young birds range

Dispersal and settlement patterns

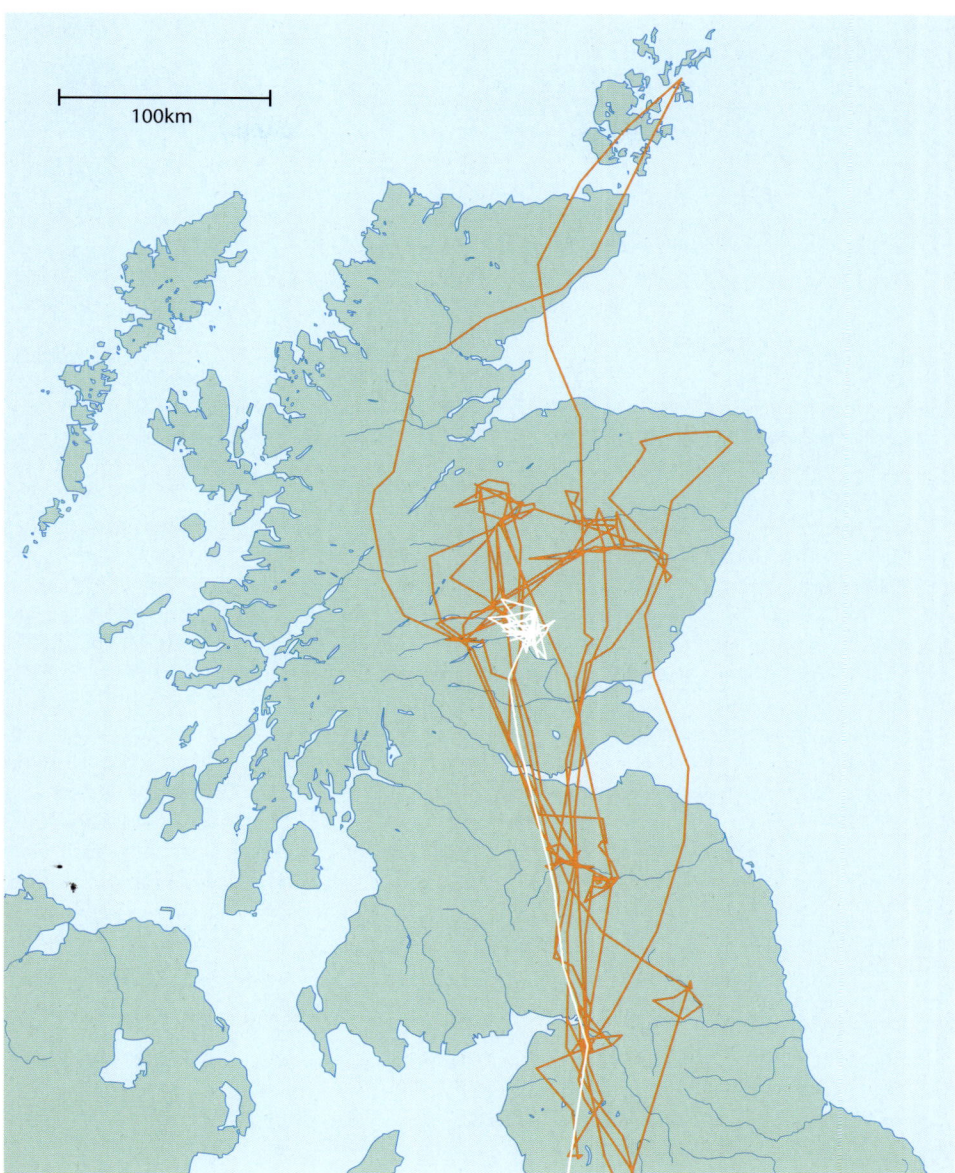

Figure 3.5. Satellite-tagged third-calendar-year male, Rothiemurchus, ranged extensively in Scotland and northern England during his first summer back in the UK (orange track), but had a much smaller summer range by his fourth summer back in Scotland (white track) (data courtesy of Roy Dennis).

when they first return. Rothiemurchus was tagged by Roy as a juvenile male at a nest on the eponymous Rothiemurchus estate in Strathspey in August 2009, as described on the Roy Dennis Wildlife Foundation website (www.roydennis.org/rothiemurchus). He subsequently migrated to Senegal and established a wintering site in a tributary of the River Gambia. The

Table 3.1. Summer home-range size of Rothiemurchus 2011–2014, based on 95 per cent utilisation distribution (UD) and 50% UD kernel density analysis

Year	Summer range: 95% UD (km²)	Core summer range: 50% UD (km²)
2011	55,004.74	11,768.06
2012	16,891.63	1,836.09
2013	5,376.34	797.04
2014	241.49	13.17

young male returned north to the UK for the first time in spring 2011, departing his wintering site on 9th May and finally arriving in Scotland on 2nd June. In a sign of the extent of explorations to come, he flew 400km north to Orkney and then back to the mainland in Caithness on 3rd June. He then spent a week exploring the Scottish Highlands to the west and south of his natal area, before heading further south and then crossing the English border into Northumberland and then Cumbria.

On 14th June, Rothiemurchus returned north again, flying 266km to Aberdeenshire and then a few days later to Loch Tay in Perthshire. Finally, on 23rd June, he overflew his natal nest, and other breeding Ospreys in Strathspey. He did not linger in the area for long though, returning south to the Scottish Borders on the 25th. However, on 1st July he headed back north once more, and again visited his natal nest that evening. He remained in Strathspey for a week, before returning to the Scottish Borders once more, again via Loch Tay. In what was becoming a familiar pattern, Rothiemurchus flew north to Strathspey again on 13th July and this time lingered four days, before flying 220km back to the Borders on the 18th. He then continued further south across the English border, to Dalston in Cumbria. After four days in England, the young male headed north to Strathspey once more, flying over his natal nest and other local breeding Ospreys again on 29th July and then heading east to Aberdeenshire on 1st August. He moved south again on 5th August, flying over the centre of Edinburgh and then onwards to Peebles in the Scottish Borders. Next day he returned to Dalston in Cumbria, the most southerly of his favoured haunts. He remained in Cumbria for another month, before setting off on his autumn migration on 7th September.

Rothiemurchus's explorations over the course of the summer had offered a unique insight into the wanderings of two-year-old birds when they first return. Far from being random, his explorations over an area of some 20,000km² were helping him to map out the locations of other breeding Ospreys. He made at least four visits to his natal site, but also visited breeding Ospreys in Caithness, Aberdeenshire, Perthshire, the Scottish Borders, Northumberland and Cumbria. Satellite tracking in North America has revealed similar patterns of behaviour among two-year-old birds (Poole 2019).

Rothiemurchus's transmitter continued to provide data for a further three summers. Although he did not breed during that period, his summer range became smaller each year and in 2014 it was clear he had settled in Perthshire in an area he had first visited in June 2011. The gradual reduction in his summer range is shown in Table 3.1 and Figure 3.5. An analysis using fixed kernel density contours and least-square cross-validation factors demonstrated that Rothiemurchus's 95% utilisation distribution (UD) was 55,004.74km², with a core area (50% UD) of 11,768.06² in 2011, whereas by summer 2014 his 95% UD had declined to just 241.49km² and 50% UD to 13.17km².

Age at first breeding

Ospreys reach breeding age at three years, although in some exceptional cases they can rear young as two-year-olds. First-time breeders are usually able to nest at an earlier age in expanding colonies where there is reduced competition for nest-sites and mates (Poole 1989). For example, in the expanding population in New England in the United States described earlier, most young Ospreys began breeding at either three or four years, with a mean age of 3.6 years (Poole 1984). Similarly, at Rutland Water, the age at first breeding of known colour-ringed females (n = 13) during the first 20 years of establishment (2001–2021) was 3.2 years, with one exceptional female rearing two chicks at just two years of age in 2003 (Mackrill *et al*. 2013). The corresponding figure for males was higher (5.9 years, n = 22) during this period, but this was due to the fact that 49 of the original 64 juvenile Ospreys translocated to Rutland Water were males, meaning competition among returning males for mates (but not nest-sites) was high. One translocated male did not breed until it was 14 years old, and another bred for the first time aged 10 years (Mackrill *et al*. 2013). This is discussed in more detail in chapter 8.

Reduced age at first breeding has been recorded following other Osprey reintroductions. Between 1984 and 1995, 143 juvenile Ospreys were released in east-central Minnesota in the United States. Subsequently, 306 nesting attempts were recorded between 1986 and 2005 from which 472 chicks fledged (Englund and Greene 2008). Age at first known successful breeding was 3.36 years for females (n = 11) and 4.39 for males (n = 38). Furthermore, two two-year-old males fledged chicks successfully in 2003. Similarly, in Italy, where 32 Corsican Ospreys were released between 2006 and 2010, mean age of first breeding among the first three males to breed was 3.6 years (Monti *et al*. 2014). These three males all paired with unringed females whose age could therefore not be verified.

Ospreys attempting to breed in established colonies often have to wait longer. Poole (1989) reported that among Ospreys breeding around Chesapeake Bay on the East Coast of the United States, where nearly all suitable breeding locations were occupied by established breeding birds, the mean age at first breeding was 5.7 years. A similar scenario is evident in Scotland, where analysis by Roy Dennis has shown that many young birds are forced to delay breeding because the population in many areas seems to have reached carrying capacity. The annual survival of breeding adults in the Scottish population is thought to be close to 90 per cent, meaning that for a colony of 10 pairs there is a requirement of only two new replacements per annum. The average productivity of pairs laying eggs in Scotland is 1.57 – meaning that, on average, 16 juvenile Ospreys will fledge from a colony of 10 pairs. Survival to two years is approximately 30 per cent and, although much lower than the corresponding figures for annual survival of adults, this is sufficient to ensure that there is always an excess of young birds in the population, with four or five new potential breeders being recruited each year. This explains why in the oldest and most studied populations in Badenoch and Strathspey, Moray and other parts of the Highlands the average age of first breeding was 4.25 years for females and 4.52 years for males, with some individuals not breeding until they were seven years of age (Dennis 2008). Roy Dennis suspects that competition for fishing resources may be the main limiting factor in these areas (pers. comm. 2022) and this could explain why Rothiemurchus moved south away from his natal area, as described above.

The impact of floaters within Osprey populations

The strong natal philopatry of young Ospreys, combined with high annual survival of established breeding adults means that as populations approach carrying capacity the proportion of non-breeding birds within the population increases. These non-breeders are often referred to as 'floaters' and were first recognised as a key component of population biology by Brown (1969). Floaters are sexually mature individuals able to enter the reproductive population when a breeding site or a potential mate becomes available. As potential breeders contributing to population fecundity and dynamics, the role of floaters commences when they are ready to reproduce (Penteriani et al. 2011).

An important concept first introduced by Newton (1988) and later discussed by Hunt (1998) is the floater-to-breeder ratio. Newton (1988) observed that for Peregrines *Falco peregrinus* a ratio of one or more floaters to one breeder could be expected for healthy populations, and that larger pools of floaters were associated with more stable breeding populations, even when breeding output was potentially impacted by floater interference (Hunt 1998). Floaters may compete with breeders and even evict them from established territories (Penteriani et al. 2011), but conversely, Kokko and Sutherland (1998) demonstrated that they also act as an important buffering pool of sexually mature individuals, which can compensate for any losses in the breeding population and thus have a stabilising effect. As such, an absence of information on floaters can potentially lead to misleading conclusions about the health and associated persistence of a population.

An increasing prevalence of floaters in Osprey colonies often leads to increased interference at established nests and may ultimately lead to a reduction in productivity if repeated intrusions cause breeding birds to remain off nests with eggs or small chicks for prolonged periods. Such behaviour has been increasingly observed in northern Scotland as the population has grown and there are fewer opportunities for young first-time breeders (Dennis 2008). A similar scenario has unfolded at Rutland Water, where the return rate of young males, in particular, is high, as described above. Small groups of half a dozen or so non-breeders – usually males – sometimes intrude at active nests, particularly in mid–late summer. The repeated *tioop* guard call given by breeding adults upon sight of an intruding Osprey becomes a familiar sound during the summer and, as described in chapter 1, sometimes the air is filled with a cacophony of *tioops* and *chips* as groups of non-breeding birds fly from nest to nest, with the floaters also calling at each other.

Intrusions in late summer are usually no more than a nuisance for breeding birds, but a more serious problem for returning breeders is increased competition for nests in the spring. As breeding opportunities become scarcer, fights over established nest-sites grow more frequent and, in some cases, can lead to serious injury or even death. The arrival dates of young birds typically become earlier with age. In the UK, returning two-year-old birds usually do not appear until late May or early June, but they will invariably arrive considerably earlier as three-year-olds – often in mid–late April – and earlier still as four-year-olds. These early returners will use knowledge gained over the previous two summers to fly from nest to nest in search of a vacancy. If an established breeder has not returned, they will take their chance. Ospreys are usually faithful to the same nest-site and mate each year, but both males and females will accept a potential new partner at their nest-site in the spring if their regular mate has not returned. Thus, the established breeders most likely to be usurped from their nest-site are those that return late, and this often has implications for breeding success even if the male is able to reclaim the nest.

On some occasions, certain individuals make repeated attempts to oust an established breeder over a number of years. 03(09) fledged from a nest near Rutland Water in 2009 and bred for the first time in 2013 at a nest 7km from his natal site, with a female from Argyll in western Scotland (Figure 3.6). This opportunity arose when the previous incumbent of the nest, a satellite-tagged male, died on its wintering grounds in Ivory Coast. Three chicks fledged that year and the pair continued to breed successfully together for the next three summers. 03(09) typically arrived back at the nest in early April each spring (mean 2013–2016 = 2nd April) but when he returned two weeks late, on 14th April in 2017, his mate had paired with a younger male, 8F(12), who fledged from the nearby Manton Bay nest in 2012. The female had already laid at least one egg and a ferocious battled ensued as 03(09) attempted to reclaim the nest. He eventually ousted the rival male, and kicked the egg out of the nest. It was early enough for the female to re-lay and two chicks subsequently fledged successfully.

History repeated itself in spring 2018. 8F(12) returned even earlier, on 21st March and again paired with the female, who returned two days later. By 11th April the female had laid the first egg, only for 03(09) to return late once more. A battle ensued again, and by 24th April 03(09) had reclaimed the nest. It was getting late in the season, but the female re-laid a clutch of three eggs and two chicks went on to fledge successfully, albeit later than usual, in mid-August.

Unperturbed and seemingly more resolute than ever, 8F(12) arrived even earlier in 2019. He was back at the nest on 20th March, and was joined by the female on 25th March. 03(09) returned a week later on 1st April, but on this occasion was unable to oust the younger male immediately. On 3rd April he was photographed some 30km to the south, having been kept away from the nest by 8F(12). It seemed that the younger male had finally got the better of him. The female was incubating at least one egg by 18th April, but perhaps having reclaimed his strength, 03(09) returned a few days later and this time was successful in ousting 8F(12). On 23rd April he was observed scraping out the nest-cup. We suspected that it would be too late to re-lay but incubation recommenced in early May and a single chick fledged successfully in early August.

Having been ousted from the nest for a third spring running, 8F(12) built a frustration eyrie in a stag-headed oak tree 2km away, but next spring, when he returned on 20th March – the same day as the previous year, and two weeks before 03(09) – he again paired with the female. It

Figure 3.6. Male 03(09) (right) with his Scottish mate at a nest near Rutland Water, prior to being usurped by a rival male (© John Wright).

was no surprise that fighting again ensued between the two longstanding rivals, and that the eggs were again lost. This time, however, 8F(12) was successful in keeping 03(09) away. The female did not re-lay following the loss of the first clutch of eggs, but she and 8F(12) remained at the nest for the rest of the summer. There were sporadic sightings of 03(09) in the vicinity, but he did not reclaim the nest at any point during the summer. It seemed that 8F's persistence had finally paid off, but this demonstrates how difficult it can be for young birds to join the breeding population. Ultimately, it was his earlier arrival each spring (mean arrival date = 23rd March) compared to his rival (mean arrival date = 5th April) that enabled him to prevail. In 2021, 8F(12) reared two chicks with the female for the first time, and 03(09) was not seen.

DISPUTES REVEALED BY SATELLITE TRACKING

While floaters may cause interference at established nests, they provide a valuable pool of potential breeders, should established birds fail to return in the spring. The high annual survival of returning breeders means that in years when vacancies do materialise, it is important for young birds to take advantage, but there can be intense competition between rivals when this occurs. This was exemplified at Rutland Water in 2020, when regular breeding male 28(10) failed to return to a nest where he had bred successfully for five years. His place was initially taken by S1(15) a male that fledged from the Manton Bay nest in 2015. Roy Dennis, Lloyd Park and I caught this male as a returning adult in 2018 and satellite-tagged him (Figure 3.7). The tag subsequently provided fascinating information on how young males enter the breeding population.

In 2020 S1(15) arrived in Rutland on 4th April and, finding the established breeding male absent, he paired with the breeding female, 2F(12). However, a number of other males were

Figure 3.7. The author with S1(15) after deploying a satellite-tag in 2018 (© Lloyd Park).

showing interest in the nest and this perhaps explains why S1(15) made round trips of 274km to Suffolk and Bedfordshire, 300km to Gloucestershire, 122km to Lincolnshire and 125km to Warwickshire between 11th and 19th April. He returned to the nest on each occasion, indicating that these flights were the result of territorial fighting with other males. When rival birds fight over nests in the spring they often disappear for hours, and sometimes days, as they chase each other, and S1(15)'s satellite transmitter appeared to be giving an insight into the distances they may cover in these circumstances. Despite the frequent intrusions and S1's prolonged absences, 2F(12) laid a clutch of three eggs and was incubating on 23rd April. The intrusions continued, however, and so did S1(15)'s flights away from the nest. He flew 245km to central Norfolk and back on 24th April and within a matter of days the eggs had been broken in fighting between S1(15) and a four-year-old male, T4(16).

S1(15)'s long flights away from the nest continued in early May after the eggs had been broken. He made a 185km return trip to southeast Cambridgeshire on 2nd May and a 232km flight to Bedfordshire on 4/5 May. Then during the morning of 12th May he flew south at speeds of over 70km/h, crossing the Thames estuary from Essex to Kent and then returning north over east London. By 17:00, when he returned to Rutland, he had flown 450km in 11 hours, but then headed north-east into Lincolnshire and then across the Wash to West Norfolk. At 23:00 that night he was still flying, having travelled another 121km, thus 571km in 17 hours of continuous flight. Next morning S1 flew 8km back to his nest in Rutland and then spent the rest of the morning there. This suggested that he had seen off the rival bird – almost certainly T4(16) – but the respite was short-lived. In two hours during the afternoon, S1 flew 142km south to London and between 17:00 and 18:00 he flew directly over the centre of the capital. He returned north again and then roosted in north Bedfordshire after flying 250km during the afternoon and evening alone. He returned north to Rutland at first light next morning, but it was notable that, unlike the previous morning, he did not return to the nest, and instead spent the day at Horn Mill Trout Farm, a favoured fishing location. He remained away from the nest for the next three days, and then during the evening of 18th May, flew 154km to northeast Suffolk. He set off back to Rutland at first light next morning, travelling a further 164km, and this time returned to the nest. Once again, however, the respite was short-lived because that evening he flew 112km to north Essex. He roosted close to the Suffolk border and then flew 124km back to Rutland after dawn the next day. However, he again returned to Horn Mill, rather than the nest-site and, in fact, there were no GPS fixes from the nest-site until 25th May. Instead, T4(16) was present there with the female. It was clear that the rival male now had the upper hand and it seems almost certain that the long-running battle between the birds was the reason for S1(15)'s long flights away.

S1(15) was back on the nest on the afternoon of 25th May and remained there for two days. He started ranging more widely again on 27th, and that evening flew 152km south to the Thames estuary, before returning north the next morning. Upon return to Rutland he immediately set off on a 142km loop through Northamptonshire, and then went back to Horn Mill, rather than the nest-site, later that day. This indicated that T4(16) may have deposed him once more – perhaps after pursuing S1(15) south to London and then through Northamptonshire. S1(15) was back on the nest on 30th but that evidently prompted another dispute with T4(16) because the satellite data showed that S1(15) made a 276km return flight to Hertfordshire that afternoon.

S1(15) spent very little time at the nest in early June and the long flights continued – 258km to Norfolk on 4/5th June and 285km to the Thames estuary on 8/9th June. He was present at the nest overnight on 11th June, but next morning he headed west, travelling 240km to the upper reaches of the Conwy estuary in North Wales. He flew further south to the Mawddach estuary the following day, and spent the afternoon there. At 07:00 next morning he began a 279km flight back to Rutland. This long journey to Wales appeared to signify that S1 had now given up on the nest, and once back in Rutland he spent all his time around Horn Mill Trout Farm. Curiously, on one of the few occasions where the satellite data showed he had returned to the nest, on 23rd June, he then flew 228km to North Yorkshire. At 05:00 the next morning he was flying just off the coast immediately north of Scarborough, and then he returned south, passing over the Humber estuary at 9:00 before arriving back in Rutland at 14:00. He then flew a loop into Nottinghamshire before returning to roost in a favoured location at Horn Mill. It was now clear that he was avoiding the nest altogether, and T4(16) was present on a daily basis with the female.

Having been ousted from the nest, S1(15) made an even more unusual flight, travelling 711km to Lower Saxony in northwest Germany, 500km of which was across the North Sea. After roosting in a wood south-west of Hamburg, S1(15) flew 162km south-west on 6th July and then a further 59km into the Netherlands the next day. He then lingered around a wetland on the banks of the River Vecht to the north of the Dutch town of Hardenberg for three days. On 11th July he headed north-east over the Wadden Sea coast and out into the North Sea once again, flying into a northwesterly headwind of 15km/h. At 18:00 BST he was 83km off the Dutch coast seemingly heading back to England, but the wind had strengthened to 20km/h and swung to the west, which perhaps explains why S1(15)'s heading became more north-easterly. At 23:00 he had flown a further 170km north-west and was flying in darkness at low altitude over Dogger Bank in the middle of the North Sea, with the wind now blowing from the west-south-west. Sadly, no further data were received from the transmitter until 14th July when three satellite fixes indicated the bird was floating in the water 346km north-east, 108km off the Danish coast. It appears that S1(15) had drowned at sea and then the body had drifted towards Denmark with the ocean currents. It seems likely that the bird was in poor condition after his long-running battle with T4(16) and simply could not contend with the strengthening crosswind which had

Figure 3.8. Satellite-tagged male S1(15)'s movements during summer 2020, when he was usurped from a nest by a rival male. He subsequently drowned in the North Sea (data courtesy of Leicestershire and Rutland Wildlife Trust).

resulted in him drifting further and further to the north. Of course, had S1(15) been able to keep the rival birds away from the nest then he would not have crossed the North Sea in the first place. It was a sorry outcome, but S1(15)'s remarkable flights, shown in Figure 3.8, helped to shed new light on how rival birds compete for nests.

Density dependence in Osprey populations

As local populations grow – such as the one at Rutland Water – competition for resources increases. Interference by non-breeding birds at nests, and increased competition for mates and nest-sites are evidence of density dependence within the population; this constitutes a negative feedback loop between population growth rate and population density (Newton 1998). Small and recovering populations are a good example of how density dependence can halt population increases in areas with a constant environment (Sutherland and Norris 2002). The actions of non-breeders, in particular, can play a key role in population regulation (Penteriani et al. 2005, Carrete et al. 2006) because direct behavioural interference in relation to factors such as mate choice, copulation and territory establishment are capable of limiting the growth of the local population (Mougeot et al. 2002).

One of the most comprehensive studies of density dependence in an Osprey population was conducted in Corsica over 31 years between 1974 and 2004 (Bretagnolle et al. 2008). Here Ospreys build their nests on coastal cliffs (Thibault et al. 2001) and make either relatively short migrations or remain sedentary at breeding sites (Monti et al. 2018a). The population was estimated at approximately 40–100 breeding pairs at the beginning of the twentieth century, but then declined rapidly due to human persecution, with only three pairs remaining in the early 1970s (Thibault et al. 2001). It gradually recovered thereafter, and increased from 3 to 31 pairs during the period of the study. Two main phases were apparent: a period of population increase between 1974 and 1989, when it grew at approximately 15% per year, followed by a phase of relative stability with the population in the study area fluctuating around 22 occupied nests. In addition, the population increase phase could be subdivided into a phase of rapid growth with a decrease in nearest neighbour distance (distance to the nearest occupied nest), and a period of clustering where nearest neighbour distance did not change, with an average spacing of 1.8km between nests. The average number of neighbours around occupied nests increased regularly up to an average of three to four pairs within 5km.

Bretagnolle et al. 2008 found that population growth was significantly and negatively related to population size and average population density, and therefore strongly indicative of density dependence. Carrying capacity of the study area was estimated to be approximately 21 territorial pairs, with the population increasing rapidly to this point but then levelling off. Most breeding parameters were also density dependent, with productivity and fledging success showing a negative relationship with population size and density. Furthermore, pairs breeding in areas of higher population density had lower hatching success and produced fewer fledged young. It was notable that after 1990 the non-breeding part of the population increased significantly, likely due to lack of nesting sites outside of the core area and the preference of young non-breeders to settle close to conspecifics. As a result, the interaction rate with other Ospreys increased six-fold between 1977 and 1997. The hatching success of individual pairs was negatively related to the frequency of territorial interactions at breeding sites during incubation, probably due to direct disturbance and indirectly via a reduction in the number of successful copulations during the pre-laying period, reducing egg fertilisation and hatching success. Hence, the increase in the number of non-breeding floaters within the population contributed to a reduction in fecundity (Mougeot et al. 2002, Bretagnolle et al. 2008).

The clear density-dependent limitation evident in the Corsican population was likely exacerbated by the fact that the local Ospreys breed on linear habitat along coastal cliffs, which limits breeding opportunities compared to populations that breed around lake and river systems (Bretagnolle et al. 2008). Nevertheless, similar examples can be found elsewhere. In Scotland, where Roy Dennis and others have monitored the population closely since the 1960s, the population appears to have reached carrying capacity in some areas. As Dennis (2008) outlined, the Osprey population in western Moray increased from a single pair in 1966 to a maximum of 13 pairs in 1993, an annual population growth rate of 13 per cent. Thereafter, the population in this colony has fluctuated between 8 and 13 pairs, indicating that it has reached carrying capacity. Similarly, the population in nearby Strathspey expanded from two pairs in 1966 to a maximum of 13 pairs in 1996 – an annual increase of 8 per cent – before levelling off at between 8 and 13 pairs. In eastern Moray the population rose from a single pair in 1985 to a maximum of eight pairs in 2000, a yearly increase of 17 per cent. It subsequently ranged between four and eight pairs. As explained earlier, it is thought that competition over fishing resources, rather than nests, may be the key limiting factor in these areas (R. Dennis pers. comm. 2022).

In another closely monitored population in mainland France, the Osprey population increased from a single breeding pair in 1985 to 38 breeding pairs in 2011. During this period, 22.2 per cent of colour-ringed individuals returned, with an average age at first breeding of 4.28 years for females, and 4.64 years for males. Breeding pairs produced an average of 1.89 fledglings annually and this was relatively stable between years (Wahl and Barbraud 2014). There was no evidence of density-dependence regulation on the number of breeding pairs or the number of chicks fledged per nest, suggesting that the population had not yet reached carrying capacity. Indeed, breeding density during the study was approximately three to four nests per 100km^2, which is lower than the corresponding figure in Corsica (five to six nests per 100km^2). Furthermore, a high proportion of breeding individuals within the population is also consistent with the absence of density dependence. There was some evidence of an effect of density on juvenile survival (0.487), which was negatively related to the number of breeding pairs. It is possible, therefore, that an increasing number of young birds were forced to emigrate when density increased and competition for nest-sites and territories became more intense (Wahl and Barbraud 2014).

When the sex ratio of non-breeding floaters is weighted towards males, interference at nests invariably becomes a problem, but if it is skewed towards females, then a situation where a male breeds simultaneously with two different females may occur. Poole (1989) recorded three polygynous trios out of 190 monogamous pairs he monitored on the East Coast of the United States in the early 1980s. More recently, polygyny has been recorded in the expanding Welsh population, most notably in 2016 when two females from Rutland Water were involved in polygamous partnerships. That year there were four active nests in Wales and breeding opportunities for females were limited due to an absence of unpaired males with nests. So, two young females from Rutland Water, both of whom had been recorded in Wales the previous year, were attracted to artificial nests close to established breeding sites. 24(10), a six-year-old female, settled on a vacant nest at Cors Dyfi and four-year-old 5F(12) did likewise on a nest at Glaslyn. In each case, the male at the established nest accepted the presence of the second female and began delivering fish and mating. Egg-laying followed and early on in the incubation period the males continued to deliver fish and even took occasional turns at incubation. However, the time they spent at the second nest began to decline as the incubation period progressed, and eventually both females were forced to begin fishing for themselves, and abandoned the eggs. This was not unexpected: it is extremely rare for the second female in a polygynous trio to produce young (Poole 1989), and this has never been documented in the UK. Both females eventually bred successfully at other nests in Wales: 24(10) at Llyn Brenig in 2018 and 5F(12) at Llyn Clywedog in 2020. Meanwhile, in 2022, when there was a greater availability of mates, three pairs bred successfully within close proximity of each other in the Glaslyn Valley.

Rarer still are cases of polyandry, where a female pairs up with two males at the same site. In one exceptional case, on Kangaroo Island in South Australia, colour-ringed males, 19 and 28, formed a polyandrous relationship with the same unringed female, who was identified by her iris characteristics, for four consecutive breeding seasons between 1993 and 1996 (Dennis 2007b). Both males were observed to bring fish to the nest and all three adults were observed on the nest-platform together, with no aggression recorded – including when young were present. Both males, one four years old and the other five years old when first observed in 1993, were from the same natal site (and parent pair), located approximately 15km away. The polyandrous partnership reared a total of five chicks over the four-year period.

Site fidelity and breeding dispersal

The difficulties encountered by some young Ospreys as they search for a first breeding site, particularly in areas nearing carrying capacity, is caused by low annual mortality and strong site and mate fidelity of established breeders. Ospreys generally exhibit very limited breeding dispersal once they have bred successfully, usually returning to the same nest and mate each spring. This means that a single site may be occupied by the same pair for 10 or more years.

The long-term colour-ringing of Ospreys in the UK has provided interesting insights into the site fidelity of established breeding birds. Furthermore, in recent years many Osprey nests

Figure 3.9. Ospreys show strong fidelity to the same nest-site and mate. Male 03(97) bred at the Site B nest at Rutland Water between 2001 and 2015, rearing 32 chicks with three different females. The change of mate occurred only once an existing partner failed to return (© John Wright).

have been equipped with high-quality cameras and attract an audience of thousands thanks to online streaming. The return of known birds to their regular nest-site is keenly awaited and celebrated each spring.

In the UK the longest a pair are known to have bred together is 22 years (Dennis 2008). The male of this pair had been breeding at same locality, albeit on two different nests, for 25 years and was thought to be at least 28 years old, while the female bred for 22 years and was at least 25 years old. They produced 38 chicks during this time, despite the fact that their eggs were robbed by egg-collectors on five occasions and broken by an intruding female twice. Another pair bred together for 13 years, producing 25 young. The female reared 32 young in her lifetime (Dennis 2008).

As Ospreys have recolonised England and Wales, similarly high levels of site fidelity have been recorded. Only eight of 31 breeding birds moved nests after their first breeding attempt (to egg stage at least) in the Rutland Water area between 2001 and 2022. Seven of these individuals moved only after a failure, and never when they had reared chicks the previous year, and the eighth moved when its regular mate failed to return. One bird, male 03(97), bred for 15 successive years and reared 32 chicks at the same nest, with three different females, between 2001 and 2015 (Figure 3.9). The changes of mate occurred only when the female from the previous year failed to return, rather than being usurped by a rival bird. Likewise, a metal-ringed female bred at the same site each year from 2009 onwards, raising 29 chicks with four different males during this period; an unringed female known as Maya, recognisable by a distinctive head pattern, bred each summer at the Manton Bay nest from 2010, rearing 34 chicks with three different males; and female 5N(04) – one of 03(97)'s offspring – reared two chicks for the first time at the Manton Bay nest in 2007. She failed on eggs the next summer and moved to another artificial nest nearby where she continued to breed every year from 2009, raising a further 36 chicks with four different males by the end of the 2022 season (chapter 8, Table 8.5). As with 03(97), the changes of mate for each female occurred only when their partner from the previous year was lost, except on two occasions when breeding males were ousted by a rival.

The longest standing partnership in the Rutland Water area during this period was between two locally fledged birds – male 11(10) and female 25(10) – who reared 27 chicks over 10 successive summers at the same site after breeding together for the first time as three-year-olds in 2013. Just one pair, 08(97) and 5N(04), bred successfully together at two different sites, but they only moved following a failure caused by repeated intrusions and interference by young males. Five of the birds that moved nests remained in the local area (maximum dispersal = 28km) and one bird, 3J(16), joined the Welsh breeding population at Cors Dyfi, as previously described; this represented an exceptional movement of 218km, given that she had already attempted to breed at Rutland Water. In Scotland, it is rare for birds to make long-distance moves between nests, especially if previous nesting attempts have been successful, but there are occasional examples. Female white/black EV bred successfully in Aberdeenshire in 2000, but then moved 80km to nest in Moray, where she reared three young (Dennis 2008). Despite this she moved to another local nest, 1.5km away, the following spring, and this time the breeding attempt failed. Meanwhile, a female that Roy Dennis satellite-tagged as an adult in Moray in 2008, known as Morven, bred at five different sites in the local area over the course of 12 years (R. Dennis pers. comm. 2022).

Ospreys returned to Kielder Forest in Northumberland in northern England in 2009. By the end of the 2022 breeding season, when the population had reached eight breeding pairs, the longest standing partnership was between male White YA from North Wales and an unringed female. They raised a total of 25 chicks to fledging, including two broods of four, in 10 successive years of breeding at two different nests (Figure 3.10). Sadly, the male disappeared as the

Dispersal and settlement patterns

Figure 3.10. An unringed female and Welsh male were the most successful pairing in the first 13 years that Ospreys bred in the Kielder Forest area, rearing a total of 25 chicks in 11 years, including this brood of four in 2016 (© Forestry England).

young were fledging in 2022, but the female continued to provision the chicks in his absence (Dailey 2022).

In Wales, the unringed female that first bred at the Glaslyn nest in 2004 (as described above) continued to return to the same site up to and including 2022, by which point she was thought to be at least 21 years old. During this period she reared 44 chicks to fledging, 26 with Rutland Water translocated male 11(10) from 2004–2014 and then 18 with an unringed male known as Aran between 2015–2022 (Glaslyn Wildlife 2022a).

The site fidelity of breeding Ospreys has been recorded in other parts of the world too. On Kangaroo Island, South Australia, Dennis (2007) reports that between 1985 and 2004, when an average of nine pairs were located each year, one pair, male L1 and female E1, bred for at least eight consecutive seasons, while male 13 and female 41 bred for 12 successive years between 1993 and 2004. Moreover, no known individuals, of either sex, were found to relocate from the territory where first identified, or to desert an established pair-bond. Meanwhile in North America Spitzer *et al.* (1983) reported that only 3% of colour-ringed birds changed nesting sites over a 10-year period in New England, and no bird nested more than 18km from its old site. Kinkead (1985) found that 73% of Ospreys ringed in his study area, in a Virginian section of Chesapeake Bay in 1983, returned to breed at the same site the following year. He noted that site fidelity may have actually exceeded this figure because 11 nest-sites, where a total of 15 individuals had been colour-ringed in 1983, were partially or completely destroyed by storms in 1984. Separation of Osprey pairs was also uncommon. Of 19 pairs ringed in 1983, 12 remained together, four were found to be missing and two new pairings were discovered the following year. The author suggested that actual pair fidelity may also have been higher than stated depending on the fidelity of the four Osprey pairs missing in 1984. Severe windstorms in April and May of 1984 destroyed many nest-sites and it is conceivable that some of these pairs remained together at new unknown sites. Thus, separation of only 11 per cent of the known pairs was verified visually in 1984.

As these various examples demonstrate, Ospreys are generally monogamous, and this is strongly influenced by the attachment of birds to their breeding site. When breeding partners fail to return in the spring they are usually replaced that year, especially in established colonies where there is a surplus of young non-breeders. This means that the same nests can be used by many generations of Ospreys. The best example of this in the United Kingdom is the famous Loch Garten nest, which was occupied by Ospreys every year between 1958 and 2018 (Dennis 2008, R. Dennis pers. comm. 2020). A total of 105 chicks were reared by a combination of at least seven different males and ten females during this time. The most successful pair were female EJ and an unringed male known as Odin, who reared 17 chicks at the site between 2009 and 2018. Similarly, successive generations of Ospreys have occupied a site in Moray, known as B01, every year since 1967, with a total of 77 chicks reared by 2020. Meanwhile, another Moray nest, known as B03, has been used almost continuously since 1968, with 73 chicks fledging by 2020 (Dennis 2020 unpub. data). It is not surprising that such sites become highly sought after by young birds attempting to breed for the first time.

Conclusions

- Male Ospreys show a high degree of natal philopatry, and usually return to breed close to their natal nest. Females often disperse further and readily join other breeding populations.
- Young birds explore extensively when they first return to breeding areas during their third calendar year. Colour-ringing and satellite-tracking research has shown that these movements are key to future nest-site choice.
- Ospreys reach breeding age at three years, although in some exceptional cases they can rear young as two-year-olds. Age at first breeding is usually lower in expanding colonies where there is reduced competition for mates, nest-sites and fishing resources.
- Density dependence may be evident as local breeding populations near carrying capacity. Increased competition for mates and nest-sites results in a higher proportion of non-breeding floaters within the population and this can lead to interference at nests and potentially a reduction in breeding productivity.
- Ospreys show a high degree of fidelity to the same mate and nest-site and, as a result, the same pair may breed together for 10 years or more.

CHAPTER 4
Nesting and courtship

The nest

The Osprey's global distribution is testament to its ability to adapt to local environmental conditions both for feeding and breeding. Ospreys build large conspicuous nests in a variety of locations, on a range of different substrates and in different habitats. The key requirements are that the nest is situated in a prominent place with expansive views across the surrounding landscape, and relatively close to suitable foraging grounds.

Tree nests

At northern latitudes, Ospreys generally breed in forested areas. Here they favour tall mature trees, either alive or dead, with adequate strong support branches at the top and easy passage in and out. The Osprey may be well adapted to catching fish, but it lacks the manoeuvrability of species that breed in dense forest. As a result, isolated trees, or those on the edge of clearings, are often favoured for ease of access. Nestlings are also less vulnerable to surprise attacks by Northern Goshawks *Accipiter gentilis* and other potential predators if nests are situated in open areas where adult birds have a better view. The height of the nest depends on the surrounding trees. If the nest is located in a block of forest, Ospreys usually select the tallest tree above the height of the surrounding canopy (Figure 4.1)

Figure 4.1. When nesting in forested areas, such as here in Belarus, Ospreys usually select the tallest trees on which to build their nests (© Denis Kittel).

In Europe the northern forests and peat bogs of Scandinavia and Russia are a stronghold for breeding Ospreys. Here the most common nesting tree is the Scots Pine *Pinus sylvestris*, with the flat top of mature trees providing a strong and stable base for the huge nest. The importance of this species is exemplified by the fact that 88 per cent of natural Osprey nests in Finland are found on Scots Pine, with the next most regularly used tree, the Norway Spruce *Picea abies*, accounting for just 3 per cent and broadleaved species such as Birch *Betula* spp., Poplar *Populus* spp., Alder *Alnus* spp. and Oak *Quercus* spp. lower still, at just 1 per cent combined. In addition, 7 per cent of nests in Finland are built on dead trees (Saurola 1997). The Norway Spruce is only suitable if the top of the tree has been broken during winter storms or due to the weight of lying snow, otherwise the uppermost branches are not strong enough to support a nest. Scots Pine is also the most commonly used nesting tree in northern Scotland, with 56 per cent of nests located in live Scots Pines (see Figure 4.1) and 13 per cent in dead Scots Pines (Dennis 2008). Meanwhile the Douglas Fir *Pseudotsuga menziesii*, a native of North America, accounts for 22 per cent of nests, with a range of other introduced conifer species, including Sitka Spruce *Picea sitchensis* and Noble Fir *Abies procera* contributing a further 2 per cent. Deciduous trees, including Birch (3 per cent), Alder (2 per cent), Oak (1 per cent) and Wild Cherry *Prunus avium* (1 per cent), are also used on occasion. Similarly, Scots Pine is the favoured tree species in Germany (Bai *et al.* 2009), although in recent decades there has been a considerable shift to the use of artificial structures, particularly pylons, as discussed later (Canal *et al.* 2017).

Elsewhere, the Osprey's ability to adapt to local availability of potential nest-sites is apparent. In central England, where there are considerably fewer pine trees than northern Scotland, Ospreys readily build nests in stag-headed Oaks (Figure 4.2) (Mackrill *et al.* 2013). Meanwhile

Figure 4.2. Dead-topped, or stag-headed, Oaks provide ideal nesting trees for Ospreys in central England (© John Wright).

in the Netherlands, recolonised by Ospreys in 2016, Poplars are employed (P. Voskamp pers. comm. 2020).

A similar scenario is evident in North America. Ewins (1997) found that of 85 nests surveyed in the forests of Ontario in eastern Canada, 80 per cent were built in conifers, mostly White Pine *Pinus strobus*, and 20 per cent were located in deciduous species, predominantly Birch *Betula* spp. and some poplars *Populus* spp. Live trees supported 47 per cent of the nests, flat-topped or dead-topped conifers supported 12 per cent, and the remaining 41 per cent of nests were on dead trees, often in swamps originally created by Beaver *Castor canadensis* activity. In Oregon's Deschutes National Forest, Ewins (1997) reports that 90 per cent of Osprey nests were located in Ponderosa Pines *Pinus ponderosa*, with a mean tree height of 35 metres and mean diameter at breast height (DBH) of 95cm. Of these nests, 30 per cent were in live trees, 21 per cent in dead-topped trees and 49 per cent on dead snags. Poole (1989) remarked that Ponderosa Pines are often favoured in western parts of the United States, such as Oregon, with Tupelo *Nyssa sylvatica* regularly used in the east.

The suitability of some tree species means that they are often used disproportionally compared to their relative abundance. For example, in Minnesota's Superior National Forest, 77 per cent of 301 Osprey nests over 31 years were built in White Pines, even though the species represented less than 0.5 per cent of trees with a DBH of greater than 10cm (Rogers and Lindquist 1993). In Yellowstone National Park in Wyoming, 39 per cent of Osprey nests were built in Lodgepole Pine *Pinus contorta*, 30 per cent in Engelmann Spruce *Picea engelmannii* and 30 per cent in Subalpine Fir *Abies lasiocarpa* (Swenson 1981). Of these, 73 per cent were taller than other trees in the surrounding landscape, thereby emphasising the Osprey's desire to nest in a prominent location.

Figure 4.3. Ospreys will sometimes nest on very low trees when they are situated in open areas or standing in water, particularly in the United States (© Ingo Arndt/NPL).

In southern parts of the United States, Ospreys nest on flat-topped mangroves, particularly the Red Mangrove *Rhizophora mangle* (Ogden 1977), and in coastal parts of North and South Carolina, flood-tolerant Cypress trees *Taxodium* spp. provide similar nesting opportunities (Figure 4.3) (Hagan and Walters 1990). Meanwhile, in parts of northwest Mexico, including Baja California peninsula and the eastern shore of the Gulf of California, where trees are scarce, Ospreys build nests on various species of tall cacti (Henny and Anderson 1979). In this region cacti were found to account for 33% of all nests in 2006, and enable Ospreys to achieve the prominent location within the landscape that they always desire (Henny *et al.* 2008b). In fact, cacti were the second most-used substrate for breeding Ospreys in the region for three successive decades (Henny *et al.* 2008b).

GROUND AND CLIFF NESTS

Where trees are absent, Ospreys will nest on the ground. Significant concentrations of ground-nesting Ospreys occur in the coastal lagoons of Baja California, in northwest Mexico (Henny and Anderson 1979, Danemann and Guzman-Póo 1992, Henny *et al.* 2008b). Ground nesting has also been recorded in Japan at sites with good all-round visibility (Shoji *et al.* 2011), and close to the shores of the Red Sea, including in Egypt where nests are located in sand dunes (Habib 2019). Ground nesting has been recorded twice in northern Scotland, but the nest was predated on both occasions (Figure 4.4) (R. Dennis pers. comm. 2020). It is for this reason that ground-nesting Ospreys are usually found on islands free of mammalian predators. For

Figure 4.4. Ospreys can nest on the ground in open areas, such as this hillock in Caithness, Scotland. Such nests can be vulnerable to predation, and this site was unsuccessful (© Tim Mackrill).

instance, 250–300 pairs of Ospreys bred on Gardiner's Island, New York, in the early part of the twentieth century, with most clustered in just a few small areas of this 13.4km² island (Poole 1989).

Cliff-nesting Ospreys are found in various locations around the world. Shoji *et al.* (2011) located 62 Osprey nests in Hokkaido, the northernmost island of Japan, and confirmed breeding in eight of 14 districts, predominantly in the west. More than half of the nests (n = 35) were on rocks or cliffs, with some 90 metres above the ground. Ospreys also breed on rocky coastal cliffs around the Mediterranean, with breeding populations located in Corsica, the Balearic Islands, Algeria and Morocco (Thibault and Patrimonio 1991; Monti *et al.* 2013). Likewise, in northwest Mexico large cliffs located close to the sea have traditionally been the favoured nesting locations, constituting 59.1 per cent of sites in 1977, 39.8% in 1992/93 and 38.3% in 2006. The cliffs sometimes consist of large pinnacles or stacks, and small sandy cliffs are also used (Henny *et al.* 2008b). Similarly, in South Australia a lack of coastal forests means that Osprey breeding habitat is predominantly limited to semi-arid open coastal landscapes. Here most nests are typically built on collapsed or remnant sections of coastal cliff, including near-shore rock-stacks (Dennis and Clancy 2014).

A small population of 14–17 pairs of Ospreys breed on Boa Vista, the easternmost of the Cape Verde archipelago, about 500km west of Senegal, at the southern edge of the Osprey's breeding range in the Western Palearctic (Siverio *et al.* 2014). The mostly flat island is characterised by a range of habitats including dunes, rocky plains, hills, low mountain ranges, coastal cliffs, sandy beaches and offshore islets. In this arid environment vegetation is limited and so

Figure 4.5. Some Osprey nests, such as this one on Boa Vista, Cape Verde, become very large when used by successive generations of Ospreys (© Pedro López).

Ospreys typically build nests on rocky outcrops, cliffs and on the ground, although often away from the coast (Figure 4.5). Siverio *et al.* (2014) identified 79 nest structures, of which 37 were occupied for at least one year between 2004 and 2007. Of occupied nests, 29.7 per cent were located on ledges on the crests of hills, 24.3 per cent on inland cliffs, 10.8 per cent on hillsides, 10.8 per cent on headlands (rocky or compacted sand), 5.4 per cent on sea cliffs and 2.7 per cent on islets/sea stacks. A further 5.4 per cent of nests were situated on palm trees, and 5.4 per cent on artificial structures.

USE OF ARTIFICIAL STRUCTURES

The wide variety of nesting substrates used by Ospreys exemplifies the species' ability to adapt to local environmental conditions, and this has been reflected in the increasing use of artificial structures in recent decades. In Europe this is most apparent in Germany, where a significant proportion of the breeding population nest on medium- and high-voltage electricity pylons. This was first recorded in 1938 when a nest was found on a pylon between Angermünde and Templin, north of Berlin. Since then, pylon nesting has become increasingly common, and by 1996, 75 per cent of the population were utilising these structures (Meyburg *et al.* 1996). This has been facilitated by the installation of artificial nests on the top of pylons, and has allowed expansion into areas, such as eastern Germany, where natural nest-sites are scarce (Canal *et al.* 2018). Pylon nesting has also been documented in other European countries, including Spain (Muriel *et al.* 2006), France (R. Wahl pers. comm. 2020) and Scotland, where mobile phone masts have also been used (Dennis 2008).

Ospreys are even better known for their ability to utilise a range of anthropogenic structures for nesting in the United States, where there has been a dramatic shift from nesting in trees in recent decades. Utility poles, mobile phone masts, floating rafts, billboards, chimneys, channel markers and buoys, and bridge superstructures have all been exploited by Ospreys in the human-dominated modern landscape (Poole 1989).

Chesapeake Bay is a stronghold for Ospreys in the United States, and is one of the places where the shift from tree nesting has been most apparent. Here the majority of Ospreys nested in trees prior to the 1950s, but there has been a dramatic shift since, with 90 per cent of 3,500 nests located on artificial structures by the mid-1990s (Watts and Paxton 2007). It is thought that this has been a key factor in the recovery of the local populations in recent decades. Many of the sites now in use are on poles erected specifically for Ospreys, but nests are also built on channel markers and duck blinds (used for wildfowling). Channel markers, which act as aids to navigation, are particularly valuable (Figure 4.6). There were a total of 1875 channel markers in 1973, 17 per cent of which were occupied by breeding Ospreys, but by 1999 the number of markers had risen to 2929 and 57.1 per cent supported Osprey nests (Watts and Paxton 2007). This increase coincided with a change of attitude towards breeding Ospreys by coastguards, who routinely removed Osprey nests – at any stage of the breeding cycle – during the 1960s and 1970s. It is clear that a shift to more Osprey-friendly policy during the late 1970s had a very beneficial impact on the breeding population (Watts and Paxton 2007). By 2010 it was estimated that there were as many as 9,000 pairs of breeding Ospreys at Chesapeake Bay, with 60–70 per cent of nests located on navigational markers (Poole 2019).

The acceptance of human-made nest structures has also facilitated the expansion of the Osprey population in southern New England and Long Island, New York, from the 1970s onwards. As in other parts of the world, a variety of artificial structures are used by breeding Ospreys in this region. For example, 17.12 per cent of nests in Rhode Island were on mobile phone masts, with a further 15.2 per cent of nests on electricity pylons and 15.2 per cent on light

Nesting and courtship

Figure 4.6. Ospreys use a range of artificial structures for nesting in the United States, including channel markers (© Lonny Garris/Shutterstock).

towers (Bierregaard *et al.* 2014). Overall, just 6.2 per cent of nests in Rhode Island were in trees, with the remaining 46.5 per cent of breeding Ospreys utilising artificial nests on poles. Mobile phone masts, electricity pylons and light towers were also occupied by breeding Ospreys in Connecticut as well as Cape Cod and Martha's Vineyard in Massachusetts (Bierregaard *et al.* 2014). Ospreys readily took to artificial nests in these areas, which accounted for 72 per cent of nests recorded across the study area in 2010. Indeed, the readiness of Ospreys to utilise artificial nests in a variety of settings across the world has been a great conservation success story, and one that is covered in more detail in chapter 8.

The shift from trees to artificial structures has also been well documented in Australia. In New South Wales almost all known nests in the 1970s and 1980s were located in tall dead or living trees, but as the population has expanded nesting attempts were made on electricity pylons and other anthropogenic structures for the first time (Clancy 2006). Now, an estimated 30 per cent of nests are found on artificial structures (Moffat 2009), aided by electrical companies who have modified structures to support nests, moved conductors and cabling, and also erected alternative poles in areas lacking natural nest-sites (Dennis and Clancy 2014).

A shift to the use of artificial structures has also been recorded in northwest Mexico, where towers, pilings, channel markers, debris washed ashore, boats (sunk and aground) and electricity pylons are all used (Henny *et al.* 2008b). The use of artificial structures markedly increased over a 30-year period, from 4.3 per cent in 1977 to 6.2 per cent in 1992/93 and 26.4 per cent in 2006. This shift has been most evident on the western side of Baja California, where 79.5 per cent of nests were located on anthropogenic substrates in 2006 (Henny *et al.* 2008b). This increase was due to an apparent shift of ground-nesting birds on small islands in San Ignacio Lagoon to electricity pylons and towers in nearby towns, and the development of the salt industry at Scammon's Lagoon beginning in 1953, which resulted in electricity pylons, channel markers and other associated structures being built in the area.

Proximity to water

As an exclusively piscivorous species, there are obvious advantages for Ospreys in breeding close to water, and this is usually what they do (Figure 4.7). Ewins (1997) reported that 93 per cent of 179 tree nests used by Ospreys in Ontario, Canada, were located within 500 metres of water, with a median distance of just 10 metres (Ewins 1997). Similarly, 83 per cent of 73 nests surveyed in Oregon, United States in 1978 were within 1km of water, and all sites were within 2km (Henny *et al.* 1978). Bai *et al.* (2009), meanwhile, found that waterbodies were the most important factor in nest-site selection of German Ospreys, and comprised 5.9 per cent of the area within a 7km radius of nest-sites. They reported that Ospreys shape their home range to encompass the main foraging sites, meaning they are not simply a circle based around the nest-site, but instead influenced by the spatial distribution of favoured fishing localities. This has been corroborated by subsequent satellite-tracking studies, as described in chapter 2. In general, proximity to a range of different foraging sites is important.

Saurola (1997) observed that optimum sites in Finland are located in trees close to the shoreline of lakes, but in areas where lakesides are disturbed by human activity, Ospreys may nest

Figure 4.7. Many Osprey nests, such as this one on an artificial nest platform beside the Yellowstone River, Montana, are located close to water, but some birds nest several kilometres from the nearest fishing sites (© Marco Restani).

several kilometres away. A similar pattern of settlement is apparent in northern Scotland. Some nests are located on the water's edge, but most are situated a few kilometres from favoured fishing grounds (Dennis 2008). Satellite tracking by Roy Dennis has shown that breeding males make daily commutes of 10–12km to favoured fishing grounds in freshwater and marine environments in Moray in northern Scotland, as described in chapter 2. Likewise, in my study area in central England, the mean distance to water of 14 nest-sites used in the Rutland Water area between 2001 and 2022 was 1.4km. Two of these nests were on poles over water, but the others ranged 0.22–2.79km to the nearest fishing site. On occasions, however, some males make much longer foraging flights. For example, one breeding male made daily visits to a trout farm 19km from its nest-site, while another bird made at least weekly visits to reservoirs 25km from its nest (chapter 2).

Nesting over water can offer valuable protection from terrestrial predators. In North America, Ospreys breeding in trees are vulnerable to predation by climbing mammalian predators such as Raccoons *Procyon lotor* which predate eggs and small young. The clear shift from tree nests to sites over water – including channel markers and poles erected specifically for Ospreys – in places such as Chesapeake Bay may have been a selective response to this threat (Bierregaard *et al.* 2020).

Islands, such as those in the coastal lagoons of Baja California, in northwest Mexico, also provide an optimum location for Ospreys to build their nests. In Los Angeles Bay, for instance, where Ospreys nest on small islands, the breeding population changed very little between 1977 (35 pairs) and 2006 (40 pairs) (Henry *et al.* 2008). In Finland, islets in lakes with tall mature trees are often favoured by breeding Ospreys (Saurola 1997).

Although it is preferable for breeding Ospreys to nest close to water in order to limit energy expenditure, there are examples of pairs breeding successfully much greater distances from favoured fishing sites. For example, male Ospreys nesting around Lake Ellis Simon – a shallow, marshy lake of 600 hectares in North Carolina, United States, where 50–60 pairs breed on cypress trees over the water – rarely fished in the lake itself, and instead travelled 14km to three different estuaries to forage (Hagan and Walters 1990). In this case, the authors noted that nesting sites were available in tall trees close to these favoured foraging sites, but concluded that the benefits of colonial nesting at Lake Ellis Simon in the over-water nests, where predation risk was reduced, outweighed the greater energetic costs of the longer foraging flights.

A similar scenario is evident at Mono Lake in eastern California in the United States, as reported by Fields and Pagel (2016). This lake is hypersaline and alkaline and lacks any fish. However, where freshwater springs meet the alkaline lake water, a precipitate of calcium carbonate, known as tufa, forms in the water column. Many of these tufa towers, which are mainly clustered around the shoreline, and can be greater than 15 metres in height, became exposed during the mid to late 1990s as freshwater inflow was diverted away from the lake. The tufa provide highly suitable locations for Ospreys to build their nests, and the first breeding pair was recorded in 1989. By 2010–2014 there were eight occupied nests at the lake, up to a maximum of 11km from the nearest fishable water, but there was no correlation between breeding success and distance to foraging sites. In this instance the lack of suitable trees closer to good foraging grounds, and the protection afforded by the tufa towers from mammalian predators, appeared to be the key overarching factors in nest-site choice (Fields and Pagel 2016).

Similarly, the breeding distribution of Ospreys in New Brunswick, on the east coast of Canada, is heavily influenced by the location of hydroelectric poles. The median distance to water of 151 nests in 1993 was 1km, but 45% were 1–5km from the nearest fishing site (Ewins 1997). It appears Ospreys actively sought out vacant poles, even if that meant breeding further from water.

Interspecific competition

While nesting on or near water does have clear benefits, interspecific competition has changed the behaviour of breeding Ospreys in some areas. Bald Eagles *Haliaeetus leucocephalus* are known to steal fish from Ospreys, and on occasion they predate chicks from nests (Figure 4.8) (Poole 2019). Studies have shown that, in some circumstances, these very large eagles may displace Ospreys from the best breeding sites close to water. In southern Florida, for example, Ogden (1975) found that a newly established pair of eagles resulted in eight of 13 pairs of Ospreys relocating at Murray Key. However, this conflict was not repeated in subsequent years, indicating that eagles may only exhibit territorial behaviour towards Ospreys in the year they first settle in a territory, becoming more tolerant in future years as they recognise neighbouring pairs – a process that is possible due to the site fidelity shown by both species. Meanwhile Gerrard *et al.* (1976) found a ratio of one Osprey to 66 Bald Eagle nests in the highest quality undisturbed habitat in Saskatchewan, Canada. They suggested that in these areas Ospreys may be competitively excluded by Bald Eagles, partly due to the earlier arrival of Bald Eagles to the area in the spring. The relationship between the two species is complex, however, because in Chesapeake Bay around 2,000 pairs of Bald Eagles and about 9,000 pairs of Ospreys share the same rich estuarine habitat. In such circumstances, Ospreys may benefit from safety in numbers, whereas lone birds may be more vulnerable (Poole 2019).

Similar interspecific interactions are recorded where the breeding range of Ospreys overlap with the White-tailed Eagle *Haliaeetus albicilla*. Shoji *et al.* (2011) reported that the distribution of nesting Ospreys in Japan is concentrated in areas where the frequency of breeding White-tailed Eagles is low. They suggest this may be due to a preference among Ospreys to limit interspecific competition and to avoid potential predation of young by White-tailed Eagles.

Figure 4.8. In some areas, Bald Eagles may displace Ospreys from the best breeding sites close to water (© Michael Cohen/Getty).

Likewise, in recent years there is evidence that Ospreys are nesting farther from water in the upper reaches of the Volga River, in the northwestern part of the Rybinsk reservoir in northwest Russia, due to increased competition with White-tailed Eagles (M. Babushkin pers. comm. 2019).

Height of nests

As already discussed, choosing a prominent location is one of the key requirements for breeding Ospreys. As such, the height of nests varies according to local environmental conditions.

In forested areas, where Ospreys select the tallest trees, nest heights can range from 15–50 metres. For instance, the mean height of nests in Deschutes National Forest, Oregon, United States, where Ospreys favour Ponderosa Pine, is 35 metres (Ewins 1997). Meanwhile in northern Scotland, nest locations often exceed 30 metres in height, with some as high as 50 metres (Dennis 2008). In other areas, nests can be much lower. Ospreys breed in emergent cypress trees at Lake Ellis Simon in North Carolina, United States, with nests situated 1–9 metres above the water (Hagan 1986). Moreover, Ospreys readily use artificial nests that are 2–5 metres in height in the wide-open expanses of the New England saltmarshes (Poole 2019). Ground nesting is widespread in areas that lack trees or alternative artificial structures, including in northwest Mexico, Japan and around the Red Sea, as described earlier (Henny *et al.* 2008b, Shoji *et al.* 2011, Habib 2019). In areas where artificial structures are used, the height of nests can range from 50 metres on electricity pylons, such as those favoured in Germany (Canal *et al.* 2018), to just a few metres when Ospreys build nests on navigational markers in the United States (Poole 2019). It is this adaptability that has enabled the Osprey to prosper in such a wide range of habitats and settings around the world.

Nearest neighbours

Numerous studies demonstrate that conspecific attraction plays a key role in shaping settlement patterns, and spacing between nests varies according to a range of different factors. The densest concentrations of breeding Ospreys are found in the United States, most notably around the lakes of south Florida (Poole 2019). Lake Istokpoga is Florida's fifth-largest lake, with a surface area of 11,207 hectares. It is shallow, with an average depth of about 1.6 metres, and more than half of the shoreline is ringed with Bald Cypress *Taxodium distichum* which offer ideal sites for Osprey nests, as described by McMillian (2013). This population, like many across North America, was badly affected by the impacts of DDT and declined to just nine pairs by 1973. However, Ospreys have recovered spectacularly in the area since then. The population increased from just over 50 breeding pairs in 1989 to more than 200 in the area monitored in previous surveys, and 300 pairs overall, in 2010. This rapid increase was thought to be related to increases in fish abundance following the accidental introduction of the invasive aquatic plant Waterthyme *Hydrilla verticillata* in 1979. The Osprey nests are now often found within 10–15 metres of each other and, in the most densely populated areas, two or three nests are sometimes located on the same tree (Poole 2019).

As the example from south Florida demonstrates, the availability of safe nest-sites close to areas of high prey availability can facilitate establishment of dense Osprey colonies. It is for this reason that the provision of artificial nests enables Ospreys to breed at similarly high densities close to optimal foraging sites, even in places where natural nest-sites are lacking. Some of the highest densities of breeding Ospreys are now found on the saltmarshes of New England on the East Coast of the United States, where large numbers breed on artificial nests sited at regular intervals across the marshes (Figure 4.9). The erection of nests 25–50 metres apart has resulted in colonies of 50–100 pairs becoming established in just a few square kilometres due to high prey availability in the adjacent estuaries (Poole 2019). In these areas Ospreys defend the immediate vicinity of the nest, but share common fishing grounds (Bierregaard *et al.* 2020).

Ospreys have also been recorded nesting at very high densities in places where they breed on the ground. Historically 200–300 pairs nested on Gardiner's Island, a small island of 13.4km² at the east end of Long Island, New York. Along one section of beach there was a succession of 22 nests, spaced at intervals of 10–200 metres (Poole 1989). Clusters of ground-nesting Ospreys also occurred on the west coast of the Baja Peninsula in northwest Mexico, most notably on two small islands in the San Ignacio Lagoon, where 143 occupied nests were counted in 1989 (Danemann 1994). However, by 2006 this ground-nesting colony was no longer present, with the birds choosing to breed on artificial structures instead, likely as a result of the combined effects of predation and disturbance. This shift is exemplified by the fact that in 1992, an estimated eight pairs were nesting on artificial structures in the region, but by 2006, the number had increased to approximately 175 pairs (Henny *et al.* 2008b).

The use of artificial structures has also had an impact on the spacing of breeding Ospreys in Europe. Although Ospreys tend not to breed here in the same high densities recorded in North America, the increasing use of electricity pylons in Germany has greatly facilitated the expansion of the species to different regions, and in some areas active nests are located on neighbouring pylons within a few hundred metres of each other, particularly if they are close to optimum fishing grounds (Bai *et al.* 2009, Schmidt-Rothmund *et al.* 2014). In such instances, substrate type has been shown to strongly influence local density, with pylon nesters breeding closer together than birds using trees (Bai *et al.* 2009).

Figure 4.9. The provision of artificial nests on the saltmarshes of New England has proved highly successful, and low nests can be easily checked by Osprey fieldworkers (© Alan Poole).

Osprey nests are usually more widely spaced in northern boreal forests. For example, the breeding density of Ospreys in northern Quebec, Canada, was reported as 0.2–2.1 pairs per 100km^2 (Bider and Bird 1983). Similarly, in southern Finland, Solonen (1993) recorded one Osprey nest per 100km^2. Modern forestry has severely limited the suitability of many areas for breeding Ospreys, and at least 47–49% of Finland's Ospreys now breed on artificial nests, with the figure closer to 90% in some regions, as discussed in chapter 8 (Saurola 2005). In the boreal forests of Värmland, in central Sweden, mean distance between nests was 7.8km in 1996 and 6.2km the following year (Widén and Richardson 2000). In southwestern Sweden, Eriksson (1985) reported an average nest spacing of 7.9km. There are some regions, though, where the breeding density is higher. Large waterbodies such as Lake Asnen and Lake Helgasjön in Småland, where the boreal and nemoral (temperate biome characterised by broadleaved forest) zones meet, provide excellent breeding sites, and consequently support much denser concentrations of breeding Ospreys. Widén and Richardson (2000) reported that 40–50 pairs bred an average of 1.2km from each other around Lake Asnen, while the mean distance between the nests of 15 pairs at Lake Helgasjön was 1.6km. In Scotland nests are usually spaced a few kilometres apart, but as numbers have increased some pairs have nested successfully less than 1km apart (Dennis 2008).

In some coastal areas the distribution of Ospreys is limited by the availability of suitable sea cliffs. Bretagnolle *et al.* (2008) reported that the average spacing between nests on cliffs in Corsica was 1.8km. Meanwhile in Boa Vista in the Cape Verde archipelago, where Ospreys breed in a range of localities from costal cliffs to more remote inland locations, mean nearest neighbour distance was 3.1km, but ranged from 241 metres to 12.7km, with 45 per cent of nests situated within 2km of their nearest neighbour (Siverio *et al.* 2014).

Nest-building

The fact that Ospreys readily take over existing nests means that many young birds do not actually build a nest from scratch. When they do, it is rarely in the spring of the year in which they breed for the first time and instead is more likely to have been undertaken during the preceding summer. Lone males often build their first nest as three-year-olds, once they have chosen a suitable site. Construction can take several weeks, depending on location and substrate. Sticks are broken from nearby trees or picked up from the ground and then carried to the chosen locality in the bird's talons. Watching an Osprey forcibly snapping branches off trees by grabbing them with their talons makes for a spectacular sight – and sound – especially as the larger sticks that are used to form the base are typically 2–4cm in diameter and up to a metre in length (Figures 4.10 and 4.11). On some occasions, birds actually alight on the outer part of branches and then use their weight to break them (Jamieson *et al.* 1982). In New South Wales, Australia, the mean length of sticks brought to Osprey nests was 48cm, with a maximum length of 1.16 metres, and diameter of 2.8cm (Clancy 2006).

The most problematic aspect of nest-building, particularly for inexperienced individuals, is usually the initial phase. This can prove especially difficult if the bird is building the nest in an upper fork of a stag-headed deciduous tree. Many sticks fall through the branches and down to the ground until the first few finally lodge in place, thereby creating a stable base for the rest of the structure. Within a matter of weeks, and after hundreds of flights to collect sticks from the local area, the male will have created a structure that is recognisable as an Osprey nest. As the nest grows, the bird begins to switch its attention to material with which to line it. Clumps of grass, moss, dried algae and other aquatic vegetation, or even seaweed, are snatched from the ground and then carried to the nest in the bird's talons. If a young male is successful in

Figure 4.10. Ospreys forcibly break sticks off trees or pick them off the ground before carrying them back to the nest (© John Wright).

Figure 4.11. Ospreys can make many trips per day when collecting sticks early in the breeding season (© John Wright).

attracting a female, then his forays to collect nesting material may become even more frequent, spurred on by the presence of a potential mate at the nest. A young pair may spend several weeks together at a newly established nest-site and then, assuming both survive the winter, return to breed the following spring.

Most established breeders come back to the same nest-site each year, and it is often necessary for them to spend several days repairing damage that has occurred during winter storms (Figure 4.12). Males collect more sticks than females, but once a stick has been dropped onto the nest the female will often take over, meticulously interweaving each stick to form a remarkably strong structure (Green 1976, Jamieson *et al.* 1982, Poole 1989). Levenson (1979), who studied breeding Ospreys in California, found that nest-building flights were most commonly undertaken in the pre-incubation stage, but never accounted for more than 4.1 per cent of daylight time, thereby exemplifying the benefits of using the same nest over successive years.

Another task when an Osprey returns to its nest in the spring is to scrape a depression in the centre, to form the nest-cup. The birds achieve this by leaning forward and then kicking backwards with their powerful feet, sending a shower of old lining behind them in the process (Figure 4.13). This depression is then lined with clumps of fresh moss, grass or seaweed, depending on what is available close to the nest (Figure 4.14). As egg-laying approaches, the female, in particular, makes multiple trips to collect nest-lining material.

A newly built Osprey nest is typically 70–100cm in diameter and, depending on the substrate, 10–50cm deep. However, because some nests are used for successive years by multiple generations of Ospreys, they can become very large indeed. The largest nests, which are usually located on solid substrates such as rocks or on the ground, can measure anything up to two metres in diameter and up to four metres in depth (Bierregaard *et al.* 2020). A notable feature as the breeding season progresses is that nests become flatter. As chicks near fledging, the adult

Figure 4.12. Ospreys are usually faithful to the same nest-site each spring, and may spend several days repairing the nest after winter storms (© John Wright).

Figure 4.13. One of the first tasks for returning Ospreys in the spring is to scrape out a shallow depression in the nest to protect the eggs (© John Wright).

birds add further nest lining in order to build up the centre of the nest and provide a solid base for the young to propel themselves into the air for the first time. A favoured material for building up the centre of the nest in pastoral areas of both Europe and the United States is dry cowpats, but typically the birds will bring in further clumps of grass, moss, aquatic vegetation and seaweed (Poole 1989, Dennis 2008, Mackrill *et al.* 2013).

The fact that Osprey nests can become such large structures means that they are sometimes utilised by other species, even when occupied. We recorded Tree Sparrows *Passer montanus* breeding at three different Osprey nests (thus 21.4% of breeding sites), sometimes in multiple years, in the Rutland Water area between 2001 and 2002. Ewins *et al.* (1994) recorded four species – Common Grackle *Quiscalus quiscula*, Tree Swallow *Tachycineta bicolor*, European Starling *Sturnus vulgaris* and House Sparrow *Passer domesticus* – breeding within or immediately beneath occupied Osprey nests in the Great Lakes Basin of Ontario and Michigan in North America, while Reese (1977) reported that both House Sparrows and Common Grackles also used occupied Osprey nests at Chesapeake Bay.

Osprey nests attract a range of invertebrates too, particularly as the same nests are used each year. Philips and Dindal (1977) reported that at least 20 arthropod species of six families occur in Osprey nests, while Monti *et al.* (2019) used cameras to record evidence of the nocturnal activity of insect fauna in three Osprey nests in Italy. They observed larval activity on two of the three sites throughout the breeding season, including larval swarms covering the whole of the nest and, at times, the back of the incubating female and on nestlings' bodies. They were later identified as the beetle *Dermestes frischii*, the larvae of which are known to feed on a wide range of material including dead tissue, wool, skin, food boluses (a semi-solid mass of food and saliva) and carcasses. It is possible, therefore, that these larvae contribute to the cleaning of the nest, by accelerating the degradation of fish remains. However, a single nestling at each of the parasitised nests had abdominal skin lesions consistent with the effects of dermestid larvae,

Nesting and courtship

Figure 4.14. Ospreys line their nest with clumps of grass and other soft materials, which they collect from the ground nearby (© John Wright).

previously recorded on Snail Kites *Rostrhamus sociabilis*, Wood Storks *Mycteria americana* and Great Blue Herons *Ardea herodias* (Snyder *et al.* 1984). Both affected nestlings survived, but it seems possible that larval infestations of this type could have more serious implications in some cases.

Breeding phenology

The timing of breeding activity varies according to geographic region. Ospreys nesting in temperate areas usually return to breeding sites during spring, but those in tropical or subtropical latitudes begin breeding in winter. For example, breeding activity usually takes place from November or December onwards in Cape Verde (Palma *et al.* 2004) and the Red Sea (Fisher 2001), and from July to November in Australia (Clancy 2006).

In temperate regions, the onset of breeding varies according to latitude. Within Europe, Ospreys return to nest-sites in the Mediterranean (Monti *et al.* 2018a), and central France (Wahl and Barbraud 2014) during February, and to central and southern England, as well as Wales, during mid-March (Figure 4.15). Further north, most Scottish Ospreys arrive in early April, although some early birds may be back at nests by late March, when they may be met by snow (Figure 4.16) (Dennis 2008), while Swedish Ospreys typically arrive in mid-April, some 50 days after birds from the Mediterranean (Monti *et al.* 2015). The last birds to return to their nests are those that breed in Lapland, in the north of Finland, who delay their return until mid-May (Saurola 2020). Ospreys in Finland now begin breeding an average of three days earlier than they did in the 1970s, likely due to climate change. There is also considerable latitudinal

Figure 4.15. Male 03(97) was usually the first Osprey to return to Rutland each year, here arriving at his nest on 19th March 2012 (© John Wright).

Figure 4.16. Ospreys migrating to nests at northern latitudes often return to snow in the spring (© John Wright).

variation within the country, with southern breeders initiating breeding an average of 17 days earlier than those in Lapland (Saurola 2021).

An even greater variation is evident in North America, with a helpful summary provided by Bierregaard *et al.* (2020). The population in southern Florida comprises both residents and full migrants. Residents lay eggs from late November to January, but migrants arrive later and begin incubation between January and early March. Further north, the first breeders usually arrive at nests in North Carolina in late February (Hagan and Walters 1990), at Chesapeake Bay in early to mid-March (Reese 1977), and in southern New England in mid to late March (Poole 1984). As in Europe, the most northerly breeders are the last to return, with birds arriving at nests in southeastern British Columbia in mid- to late April and in east-central Labrador in early May (Wetmore and Gillespie 1976).

In migratory populations there is a tendency for males to arrive a few days before their mate, but older females may arrive first (Dennis 2008). An analysis of known arrival dates of breeding birds in the Rutland Water demonstrated that females arrived first on almost twice as many

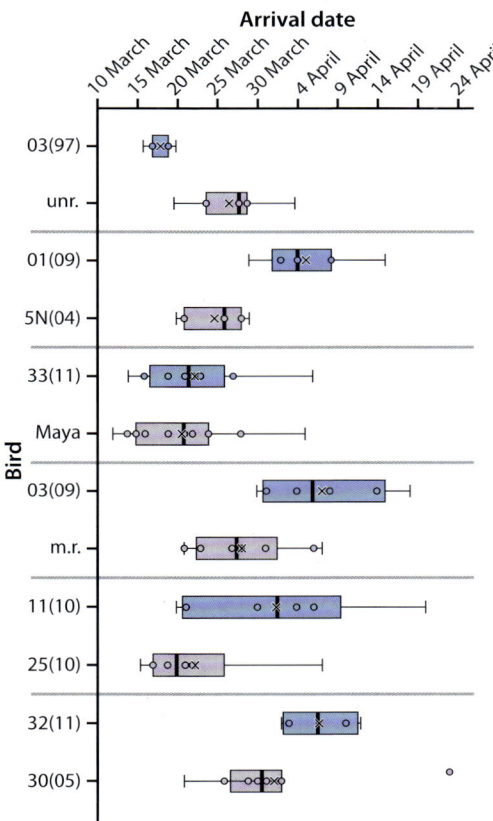

Figure 4.17. Box plot showing annual arrival dates of known male and female Ospreys that bred for a minimum of five years between 2009 and 2022 in the Rutland Water area (where accurate data available). Blue shading = male, grey = female, unr. = unringed, m.r. = metal ringed; bird ID refers to colour ring number with year fledged in brackets; birds that formed a breeding pair are shown together (Tim Mackrill).

The Osprey

Table 4.1. Order of return of breeding pairs to nests in the Rutland Water area

	Relative age of breeding birds			
	Male older	**Female older**	**Same age**	**Total**
Male arrived first	18	1	1	**20**
Female arrived first	7	18	14	**39**
Same date	2	4	0	**6**

occasions as males, usually in cases where the female was older, or if the two birds were the same age. If the male of a pair was older then it would invariably arrive first (Table 4.1). There is an increasing trend for northern European Ospreys to winter in Iberia (Martín *et al.* 2019), which may enable birds wintering there to make an earlier arrival on the breeding grounds, compared to those migrating north from more distant wintering sites in sub-Saharan Africa. Ospreys tend to arrive earlier as they get older, and to be fairly predictable in their arrival date. Figure 4.17 shows annual arrival dates of known males and females to established nests in the Rutland Water area.

Figure 4.18. Unpaired male Ospreys advertise their presence with a spectacular aerial display (© John Wright).

Courtship behaviour

Once back on territory, male Ospreys immediately set about repairing any wind damage that may have occurred to the nest during the winter, and wait for their partner to return. Unpaired males advertise their presence at the nest with spectacular aerial displays (Figure 4.18). They climb up to a height of many metres, often carrying a fish held prominently below them, before suddenly folding their wings and dropping sharply. They may drop a hundred metres or more before pulling out of the dive and rising again with fast wingbeats, all the while uttering the distinctive *eep eep eep* display call, which is sometimes barely audible such is the altitude of the bird. These calls, which are so evocative of spring in an Osprey colony, are more frequently heard in areas with a greater prevalence of young, unattached birds. Roy Dennis comments that these display flights are now much less common in northern Scotland than they were in the 1960s and 1970s when the population was just becoming re-established (Dennis 2008). Likewise, the sky dance display was found to be relatively rare in Corsica, probably due to a high proportion of established pairs and associated site and mate fidelity (Bretagnolle and Thibault 1993).

If a male is successful in attracting a female to the nest, he may continue to display high above for 15 minutes or more before eventually landing beside her. At this point he will often employ a defensive posture, mantling – spreading his wings and then holding them low and outstretched – and facing away from the female, whether holding a fish or not. If he does have a fish then the female may inch her way towards him and gently attempt to pull the fish in her direction with her bill. Some males will allow the female to take the fish at this point, while others are much more reluctant, particularly early in the courtship phase. I have watched some males teasing newly arrived females with fish, repeatedly landing on the nest but then carrying the fish off again. Established pairs also perform the mantling display when they are reunited at a nest in the spring (Figure 4.19).

Figure 4.19. Established pairs perform a mantling display when they are reunited in the spring. Male Ospreys carry out the same display when they are joined on their nest by a newly arrived female. Here 33(11) (right) and his mate, known as Maya, are mantling after their return to the Manton Bay nest at Rutland Water (© Leicestershire and Rutland Wildlife Trust).

Fish plays a central role in courtship, and virtually as soon as a new female arrives at a nest she begins soliciting food by uttering a distinctive repetitive call. This food solicitation call is only given by females and has three versions, as described by Bretagnolle and Thibault (1993). These differ in temporal and frequency parameters, but represent a continuum, as the female usually gives them successively according to her motivation. In essence, the hungrier she is, the louder and more intensely she calls. Mougeot et al. (2002) found that mean female solicitation rate in Corsica was significantly negatively related to the average male courtship feeding rate and to the mean size of fish delivery by the male.

It is thought that courtship feeding is a key way in which females judge the suitability of potential mates (Poole 2019). Males undertake virtually all of the foraging during the breeding season, and so a male who provides an ample supply of fish during courtship is also likely to provision chicks well. This was tested by Green and Krebs (1995) who studied courtship feeding rates in British Colombia in western Canada. They found that courtship feeding rate was positively correlated with male provisioning rate when chicks were one-to- two weeks old, as well as mean brood growth rate. They concluded, therefore, that courtship feeding rate likely enabled female Ospreys to predict the quality of subsequent paternal care. Secondly, Mougeot et al. (2002) found that copulation rate and successful copulation rate were both positively associated with the amount of food provision to the female. Thus, the more frequently the male delivered food to the female, the more often the pair copulated, and did so successfully. Poole (1985), meanwhile, found that well-fed females in southern Massachusetts were less likely to solicit food from other males or to copulate with them.

Spring is an exciting time in Osprey colonies, and small, closely monitored populations with a high proportion of ringed birds provide an excellent opportunity to follow how females select their breeding location. Female 2AF(16) returned to the Rutland Water area for the first time in May 2018. We followed her progress closely that summer and noted that she spent time with three different unpaired males. With this in mind, Lloyd Park and I installed trail cameras at each site the following spring, to enable us to monitor the nests remotely. We were particularly interested to know how much 2AF(16) alternated between sites before settling on one in particular. Two of the males, 51(11) and S1(15), returned on 3rd and 4th April respectively, and immediately set about rebuilding their nests. 2AF(16) arrived on 14th April, and the third male in the trio, 4K(13), was back the following day. 2AF(16) initially joined 51(11), but next morning she was perched on S1(15)'s nest, with the male displaying high above. Then, later that afternoon, the trail camera captured her with newly returned 4K at his nest, 28km to the north-west. This pattern continued for several more days, with all three males readily providing fish. By early May, 2AF(16) was beginning to spend longer periods with 51(11), but continued to visit S1(15), whose nest was located just under 7km to the east. She eventually settled with 51(11) and laid her first egg on 9th May, 25 days after returning. The trail camera revealed that she returned to S1(15)'s nest two-and-a-half weeks into the incubation period, on 28th May. She accepted a fish from the male and then immediately returned to her own nest, where 51(11) had been incubating in her absence. Two chicks went on to fledge successfully and 2AF(16) and 51(11) continued to breed together at the site for a further two years, before 51(11) failed to return in 2022.

Unsurprisingly, there is clear difference in the courtship behaviour of established breeders and those pairing up or breeding for the first time. Although there is little obvious courtship between established pairs when they return in the spring, it is nevertheless wonderful to witness the moment they are reunited after a winter apart. The male will sometimes display when his mate first returns, or as he approaches the nest if she is already present. I remember on one occasion, in 2004, being at Site B at Rutland Water when female 05(00) returned. The male, who had been waiting on the nest with a partly eaten fish, saw her approaching before I did, giving the shrill *eep eep eep* call as he took off and quicky gained altitude. The female headed straight for the nest and then watched for ten minutes or more as the male displayed high above her. He eventually landed on the nest and presented the fish to his newly returned mate. It was a thrilling encounter to watch.

Copulation

The male of an established pair begins provisioning his mate as soon as they are reunited in the spring. They also start copulating immediately, and typically 10–20 times per day (Birkhead and Lessells 1988, Widén and Richardson 2000, Mougeot *et al.* 2002). When a female Osprey arrives at a nest in the spring – whether her own long-established breeding site or a prospective new one – she will invariably spend much of her time perched on the nest, soliciting food and arranging sticks and other nesting material. This means that most copulation attempts occur there (Widén and Richardson 2000). The male flies towards the female and then alights gently on her back with talons closed and wings held open. If a female is receptive, she will lean slightly forward and raise her tail as the male approaches, enabling him to position himself to ensure cloacal contact, flapping his wings to hold the position for just seconds before taking off again (Figure 4.20). If the female continues to lean forward with her tail raised for a few seconds after the male has departed, it is a sure sign that the copulation attempt has been successful. Experienced pairs have a high degree of success, but inexperienced birds often do not achieve cloacal contact because they do not achieve the correct position, or because the female is less receptive (Poole 1985).

Some females who join unpaired males at nests in the spring appear to have no intention of breeding at that locality and instead remain simply to solicit food. The broad distribution of Ospreys means that females from northern populations often pass through other breeding areas on spring migration. It is relatively common for females who already have an established

Figure 4.20. Ospreys usually copulate 10–20 times per day prior to incubation. Frequency generally peaks in the days leading up to egg-laying, when the female is most fertile (© Montgomeryshire Wildlife Trust).

nest further north to linger at a nest with a male to avoid having to hunt for themselves. A sure sign that a female is behaving in this way is that she readily accepts fish from the male, but aggressively rebuts any copulation attempts. We have recorded numerous females behaving in this way at Rutland Water and, in 2020, Rutland female 2AF(16), described above, lingered on a nest with an unpaired male at the Urdaibai Biosphere Reserve in the Basque Country in northern Spain before continuing north. She accepted fish from the male but did not copulate during her two-day stay on 30th and 31st March. She then resumed her journey back to Rutland, and went on to rear two chicks at the nest she had used the previous summer. Females may behave in a similar way if they return to their nest-site before their mate. We have recorded numerous instances of Rutland Water females soliciting for food on neighbouring nests prior to the return of their established mate. This behaviour stops as soon as their regular partner is back in situ.

When a male Osprey returns to the nest with a fish, he will usually take it to a favoured perch to eat. If a female is present, she will solicit for food on the nest, calling with increased volume and intensity as she becomes hungrier. Males in established pairs usually eat the head of the fish before delivering the remainder to their mate on the nest. Poole (1985) found that older, more experienced males shared their food very equitably, whereas younger males (<5 years) were more reluctant to share their catch. This is particularly the case early in the breeding season, when inexperienced males sometimes consume a whole fish while a newly arrived female begs vociferously on the nest. This reluctance to share fish is perhaps not surprising in view of the behaviour of some females, but courtship feeding undoubtedly plays a key role in the formation of new Osprey pairs and so males who do not share their fish are less likely to keep hold of a prospective mate (Poole 1985).

While males may sometimes be reluctant to share food when a new female arrives, they usually try to copulate immediately, even though such attempts are unlikely to be successful. Poole (1985) found that young pairs were less likely to copulate successfully than old pairs (37 per cent vs 72 per cent), usually because the female is not receptive. If the female keeps her tail down it is a sign that she does not wish to copulate. Sometimes this results in the male landing on her back and resting there for a few seconds before taking off again, but on other occasions a female will aggressively rebuff the male as he approaches, raising her wings and pecking at the male as he attempts to alight on her back.

As with courtship feeding, it seems that frequent copulations, even if unsuccessful, are part of the pair-bonding and mate-assessment process. Mougeot *et al.* (2002) found that new pairs in Corsica copulated more frequently, but less successfully, than experienced pairs, particularly early in the pre-laying period. New pairs copulated a mean 329 times prior to egg-laying, with a success rate of 49 per cent compared to 235 times at a success rate of 72.1 per cent for pairs that had bred together before. The Corsican Osprey population is only partially migratory, with a longer pre-laying period than Ospreys in fully migratory populations. This perhaps explains why both established and new pairs in that population copulated more times than elsewhere. Birkhead and Lessels (1988) found that, on average, Osprey pairs monitored over a 10-year period at the Loch Garten nest in northern Scotland attempted to copulate 160 times per clutch (range: 88–338) over a 14-day period, with a success rate of 39 per cent. The corresponding figure in Sweden was about 150 times, with a success rate varying from 64–69 per cent (Widén and Richardson 2000).

Copulation frequency generally peaks in the days leading up to egg-laying, when the female is most fertile (Birkhead and Lessels 1988, Mougeot *et al.* 2002). Males guard their mate throughout the pre-laying period, and particularly during the period when she is most fertile – usually only leaving the nest-site to forage. Mougeot *et al.* (2002) found that male attendance at nests in Corsica increased significantly in line with territorial intrusion frequency by other Ospreys, suggesting they spent more time at the nest to guard their mate and thus their own paternity. The Corsican study demonstrated that males guarded females to the detriment of food provisioning of the female: the more time a male spent in the breeding territory, the less frequently he delivered fish. The authors suggest that these reduced provisioning rates may explain why pairs that faced more territorial intrusions copulated less frequently, since both attempted and successful copulation rates were positively associated with the amount of food the female received.

EXTRA-PAIR COPULATIONS

Extra-pair copulations are relatively rare in Osprey colonies, but are more likely to arise in areas of high population density and typically occur while the male is away from the nest. In Sweden, three successful extra-pair copulations were recorded at three different nests in areas of high population density (average inter-nest distance 1.2–1.6km), but not in areas of low density (average inter-nest distance 6.2–7.8km) (Widén and Richardson 2000). These copulations accounted for 3 per cent of total copulations observed in the high-density area (n = 91) and all occurred two to five days before the start of incubation, while the regular male was away. In each case an intruding male rapidly approached the nest, landed on the female's back, copulated and then flew straight off. The extra-pair copulations were not associated with delivery of food or nesting material. Birkhead and Lessells (1988), meanwhile, reported that extra-pair copulations occurred at a low frequency at the Loch Garten nest during a 10-year period when the population density of breeding Ospreys in the area was low. They recorded 14 extra-pair copulation attempts over that period, which accounted for 0.3 per cent of total copulation attempts. Mougeot et al. (2002) recorded few instances of extra-pair copulations during the study of the pre-laying behaviour of Ospreys in Corsica, where average inter-nest distance was 1.5–2km. Four extra-pair copulation attempts were observed, three of which were unsuccessful due to rejection by the female. The one successful extra-pair copulation was between a breeding female and a neighbouring male whose nest was situated 500 metres away, and within sight of the cuckolded male's nest. It occurred while the female's regular mate was away.

Englund and Greene (2008) monitored the reintroduced Osprey population in eight counties surrounding Minneapolis/St Paul in Minnesota in the United States, recording the behaviour of colour-ringed birds at a total of 306 occupied nests. From 1999 to 2005 they observed copulation with more than one partner in a breeding season 14 times at nine different nests, with most extra-pair copulations occurring before incubation and half occurring before the return of the previous year's mate. This represented 10.52 per cent of all active nests. The most interesting extra-pair relationship occurred at Auburn Lake in 2004, where a female and her 20-year-old mate had failed on eggs for four successive years. Copulatory behaviour was observed soon after the female's return, but no cloacal contact was confirmed. She was subsequently observed copulating multiple times at a second nest, located 2km away, and remained there for over a week. She finally returned to her established nest and mate at Auburn Lake on the day she laid the first egg. The chicks that subsequently hatched – but were later predated – were almost certainly not sired by the cuckolded male. The following year she moved to the second nest and paired with the male responsible for the extra-pair copulations, despite the fact that her established mate had returned to the Auburn Lake nest. All females participating in extra-pair copulation had mates (16.2 years, n = 7) that were significantly older than the mean age of males within the population (8.3 years, n = 67), indicating that this behaviour may be used as fertility insurance.

Conclusions

- Ospreys build large conspicuous nests in a variety of habitats, on a range of different substrates according to local environmental conditions. They prefer a prominent location with expansive views across the surrounding landscape, situated relatively close to suitable foraging grounds.
- Ospreys tend to select the tallest available tree (usually 15–50 metres) or one that is isolated from others. Tree species varies according to local availability but flat-topped pines are often used at northern latitudes.
- A range of artificial structures, particularly electricity pylons, mobile phone masts and channel markers, are used by nesting Ospreys in many locations across their global range. Ospreys also take readily to human-made nests.
- The height of the nest varies according to the surrounding vegetation and nests are often constructed close to the ground in open habitats. Where trees or other structures are absent, Ospreys readily nest on the ground or on cliffs.
- Most Ospreys build their nests within a few kilometres of favoured fishing sites.
- The density of breeding Ospreys varies significantly, according to a range of factors. The greatest spacing between nests is usually recorded in northern boreal forests, but Ospreys can breed in high densities in the most suitable habitat. In some extreme cases, such as in Florida, more than one pair may even breed on the same tree.
- Lone males build nests but prefer to take over existing sites. A newly built Osprey nest is typically 70–100cm in diameter and, depending on the substrate, 10–50cm deep. Once established, the same nest is usually occupied each year and can become very large.
- Ospreys that breed at more southerly latitudes generally return to nests earlier than those from farther north.
- Food provisioning by the male plays a key role in courtship, and is thought to be a crucial way for females to judge the suitability of potential mates.
- Ospreys usually copulate 10–20 times per day, and frequency generally peaks in the days leading up to egg-laying when the female is most fertile.
- Extra-pair copulations sometimes occur, usually in areas where Ospreys breed in high densities. Males guard their mate throughout the pre-laying period to prevent this.

CHAPTER 5
Incubation and chick-rearing

Watching an Osprey nest over a complete season is an immensely rewarding and enjoyable experience. Once a pair of Ospreys is established at a nest, the breeding season follows a predictable cycle from incubation through to the fledging and subsequent departure of any young. Thanks to the establishment of Osprey watchpoints and cameras in many locations, it is possible for anyone to follow the fortunes of an Osprey family through an entire breeding season, even from the comfort of your own home.

Laying dates

Ospreys in migratory populations usually begin incubating 10–30 days after arriving at their breeding territory, and so laying dates follow the same latitudinal trends as the arrival of birds at their breeding sites, as described in the previous chapter. This means that southerly breeders may lay their eggs 2–3 months earlier than those from more northerly latitudes. Within Europe, the earliest breeders in the Mediterranean begin incubating during March and early April (Monti 2012), while those in central France follow soon afterwards, during late March or the first two weeks in April (Wahl and Barbraud 2014). The earliest laying date we have recorded at Rutland Water is 28th March, but most Ospreys in the UK lay from mid-April to early May (Dennis 2008, Mackrill et al. 2013). Finnish Ospreys begin incubating during May, but the most northerly birds in Lapland delay egg-laying until June (Sivonen 2014, Saurola 2021).

Likewise, in North America the most southerly migrant breeders in Florida are the first to begin incubating during January or February (Bierregaard et al. 2020). Elsewhere, average laying dates are very similar to Europe. Most Ospreys returning to nests in North Carolina lay in early–mid April (Hagan and Walters 1990), and those breeding in the stronghold of Chesapeake Bay typically start incubating during April (Reese 1977). Further north, the peak of egg-laying occurs during mid–late April in southern New England (Poole 1984), and not until mid-May to mid-June in Labrador (Wetmore and Gillespie 1976).

Resident populations of Ospreys in the Northern Hemisphere lay their eggs much earlier. Breeding Ospreys in Cape Verde and the Red Sea lay eggs in December (Palma et al. 2004, Habib 2019), as do resident birds in South Florida in the United States (Ogden 1977, Bierregaard et al. 2020). Those in the Canary Islands may lay as early as February, but usually in March (Siverio 2006). In Australia, meanwhile, Ospreys lay their eggs in the austral winter. Those in New South Wales lay from June to August (Clancy 2006), while those in South Australia usually do not lay until September (Dennis 2007a).

Once reunited at their nest-site, established breeders settle into their usual routines very quickly, and usually commence laying earlier than younger birds (Poole 1985, Dennis 2008). Experienced females typically lay 10–14 days after the first copulations, but in one exceptional case, in 2014, the regular breeding female at the Manton Bay nest at Rutland Water laid her first egg just three days after she was observed copulating with a new male for the first time, having waited 20 days for her regular mate to return. In these circumstances, it seems likely that the appearance of a prospective mate, coupled with the time she had been present, were sufficient to trigger the hormonal changes required for the release of the first ovum into the oviduct. Successful fertilisation is dependent on the male's sperm reaching the ovum soon afterwards, and the fertilised egg then slowly descends the oviduct over a period of one–two days before it is laid (Poole 2019). Unfortunately, it was never possible to prove if the eggs were fertile because they were subsequently kicked out by another male, who usurped 28(10) and paired with the female. Even more intriguingly, an unpaired female, CJ7, laid a clutch of three eggs at Poole Harbour in Dorset in 2020. She had been present at the nest the previous summer with a two-year-old male, but he did not return and the female did not copulate with any other males in the 28 days she was present before she laid the first egg. It has been suggested that courtship feeding and early copulations trigger the hormonal changes necessary for egg-laying, but this evidence suggests that simply being present at a prospective nest in the spring can be sufficient. In 2021, CJ7 again returned to the nest and this time laid five unfertilised eggs over an 18-day period between 27th April and 14th May. As in the previous year, she did not attempt to incubate them, and each was eventually stolen by Ravens *Corvus corax*.

Younger pairs and first-time breeders are nearly always later laying because they usually arrive after more experienced breeders, and there is a longer courtship period among new pairs. This variation was well illustrated by a study in southern New England in the United States, where the earliest eggs were laid in mid-April and laying continued through to late May (Poole 1984). All of the females that began incubating during May were aged five years or under, with older birds laying eggs in April. Furthermore, pairs that had bred together in previous years laid significantly earlier than those breeding together for the first time. Similar variation is evident at Rutland Water where in some years newly established pairs have begun incubating as the earliest clutches begin to hatch. For example, in 2022 first-time breeders T7(16) and a metal-ringed (likely Scottish) female began incubating on 16th May, one week after the first chick hatched at the Manton Bay nest where an unringed female and 33(11) were breeding together for the eighth successive year. Overall, the latest known egg-laying date in the UK is 23rd May (Dennis 2008), almost two months later than the earliest. Site fidelity and monogamy thus tend to result in earlier breeding.

Clutch size

Osprey eggs are oval shaped with a creamy base that is beautifully marked with differing amounts of reddish-brown blotches, spots and streaks (Figure 5.1). They are usually more heavily marked and patterned at the larger end of the egg. On average, eggs laid by Western Palearctic Ospreys measure 61.6mm in length and 46.4mm in width (Harrison and Castell 2002), while the mean in North America, reported by Bent (1937), is 61.0 × 45.6mm. In Australia the average size is 61.4 × 45.3mm, although eggs in southwestern Australia are bigger than elsewhere (Olsen *et al.* 1993). The markings usually vary sufficiently within clutches to make it possible to identify individual eggs, and there is a tendency for the first egg to be larger than others (Poole 1982).

Osprey clutch size varies from one to four eggs, with three being the norm in both migratory and non-migratory populations, although the latter tend to have slightly smaller clutches on average (Bierregaard *et al.* 2020). Younger females may lay only two eggs in their first breeding year and three or four thereafter (Dennis 2008). In Scotland 87 per cent of 495 clutches consisted of three eggs, 10 per cent comprised two, 2 per cent had a single egg and 1 per cent had four eggs (Dennis 2008). Poole (1985) found that the mean clutch size of Ospreys breeding on the Westport River estuary in southeast Massachusetts was higher than average at 3.23, with 29 per cent of nests containing clutches of four eggs. This, however, is exceptional, and four-egg clutches are relatively rare. In 2019, two of 10 breeding females in the Rutland Water area laid four eggs and, exceptionally, all four chicks fledged in each case. Poole (1985) found that clutch

Incubation and chick-rearing

Figure 5.1. Three unhatched Osprey eggs recovered under licence from sites in the Rutland Water area during ringing in 2016 (© John Wright).

size was fairly constant year-on-year for individual females, and thus there may be a tendency for some females to lay clutches of four. An unringed female breeding at the Manton Bay nest at Rutland Water laid four eggs in 2017, 2019 and 2020 (23.1 per cent of her clutches for 2010–2022), while a female breeding in Kielder Forest laid four eggs for the first time in 2016 and again in 2019, 2020, 2021 and 2022 (35.7 per cent of clutches) (Dailey 2022) (Figure 5.2).

Clutches laid later in the breeding season, which invariably involve younger, more inexperienced breeding pairs, also tend to be smaller. Poole (1984) identified a clear decline in clutch size with laying date in southeast Massachusetts. Mean clutch size was very high, at 3.42 ±0.10

Figure 5.2. An unringed female at a nest in Kielder Forest, Northumberland, laid clutches of four eggs on five occasions between 2016 and 2022 (© Forestry England).

SE (n = 31) during the first week of colony laying, when more experienced birds were most likely to be producing eggs, compared to 3.14 ±0.14 SE (n = 14) four or more weeks after the initiation of laying in the colony. Similarly, Steeger and Ydenberg (1993) found that clutch size of Ospreys breeding in the Kootenay region of British Columbia declined by 0.05 eggs per day in 1987 and 0.02 eggs per day in 1988. Furthermore, pairs that started breeding after the median laying date of 8th May experienced significantly greater partial and total clutch failure than those that laid beforehand. In some cases, delayed laying was caused by competition with Canada Geese *Branta canadensis*, which occupy Osprey nests early in the spring. Canada Geese are strong, powerful birds and although Ospreys will divebomb them repeatedly, the geese are usually able to withstand these attacks if they are incubating eggs. Faced with such a situation, Ospreys either had to wait for the geese to vacate the nest or build a new one – unless there was a vacant nest nearby. Steeger and Ydenberg (1993) surveyed 83, 108 and 124 Osprey nests in 1988, 1989 and 1990, respectively. They found that geese occupied 45 (54 per cent), 47 (44 per cent) and 52 (42 per cent) of these nests, distributed throughout the study area regardless of location or substrate. Ospreys used a range of sites in the area, including trees, pilings, bridges and electricity poles. Ospreys without an alternative nest were disproportionately affected, with laying date delayed by 9.4 days and clutch size reduced by 0.2 eggs.

Geese are also potential competitors for nesting Ospreys in the UK. At Rutland Water nests have been occupied by Canada Geese (Figure 5.3), Greylag Geese *Anser anser* and Egyptian Geese *Alopochen aegyptiaca* prior to the return of Ospreys each spring (Mackrill *et al.* 2013). This has been a particular issue at the Manton Bay nest, which Egyptian Geese, in particular, attempt to take over each year. It is noteworthy that the spring arrival dates of the Ospreys breeding at the nest have got progressively earlier over the years, and this is probably due to competition with the geese. The same unringed female bred at the nest each year between 2010 and 2022; during that period her average arrival date was 20th March, but from 2018 she arrived between 12th and 19th. This earlier arrival helps to ensure that the Egyptian Geese do not have time to become established on the nest. Once a pair of Ospreys are back at their nest, they

Figure 5.3. Canada Geese compete with Ospreys for nests in both their native North America and in Europe (© John Wright).

Incubation and chick-rearing

Figure 5.4. Egyptian Geese usurped a pair of first-time breeding Ospreys from a nest at Rutland Water in 2016, and are regular competitors for the Manton Bay nest (© John Wright).

aggressively defend it against any goose intrusions, and are capable of keeping them away. The one instance when this was not the case was in 2016, when two first-time breeders, 51(11) and 3J(13), attempted to breed on another artificial nest on the nature reserve. 3J(13) laid an egg on 5th May, but she and her mate subsequently roosted away from the nest – a sure sign of inexperience. This allowed a pair of Egyptian Geese to claim the nest and then keep the Ospreys away, meaning that the breeding attempt failed at a very early stage (Figure 5.4). Elsewhere in their European range, including in Spain and Germany, White Storks *Ciconia Ciconia* may compete with Ospreys in a similar fashion (Bai *et al.* 2009).

Ospreys lay their eggs at intervals of one to three days but incubation usually begins with the first egg – hence, there is usually an age gap of several days between the oldest and youngest in the brood, particularly in the case of clutches of four. First eggs also tend to be slightly larger than subsequent ones which, when combined with hatching asynchrony, gives young from the first and second eggs a competitive advantage over younger and smaller siblings (Poole 1982). Nest cameras, such as the one at Cors Dyfi in Wales, have shown that some pairs only incubate intermittently until the second or even third egg is laid, and this can have the effect of somewhat reducing the age gap between the oldest two chicks (Dyfi Osprey Project 2020b).

Incubation

Even when viewing a nest from a distance, an incubating Osprey is easy to spot: it sits low in the nest, often with its white head just visible above the edge. In recent years, high-definition nest cameras have provided a unique insight into the comings and goings at Osprey nests, and it is wonderful to watch the great care with which an Osprey settles on its eggs. As it shuffles up to the eggs, the Osprey is able to draw its razor-sharp talons inward, thereby negating any chance of piercing the delicate eggshells. Initially, the bird will stand over the eggs, moving them into position, before hunkering down on top of them with a distinctive rocking motion.

Male and female Ospreys have very clearly defined roles during the incubation period.

Females usually undertake around 70–80 per cent of the incubation, with the male taking over in order to give her a chance to feed and rest. During the early years of breeding at Rutland Water when we were concerned about the ongoing threat of egg-collectors, we guarded nests round the clock during the incubation period and this enabled us to keep detailed records. Female 05(00) raised a total of 17 chicks at the Site B nest between 2003 and 2008. During that period she undertook a mean 73.6 (±2.3) per cent of incubation. The unringed female who replaced her did a larger share – 82 per cent – in 2009, while female 5N(04), who bred at the Manton Bay nest in 2007 and Site N in 2009, undertook 73 per cent and 89.4 per cent of incubation when she bred in 2007 and 2009.

The male's primary role during incubation is to supply food and guard the nest. The food solicitation call of the female usually prompts the male to depart in search of a meal. At Rutland Water we found that male Ospreys usually provide one or two fish per day during the incubation period (Mackrill *et al.* 2013), typically soon after first light and again in mid–late afternoon. Females incubate quietly while their mate is away fishing, but as soon as he appears on the horizon she will begin calling again. We quickly learnt when monitoring Osprey nests that an incubating female would always see her mate returning with a fish before we did, and her food solicitation call thus offered an early warning of his impending return. Once back at the nest male Ospreys usually take their catch to a favoured perch, where they eat the head and upper part of the fish (Figure 5.5). They then deliver the remainder to the female, and this prompts a change in incubation duties. The male alights gently at the side of the nest, prompting the female to rise up carefully and take the fish from her partner. She then carries it to a nearby perch to eat, while the male settles onto the eggs (Figure 5.6). Ospreys are fastidious incubators and established pairs will leave their eggs uncovered for no more than a few seconds at change-overs. In contrast, inexperienced first-time breeders can be quite comical to watch, often taking some time to settle on the eggs as they attempt to position themselves correctly. At other times, males will instigate a change-over by flying to the nest. This often prompts the female to stand up, but on other occasions the male will actually nudge the female and almost lever her up and off the eggs such is his volition to take over incubation duties.

Figure 5.5. Male and female Ospreys have clearly defined roles during the incubation period. The female undertakes approximately three-quarters of the incubation, while the male catches all of the fish that he and his mate require. Here, a male Osprey at Rutland Water can be seen eating a Roach while his mate is incubating. Males usually consume the head of the fish before taking the remainder to their waiting mate, and then taking over incubation while the female eats (© John Wright).

Some males appear more eager to incubate than others, and this is reflected in the findings of two North American studies where the proportion of time different males spent incubating varied considerably. Grover (1984) found that the contribution of breeding males in Montana varied from 26 per cent to 57 per cent, while Poole (1984) reported a variation of 13 per cent to 66 per cent at nests in New England (Poole 1984). It is not clear what causes this variation, but

Incubation and chick-rearing

Figure 5.6. Male Ospreys take over incubation to give their mate an opportunity to feed and rest (© John Wright).

having spent many hours watching Osprey nests during incubation, it appears to me that some males are simply more motivated than others. One thing that usually is certain is that the female always does the night shift, taking over from her mate at dusk and then sitting tight through the night until another change-over soon after dawn. Watching nests through an image-intensifier night-sight at Rutland Water revealed how fastidiously females incubate during the hours of darkness, moving only to turn the eggs, reposition themselves or to leave the nest for the briefest of periods in order to defaecate.

During the incubation period, very little else happens at Osprey nests. When he is not away fishing, the male will perch quietly in a favoured spot near the nest and take the opportunity to preen. Ospreys have a gland on their rump that produces oil to aid waterproofing, and breaks between fishing trips and incubation duties provide an ideal opportunity to preen this into their feathers. If you watch the male carefully, though, you will see that he is constantly alert, interrupting preening at regular intervals to glance upwards and scan the horizon. If he sees another Osprey approaching, he will utter the distinctive guard call, a short, sharp *tioop* which echoes around the nest-site and warns the intruding bird not to come too close. If the intruding bird is some distance away then this call acts as a warning not to approach any closer. If this initial warning is not heeded then the call increases in both frequency and intensity, to become the 'excited call' as the intruder approaches, as described by Bretagnolle and Thibault (1993), and reviewed in chapter 1. At the same time a subtle change in posture is often evident, with the male taking a more upright stance with extended neck, erect crest feathers and wings held slightly open. If the intruder lingers, the male usually flies to the nest and takes up a much more defensive posture with body axis more or less horizontal, wings mostly opened and carpal joint held low. As the intruder nears the wings are shaken and the tail is raised and contracted (Figure 5.7). At this point the female will often stand up and take on the same posture as the pair defend the nest. If the intruder is persistent then the male will take flight and attempt to chase it off, often uttering the 'screaming' call which typically follows the guard and excited calls (Figure 5.8).

The Osprey

Figure 5.7. Breeding Ospreys (left) take up a defensive posture on the nest if an intruding bird (right) is particularly persistent (© Montgomeryshire Wildlife Trust).

Figure 5.8. A key role of the breeding male is to defend the nest against intruders. Persistent intruding Ospreys are eventually chased off if initial warning guard calls are not heeded (© John Wright).

The female is more vulnerable while her mate is away from the nest, and intrusions by non-breeders can be problematic when the male is absent. Young non-breeders can be particularly persistent and repeated intrusions during incubation can sometimes lead to reduced brood size, or even nest failure. Of the 113 breeding attempts we recorded in the Rutland Water area between 2001 and 2022, frequent intrusions by rival males have resulted in at least three cases of brood reduction, and one nest failure. Although Osprey eggs are fairly robust, they can become chilled if the female is kept off the nest by an intruding bird. Leaving the eggs unattended in order to chase away an intruder also increases the risk of an opportunistic corvid predating them. In addition, a further four nest failures were attributed to territorial disputes between rival males.

Loss of eggs early in the incubation period is not necessarily terminal because females are able to re-lay if eggs are lost in the first 1–3 weeks, although replacement clutches are invariably smaller (Bierregaard et al. 2020). A remarkable example of the ability of females to lay multiple eggs occurred at the Glaslyn nest in North Wales in 2015. The same pair of Ospreys had bred at the site since 2004, rearing a total of 26 chicks during 11 years of consecutive breeding. When the male failed to return in 2015 the female attracted the attention of two unpaired males. She subsequently laid a clutch of three eggs, despite only sporadic copulations with the two males and no regular mate. These eggs were almost certainly unviable and the female did not attempt to incubate them. One of the males was subsequently found electrocuted near a favoured fishing site, but a new male arrived at the nest on 30th April, and was accepted by the female. She laid a further two eggs soon after his arrival, but again did not attempt to incubate them – indicating that, again, they were not viable. She continued to copulate with the male, and then on 18th May she laid her sixth egg of the season. This time she and her new mate took turns to incubate it and, five days later, she laid a seventh egg. Both subsequently hatched, including the seventh egg which at one point, midway through the incubation period, was left exposed for 11 hours after being accidentally knocked out of the nest-cup by the adult male. The two chicks went on to fledge successfully and both began their first migration south on 25th September that year (Glaslyn Wildlife 2022c).

Replacement clutches are usually of reduced size, but there are sometimes exceptions. As described in chapter 3, a regular breeding female at a nest near Rutland Water paired with a new male, 8F(12), in 2018 and laid a clutch of three eggs. When her established male, 02(09) eventually returned, very late, on 18th April he usurped 8F(12), and kicked the eggs out of the nest-cup. The female eventually laid a replacement clutch of three eggs just under four weeks later, and all three chicks subsequently hatched.

Hatching

As hatching approaches, the breeding females become noticeably restless, perhaps sensing movement within the eggs. High-definition cameras allow us to monitor breeding Ospreys, and to record the exact duration of the incubation period, because individual eggs can be recognised on camera according to differences in markings. The first egg hatches approximately 36–42 days after it was laid. Detailed records kept by observers from the Dyfi Osprey Project and Glaslyn Wildlife in Wales showed that the first egg hatched after a mean 38.7 days at Cors Dyfi (n = 9) (Dyfi Osprey Project 2022b) and after a mean 37.1 days at Glaslyn (n = 15) (Glaslyn Wildlife 2022b); the corresponding figure at Rutland Water was 37.7 days (n = 21). Meanwhile, the average duration of incubation of breeding pairs in southeastern British Columbia, Canada was 38.5 days (range = 36–42 days) (Steeger et al. 1992).

The first very fine cracks may appear up to two days before a chick hatches, but the most noticeable sign of impending hatching, at least if you are viewing on a nest camera, is the distinctive pip hole that appears in the egg as the chick uses its egg tooth – the hardened tip of the upper mandible – to break through the eggshell. The chick will sometimes emerge less than an hour after this hole is first evident, but sometimes it can take 12 hours or more (Figure 5.9). The female continues to incubate the eggs even as the first chick begins to hatch, but it is noticeable even if you watch the nest from a distance that she appears to be sitting higher in the nest than usual. She will also be restless, frequently standing up and looking down into the nest-cup. These subtle clues are signs that the first egg is hatching, if watching a nest from a distance and without the aid of a camera.

Second, third and fourth eggs typically require fewer days of incubation before they hatch. Mean duration of hatching at Dyfi = 38.7 days for egg one, 37.1 days for egg two and 36.1 days for egg three (Dyfi Osprey Project 2022b); Glaslyn = 37.1 days for egg one, 35.7 days for egg two, 34.9 days for egg three (Glaslyn Wildlife 2022b), but the gaps between hatching are influenced by laying interval and the onset of incubation. The largest gaps occur in clutches where incubation began with the first egg, but greater hatching synchroneity ensues if full incubation does not begin until the second egg is laid. At the Cors Dyfi nest there was a mean 4.47 days (n = 2) between hatching of three egg clutches when incubation began with the first egg, but just 1.97 days (n = 3) when full incubation did not begin until the third egg had been laid (Dyfi Osprey Project 2022b).

Ospreys incubating unviable or unfertilised eggs will continue to sit well past expected hatching date. At Rutland Water, a pair continued incubating an unviable egg for over 60 days in 2008 before eventually giving up. Similarly, the first pair of Ospreys to breed in mainland Spain since 1981 incubated a single egg for over 60 days in 2005. This pair subsequently accepted two Osprey chicks that were translocated from northeast Germany in order to encourage site fidelity of the pair and to facilitate ongoing efforts to restore breeding Ospreys to Andalucía (Muriel *et al.* 2006). This is described in more detail in chapter 8.

Figure 5.9. Nest cameras, such as the one at Cors Dyfi, provide remarkable close-up images of Osprey chicks hatching (© Montgomeryshire Wildlife Trust).

Incubation and chick-rearing

Figure 5.10. Osprey eggs hatch in the order they were laid, meaning the first chick has an advantage over its siblings (© John Wright).

Osprey chicks are semiprecocial when they hatch; their eyes open within a few hours and they are covered in short, thick buff-coloured down (Figure 5.10). It is an immense privilege to witness the first few hours of a young Osprey's life, to see it slowly gaining the strength needed to hold its head up and beg for food from the female. Most chicks are strong enough to be fed within a few hours of hatching, and female Ospreys show great tenderness towards their vulnerable young. It is particularly interesting to watch first-time breeders tending to their newly hatched chicks for the first time, the power of instinct guiding their every move. During incubation, a fish delivery by the male signals a change-over in sitting duties; but as soon as the first chick hatches, there is a clear shift in behaviour. Rather than taking the fish away to eat on a favoured perch, the female remains on the nest. She takes the fish from the male and then slowly inches her way to the side of the nest, where she tears off tiny morsels, starting at the anterior end, and delicately offers them to her freshly hatched chick (Figure 5.11). It can take first-time breeders a few hours to master the required technique, which involves leaning forward deep into the nest-cup when chicks are newly hatched and very small. The female is careful to discard bones and other hard parts of the fish, or to eat them herself. The chick will readily take food within a few hours of hatching, but only in very short stints of between five and 10 minutes. Once the chick has had its fill, the female will leave the fish on the side of the nest and settle back down on the chick and any remaining

Figure 5.11. Females tear up tiny pieces of fish and delicately offer them to newly hatched chicks (© Montgomeryshire Wildlife Trust).

unhatched eggs. The hatching of the first chick heralds a change in the roles of the adult birds: the male will now concentrate solely on catching fish and protecting the nest, leaving the female to brood and care for the young and to continue incubating the remaining eggs which may not hatch for several days. If there is a fish in the nest then the female will stand up at regular intervals – often several times an hour – to offer fish to the chick.

Newly hatched chicks weigh approximately 50g, and can gain a quarter of their initial body weight each day for the first five days after hatching, meaning there can be a considerable size difference between the oldest and youngest in the brood, particularly if there are four chicks (Steidl and Griffin 1991). At the Manton Bay nest at Rutland Water, where the female laid four eggs over an 8–9-day period in 2017, 2019 and 2020, hatching occurred over the course of one week on each occasion. By the time an Osprey chick is seven days old it may weigh in excess of 125g, meaning that the oldest of brood could be double the body weight of the youngest in these circumstances.

Unsurprisingly, the oldest chicks tend to be dominant at feeding time, principally because they are larger and more mobile than their smaller siblings. Therefore, the smallest chicks only survive if the male is able to provide sufficient fish. The largest chicks are invariably fed first, but the female will continue to feed any smaller, weaker and subordinate siblings after the largest of the brood is satiated. She then feeds herself last. The provisioning rate of the male is thus crucial at this early stage.

Provisioning rates

The sight of a newly hatched chick prompts a distinct step-change in the frequency of fishing trips by the male, and he immediately doubles or even triples the number of fish he delivers to the nest each day (Figure 5.12). Green and Ydenberg (1994) undertook a study of the energy expenditure of males provisioning young in the Kootenay region of British Columbia. They

Figure 5.12. Males increase their fishing effort once chicks have hatched, delivering fish to the nest at regular intervals during the day, which their mate then feeds to the young (© Leicestershire and Rutland Wildlife Trust).

found that males with dependent offspring spent a mean 1.34 (±0.38) hours, or approximately 8 per cent of their active day, foraging. Fish capture rates increased significantly when chicks were aged between 1–2 and 3–4 weeks, which coincides with the period when the greatest gains in body mass are made (Steidl and Griffin 1991). The male's intake rate did not increase during this period, suggesting that nestlings obtained more food as a result of the increased hunting yield.

The mean daily energy expenditure (DEE) of males in the study was 1,248 kJ/day. This was independent of brood age, but males provisioning broods of three chicks spent significantly longer in flight and consequently had higher energetic expenditure than males provisioning one or two young. Furthermore, they did not consume more food themselves, indicating that they likely lost a greater proportion of their body mass over the course of the breeding season than those rearing smaller broods. Nevertheless, the mean DEE of males provisioning broods of three young (1,336 kJ/day) remained below the expected maximum DEE for a male Osprey weighing 1,428g (Karasov 1990). Weather conditions in the study were fairly benign, but in areas with more variable weather, including prolonged periods of heavy rain or strong winds, it is likely that males may be forced to increase their DEE closer to the daily maximum, or simply stop hunting altogether. Weather at hatching is also crucial to the survival of the brood (Poole 1989, Dennis 2008). Female Ospreys brood their young almost continuously for at least the first two weeks after hatching, and this is particularly important during cold and wet conditions when small chicks can become chilled very quickly. Cold and wet conditions can be fatal through the combined effect of chilling the chicks and making fishing difficult for the male. In such a situation, the youngest chicks rarely survive, and on some occasions prolonged periods of wet weather can be fatal for the whole brood. However, female Ospreys are incredibly strong and determined mothers, sheltering their young remarkably well from torrential rain and cold winds.

Green and Ydenberg (1994) also artificially manipulated some broods in their study, but male Ospreys neither increased their energetic expenditure nor reduced their food intake when provisioning experimentally enlarged broods. As a result, chick intake rates were significantly reduced and chicks gained weight at a slower rate, following experimental brood enlargement. The authors suggest that this could be interpreted as an unwillingness to pay the costs associated with increased energetic expenditure. An earlier study by Poole (1984), in which he manipulated broods of Ospreys in New England in the United States from three to five young, demonstrated male provisioning rates did increase after brood enlargement, but not sufficiently to compensate for the increased calorific demands. Under these circumstances males did not reduce their food intake, but females did and subsequently lost three times more weight than their mates. The reluctance of males to reduce their food intake in these circumstances is perhaps due to their higher daily energetic requirements. Green and Ydenberg (1994) calculated that the mean intake rate of males (50 ±29 kJ/hour) was considerably greater than that of females (29 ±10 kJ/hour). The corresponding figure for chicks was 50 ±8 kJ/hour. Likewise, Poole (1984) found that females in New England generally receive 15–20% of the food her mate catches, with males consuming twice that amount.

Growth of nestlings

Osprey chicks grow extremely quickly thanks to their protein-rich diet. Within 10–12 days of hatching, they enter the 'reptilian stage', when the emergence of their second down gives them a much darker, scaly appearance (Figure 5.13). The first rusty-brown pin feathers appear on the chicks' head and neck once they are a fortnight old, and body feathers follow soon afterwards. At this stage, the youngsters' appearance changes markedly by the day, but there is often still a

Figure 5.13. Osprey chicks enter the 'reptilian stage' once they are 10–12 days old (© Leicestershire and Rutland Wildlife Trust).

clear size difference, particularly between the oldest and youngest of the brood. Steidl and Griffin (1991) studied the growth of Osprey chicks in New Jersey in detail and determined an overall growth rate of 0.173, which was best described by a logistic curve (Figure 5.14). Nestlings entered the fastest phase of growth – the steepest segment of the S-shaped logistic curve – at approximately 10 days of age. Schaadt and Bird (1993) calculated a similar growth rate of 0.18 among 19 broods at nests in Nova Scotia in Canada and likewise concluded that growth-rates were best represented by a logistical curve.

Squabbles between siblings sometimes occur – particularly as the chicks become more mobile – but rarely amount to anything serious unless there is a food shortage, in which case any size difference may become more pronounced as the eldest dominates at feeding time. Nest cameras provide a unique insight into the dynamics in these moments, and it is easy to see how the smallest chicks in broods of three or four suffer if there is a lack of food. Larger nestlings are more mobile and usually monopolise a position close to the female at feeding time and sometimes show aggression if siblings approach the female. This aggression can range from simple pecks, shoves and threat displays with wings raised and neck extended, to more aggressive acts such as biting while pulling and twisting the skin, and hits where a chick may lean back and then lunge forward, striking a sibling with a hard blow to the head, back or neck (Machmer and Ydenberg 1998). These behaviours are not always related to food, but aggression is most likely in food-deprived nests. Smaller chicks sometimes have bald patches at the back of their head where they have been repeatedly pecked and attacked by larger, dominant siblings.

Machmer and Ydenberg (1998) studied chick dynamics at feeding time experimentally at Kootenay Lake in British Colombia, Canada. They removed a total of 10 whole broods, with nestlings ranging from 5–13 days of age, from their natal nest for a period of two days and exchanged them with a brood from one of two 'viewing' nests located on pilings in the Kootenay River, and easily visible from a nearby railroad bridge. The transferred brood were then held in an artificial nest for a period of 3.5 hours, which corresponded to the time required to empty a three-quarters-full crop. Broods were then subjected to one of two feeding treatments in order

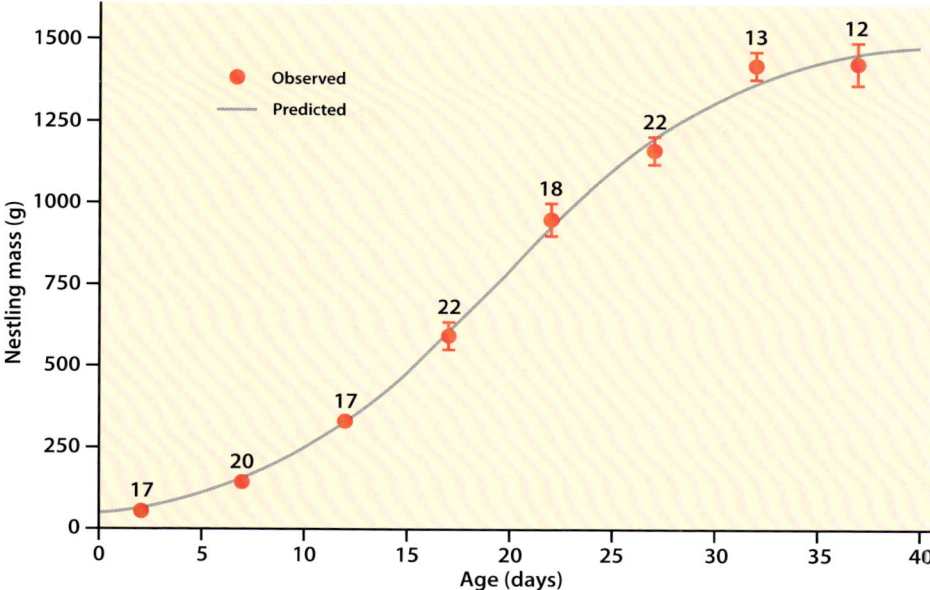

Figure 5.14. Observed values of Osprey nestling mass according to age in southern New Jersey in 1998, which conformed to predicted logistic growth. Logistic growth produces an S-shaped curve with a period of fast growth followed by a period of levelling off. The sample size, which is shown above the observed values, is the number of nestlings weighed during each interval. Error bars were too narrow to show for all points (reproduction of figure from Steidl and Griffin 1991).

to artificially manipulate their hunger level – one where the whole brood was fed to satiation and another where they were sham-fed – before being weighed and then returned to the viewing nest, along with a whole fish. The female typically initiated feeding as soon as the brood was returned to the nest. This process was repeated a minimum of four times (i.e. two feeding treatments and two sham-feeding treatments) over the two-day period before the nestlings were returned to their natal nest. As expected, meals were 45 per cent larger following the sham-feeding treatments and more aggression was recorded. The dominance of older siblings was exemplified by the fact that they received a larger share of meals when they were hungry than when they were satiated. The overall effect of hunger on aggression within the brood, however, was less significant compared with the effect of brood asymmetry. Aggression was least pronounced when there was greater brood mass asymmetry because, in this case, larger siblings are able to maintain unrestricted access to food by virtue of their size advantage. However, when siblings are closer together in terms of weight there may be a greater need for aggression to maintain dominance at feeding time. This corroborates earlier work of Forbes (1991), who demonstrated that feeding among brood members was hierarchical, with the largest of the brood eating most when hungry and leaving a smaller residual share of food for junior siblings. Furthermore, aggression was diminished with artificially exaggerated hatching intervals, suggesting that the largest sibling did not need to exert its dominance through aggression due to clear differences in size and mass.

The dominance of older siblings in food-deprived nests invariably means that the youngest chick does not survive. In their New Jersey study, Steidl and Griffin (1991) found that brood reduction (i.e. the loss of at least one nestling) was recorded in 36 per cent of broods of three and in 23 per cent of broods of two, and these losses typically occurred as chicks began to grow at the fastest rate, two to three weeks after hatching. The crucial factor in this regard was that differences in the body mass of siblings in broods of two remained relatively constant as they grew, but increased in broods of three. There was an average percentage size difference in excess of 50 per cent for broods of three, compared to 25 per cent for broods of two during the period when the greatest proportion of nestlings were lost. Indeed, percentage weight difference between the two groups was most pronounced from 15 to 20 days and subsequently decreased as nestlings approached fledging age when they require less food. The authors conclude that although brood reduction was almost certainly related to rates of fish provisioning by the male, this was not down solely to food availability or abundance, given that foraging Ospreys exploit large numbers of Menhaden that spawn in shallow estuarine waters close to nests. Instead, they posited that differential rates of brood reduction in this colony were due to differing levels of fishing ability, experience and motivation of breeding males. In Scotland, 23 per cent of 1,120 broods of fledged young analysed by Dennis (2008) contained one chick, 47 per cent had two, and 30 per cent comprised three chicks. Although brood reduction was not explicitly tested, 87 per cent of 495 clutches contained three eggs and, although not all would hatch, it is clear that brood reduction does occur on a regular basis in Scottish nests, as shown in Figure 5.15.

By three weeks of age Osprey nestlings have attained about half of their final body weight (Steidl and Griffin 1991), and are extremely mobile, though still not strong enough to walk on their feet (Figure 5.16). Instead, they shuffle around the nest on their haunches, with their disproportionately large feet giving them a decidedly comical appearance. They often manoeuvre their way to the edge of the nest and stare at the world below. By now, their wings are starting to develop, with primary and secondary feathers beginning to grow. The chicks are

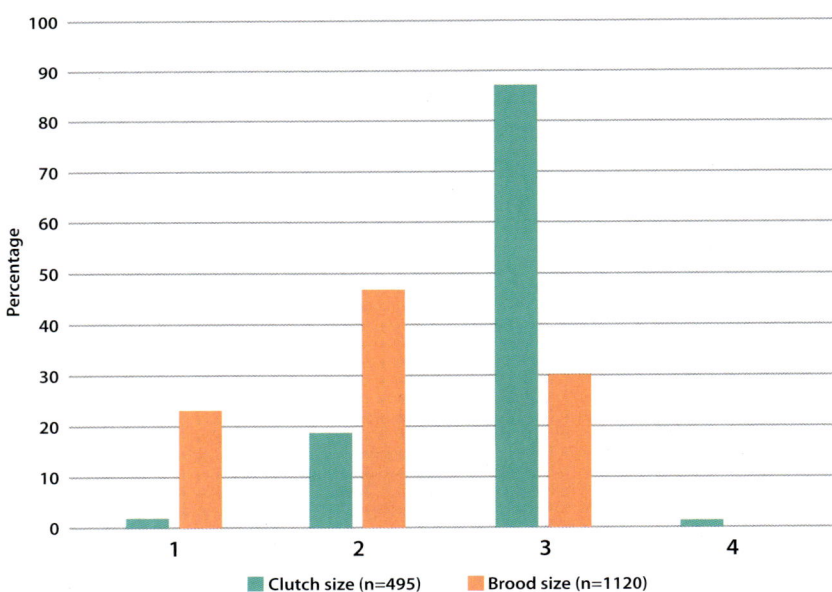

Figure 5.15. Clutch size and brood size of Ospreys monitored and analysed by Roy Dennis in Scotland (Dennis 2008).

Incubation and chick-rearing

Figure 5.16. By the time they are three to four weeks old, Osprey chicks are more mobile, with feathers rapidly replacing down (© Leicestershire and Rutland Wildlife Trust).

now so large that it becomes increasingly difficult for the female to brood them, especially if there are three or four chicks. Nonetheless, she will make every effort to shelter them from heavy rain, and shade them from strong sun.

As they get bigger, the female feeds her offspring larger pieces of fish. Green and Ydenberg (1994) quantified this change in Nova Scotia by counting the number of bites taken to consume a fish of known mass. In each case a fish weighing 200–400g was placed on the nest, and its consumption observed. The smallest chicks, aged 1–2 weeks, were fed pieces weighing a mean 0.35g (±0.16) by the female, which equated to approximately a third of the size of chunks eaten by adult birds (1.03g ±0.12). By the time chicks were three–four weeks old they were consuming much larger pieces, weighing a mean 0.87g (±0.10), with this remaining constant to fledging. The male usually leaves all feeding duties to the female, but over the years at Rutland Water we have sometimes observed males offering pieces of fish to their growing chicks, and even their mate. Rarely, some males assist with the feeding on a more regular basis, particularly as chicks in larger broods become more mobile.

Nestlings continue to grow rapidly, with the greatest gains in body mass occurring between the ages of 20–35 days, when it increases by 50–75g per day (Steidl and Griffin 1991). Their appearance continues to change too, and by the time they are a month old they are fully feathered, albeit with growing of primaries and secondaries still to do. At four to five weeks of age, the youngsters are strong enough to stand up, and this enables them to start flapping their wings for the first time. Although rather weak and feeble at first, this wing flapping grows progressively stronger each day, and by the time the chicks are 6–7 weeks of age each stroke of their almost fully formed wings is considerably more powerful and controlled. At this age the young Ospreys are no longer putting on weight, and this is reflected in a decline in food intake. This has the effect of reducing body mass as they approach fledging.

Beautiful pale fringing to their brown upperparts gives the nestlings a striking appearance, but also offers important camouflage. At the first sign of danger – usually signalled by an

The Osprey

Figure 5.17. Osprey chicks lie prone in the nest at the first sign of danger, with the pale fringing on their upperparts providing excellent camouflage (© John Wright).

intruder or alarm call from their mother – the chicks immediately lie prone in the nest, their mottled plumage making them remarkably inconspicuous from above (Figure 5.17). Eagle Owl *Bubo bubo*, Bald Eagle *Haliaeetus leucocephalus*, White-tailed Eagle *Haliaeetus albicilla*, Northern Goshawk *Accipiter gentilis* and Common Buzzard *Buteo buteo* are all known to predate Osprey nestlings at differing ages and so their plumage offers some protection against these aerial threats. Predation of Osprey nestlings is covered in more detail in chapter 9.

With her offspring now almost fully grown, there is much less space on the nest for the female, and so she usually sits on a nearby perch keeping a watchful eye over her brood. Another key difference once the youngsters reach six to seven weeks of age is that they are able to feed themselves. This means that at this stage when the male lands on the nest with his latest offering, there is often a clamour to take the fish from him. If the female gets there first, she will feed the chicks, but sometimes one of the youngsters will grasp hold of the fish, mantle over it with wings drooped and then expertly tear pieces from the flesh while holding the slippery skin in its talons. That said, it is rare for a chick to eat a whole fish in a single sitting; there is usually ample to spread between the brood if the fish is large enough.

By now, it can be possible to sex males and females, even from a distance. Females are bigger, more thick-set birds with larger bills and thicker legs. Males, in comparison, look relatively slim and sleek, and differences can be quite obvious when siblings are perched close together on the nest. Nevertheless, there are always individuals that lie somewhere in between and are more difficult to separate. Biometrics collected at ringing are usually the best way to separate young Ospreys, as discussed in chapter 1.

Fledging

Once they reach seven weeks of age, it is just a matter of time before young Ospreys fledge. During the week leading up to their maiden flight, wing flapping becomes so powerful that it begins to lift the youngster upwards; usually this is by only a few centimetres at first, but after three or four days of 'helicoptering' they sometimes hover a metre or more above the nest (Figure 5.18). This is a sure sign that the juvenile Ospreys are ready to fly, and the only barrier remaining is not a physical one. Now, it is simply a case of mustering the courage to launch themselves off the safety of the nest for the first time. Sometimes this happens purely by chance, when a sudden gust of wind results in an unplanned first flight of a helicoptering juvenile. On other occasions, a young bird will suddenly leap into the air from the edge of the nest. Irrespective of the exact mechanism of fledging, the chicks' mother will be on high alert, intently watching the every move of her offspring at this perilous time. If one of her brood takes to the air, she will often fly with it, shadowing the youngster closely during its first experience of life in the air, and ensuring that it does not attempt to fly too far.

The vast majority of young Ospreys fledge when they are seven to eight weeks of age. An analysis of fledging at the Loch Garten nest covering the period from 1968 to 1987 showed that chicks fledged at a mean age of 52.8 days (range 49–59 days, n = 46) (Bustamante 1995), while the corresponding figures at closely monitored public nests in England and Wales were similar (Table 5.1).

Poole (1989) reported that juveniles in southern New England in the United States typically made their first flight aged 50–55 days. There is a trend for Ospreys at more southerly latitudes to fledge at a slightly older age. For instance, the mean fledging age of Ospreys at nests in the Gulf of California in Mexico was 62.5 days (SD = 4.9 days) (Judge 1983), perhaps because they do not migrate (Bierregaard *et al.* 2020).

Hatching asynchrony means that young Ospreys often fledge a few days apart. Bustamante (1995) found that in broods of more than one chick at the Loch Garten nest, the second and the third chicks made their first flight a mean 1.6 days (SD = 1.4, n = 18) and 3.8 days (SD = 1.9, n = 7) after the first one. There is also a tendency for males, who are lighter than females, to fledge slightly earlier. Larger female chicks sometimes remain on the nest for a week longer than male siblings, watching the growing aerial prowess of their nest-mates but not replicating it for several days.

Figure 5.18. As they grow in strength, juvenile Ospreys begin 'helicoptering' above the nest while they prepare to make their first flight (© John Wright).

The Osprey

Table 5.1. Age at fledging, in days, of chicks at five sites in the UK

Site	N	Mean fledging age (and SD)	Source
Cors Dyfi	22	52.2 (2.0)	Dyfi Osprey Project (2022b)
Glaslyn	36	53.2 (2.0)	Glaslyn Wildlife (2022b)
Manton Bay, Rutland Water	20	53.1 (1.2)	T. Mackrill unpub. data
Kielder	31	54.9 (1.9)	Dailey (2021)
Loch Garten	46	52.8	Bustamante (1995)

The thrill of watching a young Osprey take to the air for the first time is one of the highlights of any Osprey season and it never ceases to amaze me how quickly young birds adapt to life in the air. The first flight is usually a short affair in the immediate vicinity of the nest. Quite often the fledging juvenile will complete a short one- or two-minute flight and then land back on the nest, or a nearby convenient perch. Once back on the nest they are on high alert, wide-eyed and bobbing their head from side to side as they survey their surrounds. They quickly grow in confidence and make repeated forays from the nest. One of my favourite places to watch young Ospreys flying for the first time is the Manton Bay nest at Rutland Water where I have been amazed to witness juveniles playfully twisting and turning in the air, swooping down towards the water, and even chasing Egyptian Geese just hours after making their maiden flight.

Figure 5.19. Landing can prove the most problematic element of fledging for juvenile Ospreys. Here, a juvenile male attempts to land for the first time, causing his parents to take off to give him space on the perch (© John Wright).

Landing, on the other hand, can be more problematic. Ideally, juveniles will land back on the nest, where there is a greater margin for error, but on many occasions they attempt to land in nearby trees or other perches, which is more difficult (Figure 5.19). This is particularly the case during gusty conditions or heavy rain, when the chances of missing a landing and ending up on the ground increase significantly. It is relatively common for young Ospreys to become grounded around fledging time. Landing in thick vegetation, particularly when it is wet, can be fatal for these birds, who may find it impossible to get airborne again. We have had to rescue several recently fledged juveniles at Rutland Water over the years. On one occasion we kept a waterlogged juvenile overnight to allow it to dry off before releasing it close to the nest the following morning. On that occasion it was pleasing to see the young bird fly straight back to the nest, but other times juveniles have gone missing soon after leaving the nest, never to be seen again. It is for this reason that juveniles released as part of translocation projects are fitted with miniature tail-mounted VHF radio transmitters. This means that if young birds are grounded after making their first flight they can be located, rescued and re-released.

Post-fledging

The post-fledging period – that is, the time between fledging and either migration or dispersal away from the nest – is known to have important implications for survival and natural selection among birds (Greenwood and Harvey 1982, Newton 2010). It is an opportunity to gain essential life skills and is particularly important for Ospreys and other migratory species because inexperienced juveniles undergo important behavioural and physiological changes that are essential for migration (Newton 2010). The first migration is an extremely demanding period for young Ospreys and so it is essential that they depart in good condition if they are to survive their first southward journey.

The duration of the post-fledging period varies with latitude. Migratory Ospreys fledging from nests at northern latitudes typically remain in the vicinity of the nest for a shorter period than those from sedentary populations further south, and usually in the order of 30–40 days. Ospreys fledging from the Loch Garten nest in northern Scotland between 1968 and 1987 departed the natal area a mean 30.4 days after fledging (range = 18–46, n = 35), at a mean age of 82.6 days (range 71–97, n = 35) (Bustamante 1995). Satellite-tagged juvenile Ospreys from Moray and Strathspey in northern Scotland left their nest-sites a mean 39 days after fledging (range = 18–56, n = 10) (Mackrill 2017), while the mean duration of the post-fledging period at the Cors Dyfi nest in Wales was 37.5 days (range = 31–47 days, n = 21) between 2011 and 2019 (Dyfi Osprey Project 2022c). Bustamante (1995) found that there was no correlation between fledging date and the duration of the post-fledging period over 20 years at the Loch Garten nest, but satellite-tagged juveniles that fledged later in Moray and Strathspey remained at their nests for a shorter period before initiating migration than those fledging earlier (Mackrill 2017). In north-central Florida, where Ospreys are present throughout the year, fledglings remained near their nest and took food from their parents for 8–10 weeks (Edwards 1989).

In the first week to 10 days after making their maiden flight, juvenile Ospreys rarely venture more than a few hundred metres from the nest. They make frequent short flights between favoured perches and return to the nest at regular intervals. If you visit an Osprey nest soon after young have fledged then you are invariably greeted by a cacophony of food-begging from the youngsters. This almost incessant calling betrays the fact that the juveniles are still reliant on their parents for food. The male continues to bring fish to the nest at regular intervals throughout the post-fledging period, and whenever he appears with a fish the juveniles fly to the nest immediately in order to stake their claim on the meal. Squabbles invariably ensue before

The Osprey

the bird that is successful in taking the fish either eats it on the nest, or more likely, carries it off to a favoured perch where there is less chance of it being hassled by siblings. With her offspring now less in need of her protection, the female takes the opportunity to venture away from the nest, and may supplement fish caught by the male with those of her own. At Rutland Water females often begin fishing for their offspring as the young fledge.

Most young Ospreys start to venture further afield a week or 10 days after fledging. Initially, these exploratory flights are fairly short, but as the days and weeks progress the birds begin to wander further from the nest and can be absent for periods of several hours. There seems to be

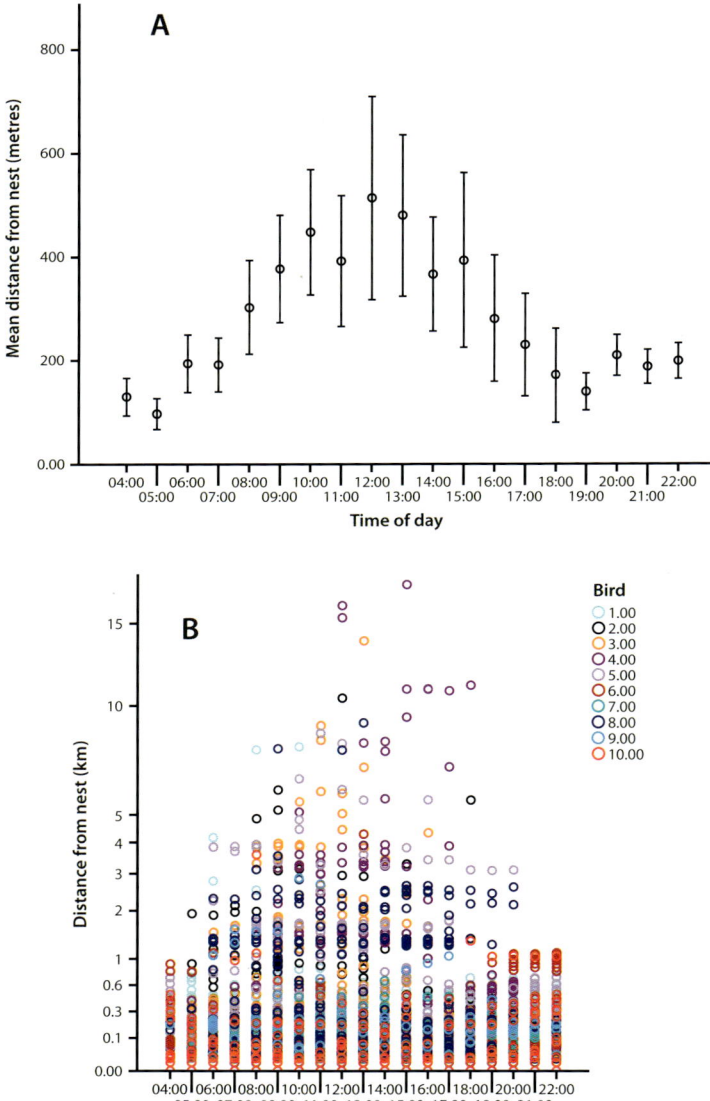

Incubation and chick-rearing

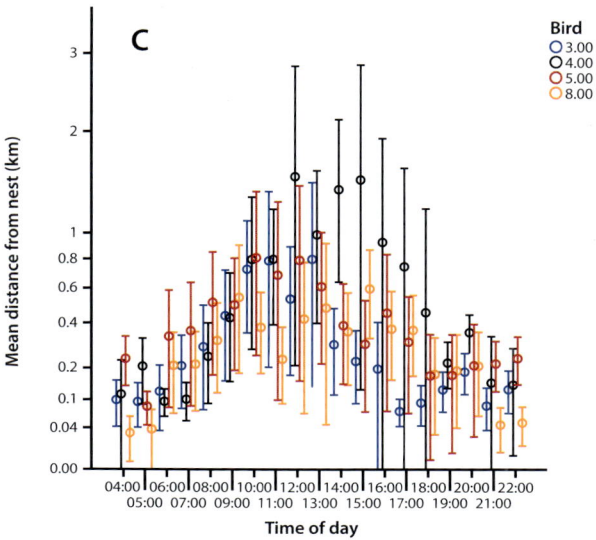

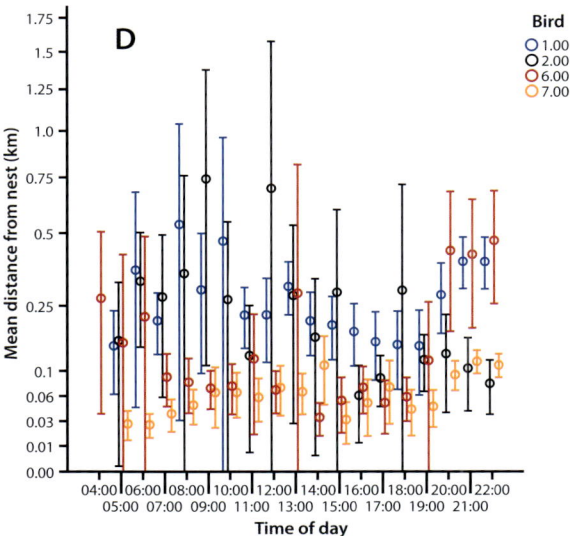

Figure 5.20. Relationship between the time of day and distance from the nest of 10 juvenile Ospreys in Scotland. (A) shows the mean distance from the nest with 95 per cent confidence intervals for all birds combined and (B) is a scatter plot of all GPS observations for each bird. (C) and (D) show the mean distance from nest of birds 1–8 according to the time of day. (C) shows the trend for birds that explored most widely (according to mean distance from the nest) and (D) shows birds that remained more sedentary. Note the scales for (C) and (D) are different to account for differing variance of data (Tim Mackrill).

considerable individual variation in the degree of exploratory behaviour prior to migration. I analysed the post-fledging behaviour of 10 juvenile Ospreys satellite-tagged by Roy Dennis in northern Scotland for my PhD. These birds remained in the vicinity of their nest for a mean 39 days after fledging (range = 18–56), with 93 per cent of all GPS fixes (generally one per hour) located within 1km of the nest. All of the birds remained in the immediate vicinity of the nest for the first week after fledging and within 1km during the second week, but there was considerable variation thereafter. One juvenile male was logged 1km or more from the nest on 101 occasions during its 53-day post-fledging period (12.8 per cent of all GPS fixes) up to a maximum distance of 9.1km, while one female did not leave the immediate vicinity of the nest at all before migrating 44 days after fledging, with a maximum distance of just 659 metres recorded. This, however, was atypical and most of the birds made longer exploratory flights from week three onwards, up to a maximum distance of 17.7km from the nest. Most of the exploratory flights were undertaken during the middle of the day, typically between 9am and 3pm, exemplified by the fact that mean distance from the nest was greatest between 12 noon and 1pm (Figure 5.20).

Monti et al. (2012) recorded a similar pattern of dispersal among translocated juveniles at Maremma Regional Park in Italy. The translocated birds remained around the release site for a mean 48.7 days (range = 3–73 days, n = 13) and most movements occurred within 1km. The birds began making longer flights, up to a maximum distance of 14.3km approximately three weeks after fledging.

There was no evidence that the post-fledging movements of the satellite-tagged birds in Scotland were associated with foraging, and instead the explorations appear to be important in the development of flight skills, as in other raptors (Bustamante 1993, 1994). They may also facilitate imprinting on the natal area. Most young Ospreys in migratory populations are dependent on their parents for food for the duration of the post-fledging period, and while it is relatively common for juveniles to spend several hours away from the nest, they return on a regular basis. The satellite data showed that the Scottish juveniles were nearly always perched close to the nest during the early morning and again in the evening. The shortest mean distances from the nest were recorded between 4am and 7am and then again from 6pm and 10pm. This coincides with periods when breeding males were likely to return to the nest with fish and is also indicative of the fact that juvenile Ospreys nearly always roost close to the nest. It was only in the nights immediately prior to migration that the Scottish juveniles began to roost elsewhere (Mackrill 2017).

One clear advantage of remaining close to the nest during the post-fledging period, particularly for juvenile Ospreys with siblings, is that it increases an individual's chances of claiming fish when they are brought to the nest by the breeding male (Figure 5.21). A key element of the post-fledging period is the need to deposit sufficient fuel for migration, and thus to depart in the best possible condition. Setting off underweight and in poor condition is potentially fatal and this may explain why some birds, such as the extremely sedentary satellite-tagged individual in northern Scotland described above, are reluctant to venture far from the nest prior to migration.

Whilst most juvenile Ospreys remain dependent on their parents for food throughout the post-fledging period, some are successful in catching their own fish prior to migrating. Fishing is innate in juvenile Ospreys and hence they do not need to learn from their parents. However, juveniles fledging from nests on or close to water have greater opportunities to practise their hunting skills prior to migration. For instance, at Rutland Water juveniles fledging from the Manton Bay nest, which is situated on a telegraph pole in the southwest corner of the reservoir, are observed diving into the water a few days after fledging (Figure 5.22), and also dragging

Figure 5.21. Juvenile Ospreys are dependent on their parents for food throughout the post-fledging phase, food-begging loudly when they are present at the nest (© John Wright).

their feet through the water in order to wash them (Figure 5.23). While practice dives are rarely successful, they provide a valuable learning opportunity. In contrast, juveniles that fledge from nests 2–3km from the nearest water usually do not begin making practice dives until three weeks after fledging, and in that sense are at a disadvantage. Even if juvenile Ospreys are successful at making a catch, they continue to return to their nest and solicit for food until they embark on their first migration. Indeed, many Ospreys from migratory populations do not catch their own fish until they have initiated migration.

The fact that young Ospreys in migratory populations spent only a relatively short period at their natal nest after fledging means that it is difficult to study the development of fishing capability. However, one notable piece of research was undertaken at Newnan's Lake in north-central Florida, where juvenile Ospreys tend to become independent from their parents 8–10 weeks after fledging, but do not migrate. This enabled Edwards (1989) to record how capture success changed over time, up to 150 days after fledging. The first successful dives were recorded 11 days after fledging and all 22 colour-ringed young monitored for the study had caught fish 20 days after making their first flight, which is exceptional compared to our observations in the UK. As expected, capture rates increased with age from 21.5–22.5 per cent 30 days after fledging, to 40.3–47.2 per cent after 90 days and 61.7–63.0 per cent 150 days after making their first flight. Capture success of young with siblings, who often foraged together, was initially greater than singleton young, suggesting some degree of social learning was important. However, by the end of the post-fledging period single young achieved the same success rates as those with siblings.

There is usually a predictable pattern of departure among Osprey families in migratory populations. Once her young are flying, the breeding female may take the opportunity to depart. Some females leave almost as soon as their offspring are on the wing, while others wait much longer and provide fish to supplement those caught by the male. A reduction in parental

Figure 5.22. Many juvenile Ospreys do not catch a fish until they begin their first migration, but will frequently make practice dives, like this individual at Rutland Water (© John Wright).

Figure 5.23. Juvenile Ospreys become increasingly skilful on the wing during the post-fledging period. Here, a juvenile female skims across the water in order to wash her feet (© John Wright).

investment has been shown to prompt juveniles of some species to disperse (e.g. Balbontín and Ferrer 2005, Vergara and Fargallo 2008), but there is no such evidence for this in Ospreys. The breeding male continues to provision the young until they set off on migration, meaning he will usually wait until the last of the juveniles has departed before beginning his own journey south. Thus, the autumn departure of males can be some weeks later than females (Mackrill 2017). The departure and subsequent migration of Ospreys is covered in the next chapter.

Conclusions

- Migratory Ospreys usually begin incubating 10–30 days after arriving at the breeding grounds. Laying dates follow clear latitudinal trends, with birds in southern populations beginning incubation two or more months before pairs returning to the most northerly nests. Sedentary southern birds lay during the winter.
- Ospreys usually lay 2–3 eggs, but clutches of four are sometimes recorded. Eggs are laid at 1–3 day intervals and incubation generally, but not always, begins with the first.
- Females undertake the majority of incubation, with her mate taking over periodically to enable her to feed or rest. The male usually provides 1–2 fish each day during incubation and spends the rest of the time perched close to the nest to guard the incubating female. Intrusions by non-breeding Ospreys can result in brood reductions, or even complete failure.
- The first egg hatches approximately 34–42 days after it was laid, and the remainder of the clutch following in the order of laying. The gaps between hatching are influenced by laying interval and the onset of incubation, but are usually in the order of 1–3 days. There can be a considerable size difference between the oldest and youngest in a brood, particularly if there are four chicks.
- Osprey chicks are fed by the female within a few hours of hatching, and she broods them between feeds. The sight of newly hatched young prompts a step-change in fishing effort by the male, who will deliver twice or three times as many fish to the nest compared to during incubation. Males do not brood young but, as during incubation, will perch close to the nest when not away fishing.
- Osprey chicks grow extremely quickly thanks to their protein-rich diet. Within 10–12 days of hatching, they enter the 'reptilian stage', when the emergence of their second down gives them a much darker, scaly appearance. This coincides with the fastest period of growth. Some sibling aggression can occur at this stage, but is generally not serious unless there are food shortages.
- Nestlings are fully feathered by the time they are a month old, but key flight feathers continue to grow. Pale fringing to their brown upperparts gives the nestlings a striking appearance, and also offers important camouflage. By six weeks the young Ospreys are no longer putting on body mass, and this is reflected in a decline in food intake prior to fledging. The young begin wing flapping as they prepare to make their first flight and, as they grow in strength, they hover, or 'helicopter' above the nest.
- Most young Ospreys fledge at approximately seven-and-a-half weeks of age. First flights are usually to a nearby perch or back to the nest itself.
- Juvenile Ospreys remain dependent on their parents for food throughout the post-fledging phase, and the male continues to provision them until they depart on migration. Some females also assist with fish deliveries once her offspring have fledged.
- Young birds begin to venture away from the immediate vicinity of the nest 7–10 days after fledging, with longer exploratory flights made after approximately three weeks, although the nest remains the focal point of activity prior to migration.
- There is usually a predictable pattern of departure among Osprey families in migratory populations, with females departing first, followed by her offspring and then the breeding male. Juveniles in migratory populations generally depart 5–6 weeks after fledging, while non-migratory birds may wait for longer before dispersing.

CHAPTER 6
Migration

One of my highlights of the year is the sight of a newly returned Osprey perched regally beside its nest, its white underside illuminated by gentle spring sun. I always think it is rather humbling to consider what the bird might have experienced in order to make a timely arrival back at its nest: traversing the Sahara, negotiating imposing mountain ranges, crossing areas of open sea. It could even have flown through the night – its path illuminated only by stars and moonlight – with the urge to reclaim its nest-site a source of great motivation to power its way north.

The seasonal migration of birds is perhaps the greatest phenomenon of the natural world. Each year almost 20 per cent of the world's nearly 11,000 bird species are compelled to make seasonal movements to take advantage of changing food availability, reduce competition and avoid predation (Alerstam 1990, Somveille *et al.* 2015). The selective advantages of these migratory movements are offset by costs associated with such travel (Newton 2008). Knowledge and understanding of avian migration increased dramatically during the past century as new methods of study emerged (Alerstam and Hedenström 1998). The first recovery of a British-ringed Barn Swallow *Hirundo rustica* from South Africa provoked widespread astonishment (Witherby 1912), yet now we can track migratory Ospreys on a minute-by-minute basis as they fly south. The Osprey is one of the most closely studied migratory species, with initial research on migration routes and wintering localities undertaken through ringing studies now greatly enhanced by advances in satellite telemetry. Following individual birds on their annual journeys has captivated the general public and brought the wonder of bird migration to life.

The seasonality of resource availability drives annual movements of migratory species, and the highly specialist piscivorous diet of the Osprey means that birds breeding at higher latitudes would not survive the cold northern winters if they did not move south. However, the selective advantages of migration are offset by the costs of the journey. Migratory flights are associated with enhanced risk of predation, diseases, exhaustion, food shortage and mortality linked with weather and wind conditions which may have short-term effects on population sizes (Newton 2008). It may also expose migrating birds to increased risk of mortality due to anthropogenic factors, such as hunting (Brochet *et al.* 2015) and collisions with wind turbines (Johnston *et al.* 2012). Klaassen *et al.* (2014) showed that mortality rate among Ospreys and two other migratory raptors – Marsh Harrier *Circus aeruginosus* and Montagu's Harrier *Circus pygargus* – was six times higher during migration periods than on the breeding or wintering grounds. Furthermore, there was a tendency for a higher mortality rate during spring migration, when mortality was most likely to occur in the Sahara. The Osprey, however, is highly adapted to combating the challenges posed by these demanding journeys and the annual survival of adults in migratory populations, which may exceed 90% from year to year (Dennis 2008), is testament to the navigational skills of experienced birds and their ability to adapt and respond to local environmental conditions encountered en route.

Migration is a particularly demanding challenge for juvenile birds because they have inferior navigational abilities (Guildford *et al.* 2011, Mueller *et al.* 2013), expend more energy when flying (Rotics *et al.* 2016), are more susceptible to adverse weather (Thorup *et al.* 2003) and are less efficient foragers than adults (Skorka and Wojcik 2008). Because of these factors, survival of juveniles during their first year is often considerably lower than annual survival of adults (Strandberg *et al.* 2010, Sergio *et al.* 2011, Guillemain *et al.* 2013). The first migration of juvenile

Ospreys is particularly demanding because, unlike species which migrate in mixed-aged flocks, or with their parents (Newton 2008), young Ospreys undertake their first migration alone, relying on an inherited programme of direction and distance known as vector summation in order to reach a suitable wintering site (Mouritsen 2003), as described later in this chapter.

Before considering the specifics of Osprey migration, it is helpful to consider three key related factors: flight method, wind and navigation.

Morphology and flight mode

A migrating bird must overcome the aerodynamic force – or drag – that opposes its movement through the air, in order to stay airborne, as described by Pennycuick (1969) and various researchers since (e.g. Alerstam 2000, Hendenström 2002, Pennycuick 2008). There are three types of drag which act on a bird in flight. Induced drag is caused by the generation of lift, while parasitic and profile drag are the result of pressure and friction acting on the bird as it moves through the air. This means that induced drag decreases with airspeed, whereas profile and parasitic drag increase as the bird travels faster in relation to the air. Profile drag is generated by pressure and friction acting on the bird's wings as it is flying, while parasitic drag refers to the drag created by the bird's body. So, a large frontal area causes a relatively high drag, whereas a slimmer, more streamlined body allows oncoming air molecules to flow past more easily, thereby reducing drag. It is easy to understand, therefore, that the morphology of a bird is integral to its ability to counteract the effects of drag during flight.

In soaring-gliding flight, birds use external sources of energy to gain altitude, which creates the power required to move forward and stay airborne. The most common method is to exploit thermal updrafts generated by differential heating of the earth's surface, which result in rising columns of warm air (Kerlinger 1989, Pennycuick 1998). The strength, spacing and vertical extent of thermals varies in different areas and on different days but as long as air mass speed within a thermal exceeds an individual's sinking speed, a bird will gain altitude with minimal energy expenditure. The bird makes no onward progress while it gains altitude, but once at the top of the thermal it glides forwards in the intended direction of migration, gradually losing altitude until it reaches another updraft. Under this scenario, the resulting cross-country speed is determined by climbing speed within thermals and the inter-thermal gliding speed.

The power required for soaring-gliding flight can also be attained through slope soaring, whereby the flying bird exploits orographic updrafts created when horizontal winds are deflected upwards by ridges and hills (Kerlinger 1989, Shepard et al. 2013). Under constant wind conditions orographic lift has the potential to provide a continuous source of lift along specific terrain features, and so onward progress may be less staggered during slope soaring compared to thermal flight (Kerlinger 1989, Pennycuick 1998). However, slope soaring birds may be forced to deviate from the desired course in order to follow the terrain feature that is generating lift (Duerr et al. 2012).

The alternative to soaring-gliding flight is flapping flight, in which wing strokes provide the necessary lift and thrust (Alerstam and Hendenström 1998). Alerstam (2000) demonstrated that the mechanical power required for flapping flight is higher than the corresponding figure for gliding flight, but that both flapping and gliding flight produce a U-shaped power curve. This is because induced drag becomes reduced with increasing flight speed, but both parasitic and profile drag increase. As such, there is an optimum travelling speed to maximise distances migrating birds can fly each day. The ability of a given species to utilise these contrasting flight modes during migration is determined by its size, morphology and ability to adjust its wings and tail at different speeds (Thomas 1996, Tucker 1987, 1998, Alerstam 2000).

The power curve (Figure 6.1) provides compelling evidence to explain the evolution of flight behaviour and daily travel distances during bird migration. Flapping flight is highly metabolically costly because the power used for forward propulsion is produced by the bird's own flight muscles, rather than an external energy source. This means that the power requirements for flapping flight increase with body mass (Pennycuick 1998). Alerstam (2000) demonstrated that given a basal metabolic rate (BMR) of 1.5, an Osprey is expected to fly at 17 ms^{-1} to maximise migration speed, whereas the optimum flight speed of the Arctic Tern *Sterna paradisaea* at the same BMR is slower (11 ms^{-1}). However, because the tern is considerably smaller and lighter it can fly at this optimum speed for 13 per cent longer, meaning that it could theoretically travel 187km per day by flapping compared to just 97km per day for the Osprey.

Soaring-gliding flight is far less metabolically costly than flapping because the majority of energy required to power flight is extracted from thermal updrafts (Alerstam 2000). As a result, more than 300 larger species of various taxonomic groups migrate long distances by this method (Del Hoyo *et al.* 1992). Wing loading, calculated by dividing body mass by wing surface

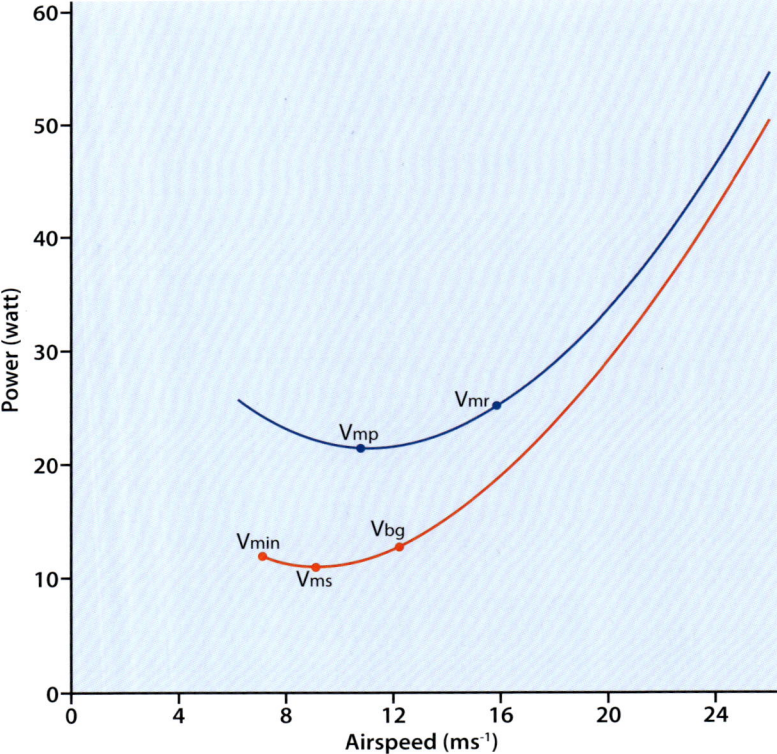

Figure 6.1. Power in relation to speed for flapping flight (blue line) and gliding flight (red line) as exemplified for the Osprey, according to the theory of flight mechanics (Alerstam 2000). Vmp = minimum power speed in flapping flight, Vmr = maximum range speed in flapping flight, Vmin = minimum (stalling speed) in gliding flight, Vms = speed of minimum sink in gliding flight, Vbg = speed of best glide in gliding flight. In flapping flight, power is generated by wingbeats, whereas in gliding flight birds exploit external sources of energy (e.g. thermal updrafts) and power thus relates to the loss of potential energy (reproduction of figure from Alerstam 2000).

area, is important in determining the extent to which migrating birds utilise this mode of flight. Species with a lower wing loading can exploit weaker lift because they have a lower sinking speed (Pennycuick 2008). Conversely, species with higher wing loading can glide faster between thermals (Kerlinger 1989). As a result, species with very high wing loadings such as Black Stork *Ciconia nigra*, White Stork *Ciconia ciconia* and Steppe Eagle *Aquila nipalensis* are extremely dependent on favourable thermal conditions to aid migration, meaning that daily travel time is restricted to periods when thermals are available (Spaar and Bruderer 1996, Shamoun-Baranes *et al.* 2003, Chevallier *et al.* 2010). In contrast, species with a lower wing loading, such as Marsh Harrier and Montagu's Harrier, are less restricted by thermal availability, even though they migrate by soaring-gliding where possible (Mellone *et al.* 2012, Liminana *et al.* 2013). It is notable, though, that even these latter 'facultative soaring' migrants (Vansteelant *et al.* 2015) rarely undertake nocturnal migration, generally only doing so

Figure 6.2. The Osprey's morphology and, specifically, its long narrow wings mean it can take a flexible approach to migration, utilising both soaring-gliding and flapping flight (© John Wright).

when migrating over unfavourable habitats or when migration is time limited (Alerstam 2006). This contrasts markedly with species that migrate solely by flapping flight. Many passerine species, for example, migrate almost exclusively at night (Gwinner 1996, Alerstam 2011).

The Osprey's long, narrow wings give it a distinctive flight profile (Figure 6.2), and this morphology enables it to adapt its flight mode according to local environmental conditions experienced on migration. Its characteristic wing shape reduces drag, which means that flapping flight is less energetically costly compared to larger, heavier species. But conversely, its wing loading ($4.9kg/m^2$) also allows it to exploit thermal updrafts and orographic lift well. This enables migrating Ospreys to take a flexible approach, both in terms of the weather conditions they travel in and the routes they use. Many migratory raptors are restricted to flight over land due to their dependence on thermal updrafts which are generally not available at sea at temperate latitudes, but the Osprey's morphology permits longer flights across the open ocean (Horton *et al.* 2014, Mackrill 2017), as discussed later in this chapter.

The influence of the wind

Wind is a ubiquitous and extremely important factor in bird migration, and like morphology and flight mechanics, it is key to understanding migratory behaviour. The flight path of a flying bird is the result of the wind speed and direction and the bird's speed and direction relative to the air, known as the triangle of velocities (Alerstam 2000). Winds can create headwind, tailwind or crosswind conditions, depending on the bird's direction of travel, and all have differing impacts on migratory flights.

Crosswinds can be particularly influential, and migrant birds may respond by either compensating, overcompensating or drifting with the wind (either intentionally or unintentionally) (Figure 6.3). This response is dependent on a range of factors, including past experience and the stage of the migratory journey. Experienced birds are able to compensate for crosswinds by increasing their airspeed (i.e. their speed in relation to the air, rather than the ground) and adapting their heading so that they maintain their preferred migratory course. However, under the rules of optimum bird migration (Alerstam and Lindström 1990), this response is expected to vary according to the stage of the journey. If an individual is migrating to a distant location and is still many flight steps away, it can gain time and energy by allowing itself to be drifted by the wind at this early stage in the journey. Yet as it approaches its intended destination it is expected to compensate to an increasing degree (Alerstam 2000). Klaassen *et al.* (2011) found that Ospreys and Marsh Harriers respond to crosswinds according to local conditions: drifting when favourable or compensating or overcompensating when approaching a goal or facing the risk of being blown into hazardous habitats. They suggest that this flexibility of response indicates that migrant raptors have excellent navigational skills and are able to estimate wind drift while aloft. Inexperienced juveniles, migrating for the first time, are more susceptible to the effects of crosswinds and less likely to adapt their flight heading as a result (Backman and Alerstam 2003). This is reflected in the migratory flights of many juvenile Ospreys on their first migration, as discussed later in this chapter.

Migrating birds may also make adjustments if they are experiencing tailwind or headwind conditions. Tailwinds enable migrants to reduce their airspeed, and thus save energy, without reducing their preferred groundspeed. Conversely, those flying into a headwind may be forced to increase their airspeed significantly in order to maintain the desired groundspeed (Liechti and Bruderer 1995).

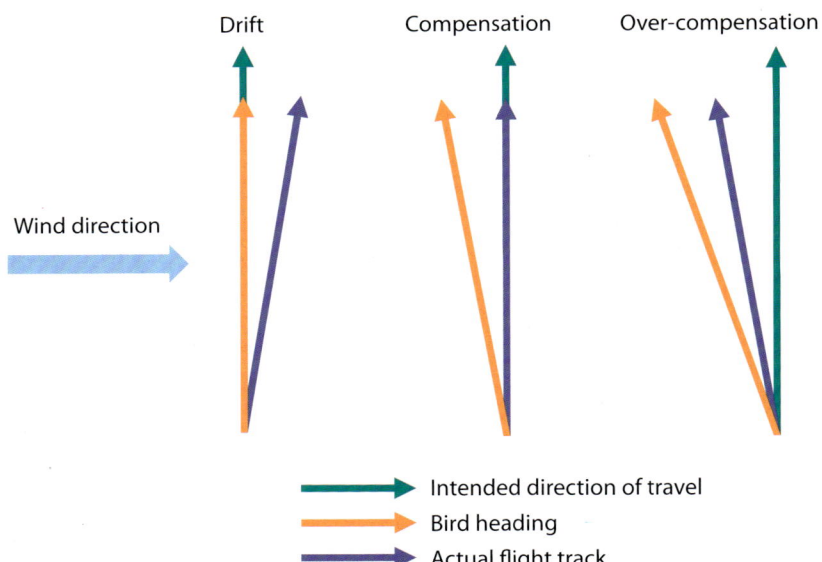

Figure 6.3. When migrating in crosswinds, an Osprey must adjust its heading correctly in order to maintain its intended direction of travel (compensation); if not, it will either drift with the wind or overcompensate, and go off course as a result (Tim Mackrill).

Navigation

Perhaps the most intriguing question in avian migration relates to how migrant birds find their way, particularly first-year birds with no previous experience. Research indicates that juveniles are usually unable to correct for displacement during their first autumn migration, suggesting that true navigation – where birds use geographical landmarks or sensory cues to guide them – is only possible among birds with experience of previous migratory journeys (Guildford et al. 2011). Instead, it is thought that many first-year birds, including Ospreys, rely on an inherited programme of direction and distance often referred to as vector summation, in order to reach potential wintering areas (Mouritsen 2003). Under this technique, first outlined by Rabøl (1978) and further explained by Alerstam (2000) migration consists of a series of flight steps, or vectors, where the orientation between each flight step varies according to a circular probability distribution around the primary (mean) direction. Any subsequent variations in orientation between different flight steps may be caused by limitations in the precision of the birds' navigational abilities and external variables such as wind drift. This supports the 'clock-and-compass concept' (Gwinner 1996), whereby migration of juveniles is controlled by an endogenous temporal/directional programme without any elements of compensation for geographical displacement or goal area navigation.

Migratory routes

Having reviewed the factors that underpin bird migration, let us now consider the migratory journeys made by Ospreys from different regions. The cosmopolitan distribution of the species means that migratory populations are found at northerly latitudes across much of the globe. Populations in Palearctic Eurasia and Nearctic America are almost exclusively migratory, with adult and juvenile birds leaving northern breeding grounds during August and September (e.g. Martell et al. 2001, Hake et al. 2001, Dennis 2008, Väli and Sellis 2016, Mackrill 2017, Østnes et al. 2019). Birds from Western Europe migrate south along the East Atlantic Flyway, with individuals from the UK, France, Germany and Sweden migrating south through Spain and Portugal. An increasing number of birds winter on the Iberian Peninsula (Martín et al. 2019), but most continue south, crossing the Strait of Gibraltar to Morocco, and then skirting around the western edge of the Sahara towards the fish-rich coastal waters of West Africa, from Mauritania south through Senegal and Guinea towards Ivory Coast and Ghana (Hake et al. 2001, Dennis 2008, Mackrill 2017). Birds from populations further east – Finland, Norway, Eastern Europe and northwestern Russia tend to use a more easterly route that takes them south through the Balkans, Middle East and then south along the great Rift Valley towards wintering grounds in eastern and southern Africa (Väli and Sellis 2016, Babushkin et al. 2019, Østnes et al. 2019). Some Finnish birds, however, use a route more typical of their western neighbours and winter in West Africa (Saurola 2021). The autumn migration routes of selected individual satellite-tagged adult Ospreys from the UK, Finland and Estonia are shown in Figure 6.4 to illustrate the different routes used.

Ospreys breeding further east in Russia migrate south to coastal Arabia, the Indian subcontinent and South-East Asia (Ferguson-Lees and Christie 2001, Babushkin pers. comm. 2019). Ospreys nesting at lower latitudes in Europe around the Mediterranean tend to remain sedentary or make short-distance migrations (Monti et al. 2018a). This means that Ospreys display a leap-frog migration, with populations breeding at higher latitudes passing through areas where more southerly breeders occur, and then wintering farther south. It means that most northern birds must negotiate a major ecological barrier in the Sahara, whereas a majority of those

The Osprey

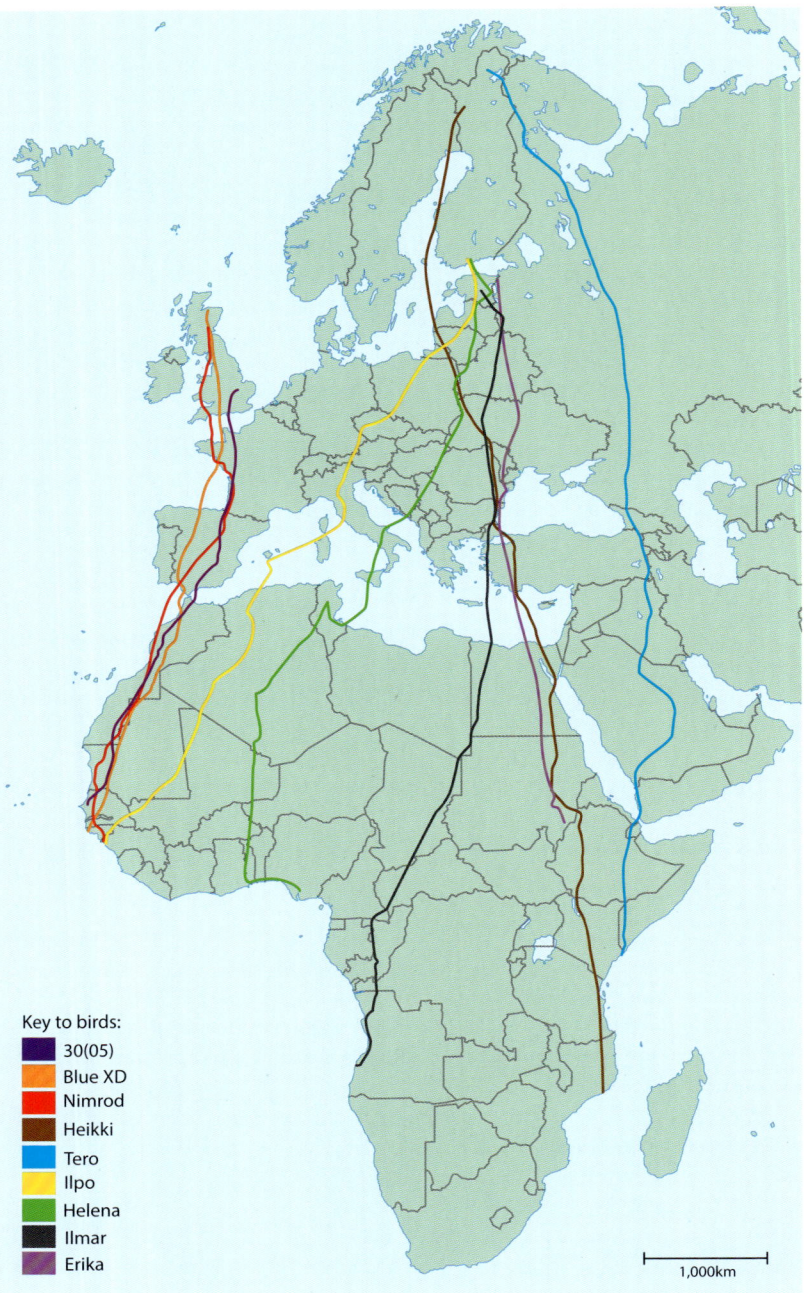

Figure 6.4. Migration tracks of selected adult satellite-tagged Ospreys from the UK, Finland and Estonia, showing differing migration routes used between Europe and Africa (sources: Saurola 2015, Väli and Sellis 2016, Mackrill 2017).

breeding at lower latitudes around the Mediterranean usually do not. The implications of crossing the Sahara are discussed later in this chapter.

In North America, the migration routes used by *Pandion haliaetus carolinensis* are similarly divergent (Martell *et al.* 2001) (Figure 6.5). Individuals that breed in eastern parts of the United States and Canada – from Florida north – tend to migrate south along the East Coast before crossing from Florida to Cuba and then continuing onwards through Haiti and the Dominican Republic before crossing the Caribbean Sea to Colombia or Venezuela. Some birds winter in these northern parts of South America, while others continue south to winter in the Amazon Basin in Brazil (Martell *et al.* 2001, Horton *et al.* 2014). In contrast, Ospreys that breed in western parts of North America, such as British Columbia, Saskatchewan and Oregon, migrate on a broad front through the western United States and then continue south through Mexico. Some remain there for the winter, while others move further south into Central America and rarely into South America (Martel *et al.* 2001, Bierregaard *et al.* 2020). Meanwhile, Ospreys from Midwestern states are known to use three distinct routes. Some fly direct across the Gulf of Mexico to the Yucatán Peninsula and then continue south through Central America to Colombia and Peru. Others make an overland journey south along the Gulf Coast of Mexico before continuing south into Central America and then into Colombia or Peru. The final group of birds head south-east and then fly south along the Atlantic coast before either overwintering in Cuba or crossing the Caribbean Sea into South America in the manner of eastern birds (Martell *et al.* 2001, Washburn *et al.* 2014, Bierregaard *et al.* 2020). As in Europe, some Ospreys

Figure 6.5. Autumn migration routes of Ospreys from North America (based on Martell *et al.* 2001).

breeding at lower latitudes, including southern coastal Florida and Baja Mexico and the Gulf of California in Mexico, do not migrate (Poole 2019).

Timing of departure

What exactly is it that triggers an Osprey to begin its migration? On the breeding grounds there is usually a predictable pattern of departure. The breeding female's key role is redundant once her young are flying, and some take the opportunity to depart relatively early. Some females in the UK may leave their nests in early–mid August (Dennis 2008, Mackrill 2017). Others remain for longer, and supplement the provisioning of the male with fish deliveries of their own. Males continue to provide food for their young throughout the post-fledging period and remain until the last of their offspring has departed. During the latter half of the post-fledging period, in particular, males are continuously harassed by food-begging juveniles while they are at the nest, and our satellite-tracking studies have shown that they often perch away from the nest during this period, seemingly to avoid this incessant food-soliciting. In some cases breeding males in the UK do not begin their autumn migration until late September, often several weeks after their mate has departed. The median departure date of 14 satellite-tagged adult males from the UK was 7th September (range 22nd August–26th September, n = 26), while the corresponding date for seven females was 12th August (range = 2nd August–13th September, n = 23) (Mackrill 2017). A similar pattern of departure has been recorded in Sweden, where the median departure of females was 9th August (range = 28th July–22nd August, n = 10) compared to 4th September among males (range = 24th August–17th September, n = 11) (Kjellén et al. 2001).

In North America, Martell et al. (2001) also found that median departure dates of females were significantly earlier than males: 19th August (range = 12th July–9th September, n = 22), and 7th September (27th August–3rd October, n = 8) on the East Coast; 31st August (range = 1st August–12th September, n = 12) and 20th September (range = 27th August–1st October, n = 13) in the Midwest. The overall mean difference in departure dates within mated pairs was 22 days (range = 7–39 days). Adults that fail to rear young typically depart slightly earlier than those with dependent young. For example, males that failed to breed successfully in Scotland and England set off a week earlier than those that reared chicks (Mackrill 2017).

Ospreys use the fly-and-forage strategy during migration, whereby birds pause to feed before, during or after a day's flight (Strandberg and Alerstam 2007, Klaassen et al. 2008). This negates the need for extensive pre-migratory fuelling as observed in some species, but birds that remain on the breeding grounds for longer in the autumn are likely to deposit fuel prior to departing. An adult breeding male Osprey that Roy Dennis and I satellite-tagged in Strathspey in northern Scotland on 6th September 2017, weighed 2,100g the day before departure on autumn migration. Adult male Ospreys typically weigh 1,500g, indicating that the bird had put on more than 500g of fat in preparation for its migration.

Weather conditions also influence the timing of departure. As part of my PhD studies on Osprey migration, I analysed the autumn departure of 14 adult Ospreys migrating south from the UK (total 34 migrations). These birds departed over a 56-day period between 2nd August and 26th September, and all began their migration on days when they were supported by tailwinds, predominantly from the north-west. This apparent wind selectivity enabled some birds to undertake very long flights at the beginning of autumn migration. Ten satellite-tracked juveniles departed when the wind was in the west, but it is less clear whether this was a factor in the decision to initiate migration. Most juvenile Ospreys in northern populations begin migration five–six weeks after fledging (Bustamante 1995, Mackrill 2017), but unlike in some species, there is no reduction in parental food provision to encourage departure. Instead, it seems likely that a combination of time since fledging, and perhaps changes in daylength and night-time temperatures, act as cues for departure.

While there is a predictable pattern of departure in autumn relating to external factors, and specifically the rearing of young, there are less obvious cues in relation to spring migration. Adult Ospreys are extremely sedentary during the winter, and satellite tracking has shown that there is no apparent change in behaviour prior to initiating migration. It seems likely, therefore, that endogenous cues, such as hormonal changes, are responsible for adult birds departing the wintering grounds (Sharma *et al.* 2018). European Ospreys usually leave their wintering grounds from early–mid March onwards. The median departure date of 17 satellite-tagged adult birds from the UK was 17th March (range = 23rd February–12th April, n = 39) (Mackrill 2017), while the corresponding figure for birds flying north to Estonia was earlier (4th March, range = 23rd February–19th March, n = 9) (Väli and Sellis 2016). In both cases, departure dates varied considerably between individuals. Most North American Ospreys depart during the same period, with no differences between the sexes (Martell *et al.* 2014). East Coast birds tend to set off slightly earlier (median date of departure = 13th March, range = 9th February–7th April, n = 39) than those from western areas (median = 31st March, range = 13th February–19th April, n = 19) (Martell *et al.* 2014). Meanwhile, birds from the partially migratory Florida population that winter in South America have been shown to embark during January (Martell *et al.* 2004).

Total migratory distance

The total distance flown by Ospreys on migration varies considerably both within and between populations. Prior to the advent of satellite tracking, ringing studies had provided the first insight into the distances individuals travel on migration. For instance, 2,977 recoveries and recaptures of Ospreys ringed in Finland up to 2005 were logged throughout Europe and the Middle East, as well as several parts of Africa, including West Africa, parts of Central and East Africa and, most notably, the southern coast of South Africa (Saurola 2005). The most northerly breeding Ospreys in Europe breed within the Arctic Circle in Finnish Lapland, and so the wide geographical spread of the recoveries demonstrated the remarkable migratory capabilities of the species very well. Satellite tracking has since added impressive detail and information on the journeys of specific individuals, including those from the Lapland population. A good example is Tero, an adult male Osprey satellite-tagged by Harri Koskinen and colleagues on 26th July 2014 at his nest near Utsjoki (Saurola 2015). Having reared two chicks, Tero set off on 18th September, heading south through Russia and Georgia before pausing for two weeks on the border of Turkey and Armenia from 11th to 24th October. He then travelled south through Iraq and across Saudi Arabia and Yemen before crossing the Bab el-Mandeb strait at the south end of the Red Sea. He spent a week refuelling in southern Eritrea, before resuming his journey on 19th November and making speedy progress south through Djibouti and Ethiopia. He finally arrived at his wintering site at Kiunga Marine National Reserve on the Kenyan coast, 285km north of Mombasa, on 24th November, some 68 days after leaving his nest-site in the Arctic Circle, a journey of over 8,500km (Figure 6.4). That, though, is not the longest journey recorded by a Finnish satellite-tagged Osprey. In autumn 2002 a bird known as Harri migrated from northern Finland to winter in South Africa, a journey of 12,500km (P. Saurola pers. comm. 2021).

Ospreys from other northern populations also travel considerable distances on migration. Four adult Estonian satellite-tagged Ospreys flew a mean 7,515km (+1,542.8km) to wintering sites in southern, Central and East Africa in a mean 44.3 days (±10.5 days) (Väli and Sellis 2016). The return journeys of three birds in spring (mean = 7,441.7km, ±1,673.2km) were completed in a shorter period of 32.3 days (±6.7 days), even though they tended to fly shorter daily distances, as they spent less time on stop-overs (Table 6.1). Meanwhile, 13 adult Ospreys from Sweden migrated a mean 6,681km (range = 5,810–10,060km) to wintering grounds, predominantly in West Africa, over an average of 44 days (range = 27–80 days). The longest migration, both in terms of distance flown and overall duration, was a female that wintered in Mozambique. She was the only individual to migrate to southern Africa (Hake *et al.* 2001).

Ospreys from the UK usually undertake a slightly shorter migration. Twenty-one adult birds tagged between 1999 and 2017 travelled a mean 5,059.8km (±679.5km) in an average of 26 days (±12.4 days), predominantly to wintering sites in West Africa, including 11 in Senegal (Dennis 2008, Mackrill 2017). The return migration of 16 of these individuals involved mean flights of 4,509.3km (±997.3km) and were completed in an average of 23.6 days (±8 days). Two of these birds wintered in Spain, rather than continuing south across the Sahara.

Ospreys breeding at more southern latitudes in Europe typically take shorter or partial migrations. Seven adult birds satellite-tagged in Corsica, the Balearic Islands and mainland Italy migrated a mean 1,622.9km (±1,121.6km, n = 12) in an average of 4.4 days (±3.5 days), mainly to wintering sites in and around the Mediterranean (Monti *et al.* 2018a). The same birds travelled a mean 1,910.4km (±1,076.9km, n = 11) in 5.8 days (±4.3 days) on the return journey in spring.

Across the Atlantic, the mean distance flown during autumn by Ospreys breeding on the East Coast of the United States, which typically winter in Brazil and Venezuela, was 5,134km (±1,964km, n = 20) in an average of 31 days (±16 days), while those from the Midwest travelled a mean 5,872km (±1,780km, n = 16) to wintering sites in Colombia and other parts of South America in an average of 26 days (±11 days) (Martell *et al.* 2001) (Table 6.1). Birds from western parts of the United States, which generally make a shorter migration to Mexico and Central America, flew a mean 3,824km (±862km, n = 16) in an average of 13 days (±4 days). As in Europe, there can be considerable variation within populations. For example, a female from Shelter Island, New York, migrated 8,720km to the Pantanal area of Brazil in 38 days, while another individual from the same population travelled only 1,890km before wintering in Florida (Martell *et al.* 2001).

North American Ospreys spent a mean period of 22 days on spring migration (range = 5–68 days, n = 63), but again this varied between populations and, in some cases, between sexes (Martell *et al.* 2014) (see Table 6.1). Females from western areas, which migrated a mean 3,341.3km (±105.5km, n = 12) back to their breeding sites in an average of 15 days (±2.5 days), spent half as long on the journey as males from the same population, which flew an average of 4,739.3km (±679.8, n = 4) in a mean 33 days (±9.5 days), although these differences may have been due to the sample size.

DAILY DISTANCES FLOWN

Satellite-tracking studies have shown that a migrant Osprey typically flies between 200km and 400km per day, but this varies according to weather conditions, stage of the journey and season. Twenty-one adult Ospreys satellite-tagged in northern Scotland and in central England flew a mean 298km (±70.4km) per day during autumn migration, while 16 birds travelled an average of 263km (±51.8km) per day during the return journey in spring (Table 6.1). I analysed a smaller subset of 14 birds with the highest resolution data for my PhD and this revealed that the greater daily distances achieved in autumn were predominantly due to more favourable weather conditions (Mackrill 2017). Mean daily distances travelled by North American Ospreys are similar to those from the UK, as reported by Martell *et al.* (2001). West Coast Ospreys travelled a mean 296km (±55km, n = 14) per day, while individuals from populations on the East Coast travelled

Migration

Table 6.1. Migration speeds of adult Ospreys from Europe and North America

Country/region	Season	No. birds	Mean distance (km)	Mean duration (days)	Mean travelling fraction (if reported)	Mean distance on travelling days (km)	Wintering region	Study
Sweden	A	11	6,680.9 ±1,204.4	44.27 ±17.7	0.70 ±0.18	249.34 ±80.68	W/E Africa	Hake et al. (2001)
Estonia	A	4	7,515.3 ±1,542.8	44.3 ±10.5	0.68 ±1.19	301.0 ±42.4	C/E/S Africa	Väli and Sellis (2016)
	S	3	7,441.7 ±1,673.2	32.3 ±6.7	0.85 ±0.21	271.0 ±11.5		
UK	A	21	5,059.8 ±679.5	26.0 ±12.4	0.76 ±0.22	298 ±70.4	West Africa, Iberia	Dennis (2008), Mackrill (2017)
	S	16	4,509.3 ±997.3	23.6 ±8.0	0.82 ±0.17	263.0 ±51.8		
Mediterranean	A	7	1,622.9 ±1,121.6	4.4 ±3.5	0.93 ±0.22	277.2 ±125.2	Mediterranean	Monti et al. (2018a)
	S	7	1,910.4 ±1,076.9	5.8 ±4.3	0.92 ±0.25	276.1 ±80.4		
USA (East Coast)	A	27	5,134 ±1,964 (n = 20)	31 ±16 (n = 22)		214 ±81 (n = 21)	South America, also Florida and Caribbean	Martell et al. (2001), Martell et al. (2014)
	S	23m	4,514.0 ±263.0	17.8 ±1.3		261.8 ±12.3		
	S	12f	5,437.3 ±448.6 (n = 11)	31.1 ±4.9		193.9 ±19.19 (n = 11)		
USA (West Coast)	A	22	3,824 ±862 (n = 16)	13 ±4 (n = 13)		296 ±55 (n = 14)	Mexico, also Central America	Martell et al. (2001), Martell et al. (2014)
	S	4m	4,739.3 ±679.8	33.0 ±9.5		181.8 ±43.74		
	S	14f	3,341.3 ±105.5 (n = 12)	15.0 ±2.5		251.8 ±29.7 (n = 12)		
USA (Midwest)	A	25	5,872 ±1,780 (n = 16)	26 ±11 (n = 17)		230 ±61 (n = 17)	South America, also Mexico, Central America and Caribbean	Martell et al. (2001), Martell et al. (2014)
	S	5m	5,952.0 (n = 1)	24.6 ±2.6		225.3 ±29.8 (n = 4)		
	S	5f	7,000.3 ±438.3 (n = 4)	31.4 ±3.6		223.8 ±18.2 (n = 4)		

Figures in brackets include repeated journeys by the same individuals from Estonia which were tracked for multiple years. m = males, f = females; travelling fraction is calculated by dividing total duration of migration by the number of travelling days. For seasons, A = autumn, S = spring.

a mean 214km per day (±61km, n = 17), and Midwestern Ospreys flew 230km per day (±61km, n = 17) (Martell *et al.* 2001) (Table 6.1). In spring, the mean daily distance flown across all populations was 237km (n = 55, range = 95–380km) (Martell *et al.* 2014).

Daily distance flown by Ospreys on migration does not vary according to the overall journey length. Monti *et al.* (2018b) compared the migration strategies of Ospreys from the Mediterranean region that undertook a short-distance migration, with those from Sweden, which complete a long-distance migration to sub-Saharan Africa. They found that total cumulative distances travelled by Swedish birds were about five times greater than short-distance migrants from the Mediterranean region. Despite this, there was no difference in daily distances flown between populations (long-distance migrants = 202.4km per day, ±41.1km; short-distance migrants = 211km per day, ±87.7km). This, combined with the fact that long-distance migrants incorporated significantly more stop-over days, meant that the overall duration of migration for the Swedish birds was considerably longer (61.0 days, ±17.8 days) than those from the Mediterranean (5.1 days, ±2.5 days).

The Osprey's ability to utilise different flight modes means that individuals can employ a highly adaptive approach to daily flights. Its ability to supplement soaring-gliding flight with flapping was referenced by Mellone *et al.* (2012), who suggest that on occasions Ospreys may employ the technique of powered glides where birds flap sporadically during inter-thermal gliding. This has the effect of increasing flight speed and, when coupled with the ability for longer flight times each day, may explain why Ospreys often achieve greater daily distances on migration than true soaring migrants, or if they were to fly by powered flight.

Juvenile performance

Migration presents a challenge for individuals of any age, but it is a particularly demanding for juveniles. Unlike some similarly sized raptors, including Honey-buzzards *Pernis apivorus* which travel in mixed-aged flocks, young Ospreys migrate alone, relying on endogenous cues, rather than experienced conspecifics, to find the way.

Although juveniles have underdeveloped flight skills, satellite tracking has shown that the daily distances flown are often comparable to adults. For my PhD, I compared the migratory flights during autumn of eight satellite-tagged juveniles from Scotland with 14 adult birds from Scotland and England (Figure 6.6). The young birds migrated towards the same region of West Africa as the adult birds and covered similar daily distances as they migrated south through Europe and Africa. Adult birds flew a mean 263.5km per day (±210.9km) through Europe, compared to 250.7km (±259.0km) by juveniles; the corresponding figures in Africa were a mean 307.0km per day (±123.1km) by adults, and 269.0km (±121.7km) by juveniles. Some juveniles made very long continuous flights over the sea, as described later in this chapter.

Despite these apparent similarities, juveniles typically took longer to complete their migrations because they often incorporated more stop-over days than adults. The mean travelling fraction of adults (i.e. the fraction of travelling days compared to total days of migration) was 0.76 (±0.20), compared to 0.59 (±0.35) for juveniles. The migration routes used by adult birds were also more direct. The journey straightness (ratio between migration distance and the great circle distance (the shortest distance between two points on the surface of a sphere (i.e. the earth's surface), measured along the surface) between the start point and end point of the migration) of the adult birds varied from 0.89 to 0.98 (mean = 0.94, ±0.03), while the value for juveniles ranged from 0.76 to 0.95 (mean = 0.90, ±0.03).

Other studies have identified similar trends when comparing juvenile and adult migrations. Juveniles migrating between Sweden and wintering grounds in sub-Saharan Africa flew similar

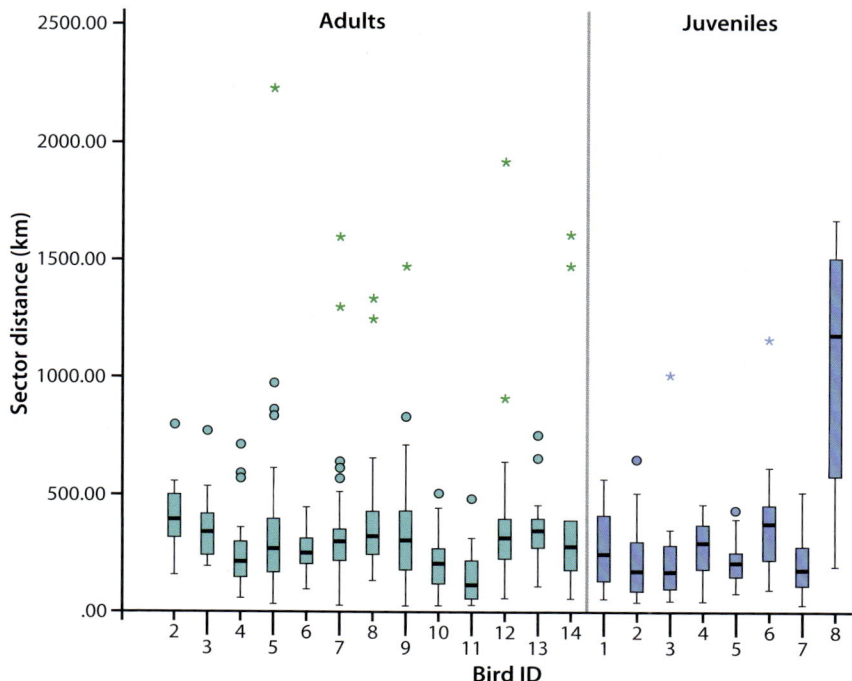

Figure 6.6. Box plot showing individual variation in daily (sector) distance flown of adult and juvenile Ospreys from the UK. If a bird continued to fly for more than one day, this non-stop flight was deemed to be a single sector and analysed in the same way as a single day. There was no significant difference in the mean distance flown by juveniles (blue) compared to adults (green), but juvenile bird 8, Stan, flew considerably further than all other birds as a result of a very unusual over-sea migration to Cape Verde (Mackrill 2017).

distances to adults on travelling days (mean 201.9km) (Monti et al. 2018b), but the same study found that juveniles undertaking short-distance migrations in the Mediterranean region flew 30 per cent less per day than adults (181.9km/day compared to 260.4km/day). Juveniles incorporated more stop-over days into their migrations than adults, stopping on average for six days more, showed greater variance in the major axis of migration and used less direct routes.

Factors affecting duration of migration

Weather has a profound effect on bird migration and, therefore, a critical examination of migration speed on both the daily and whole-journey level, is enhanced by the analysis of associated meteorological data. It is also important to consider the motivation of an individual bird. Studies have generally shown that birds migrate faster in spring than autumn (Nilsson et al. 2013) because of the selective advantages of returning to the breeding site earlier (Kokko 1999, Newton 2008); in the case of Ospreys this relates to the reclamation of an established nest-site (as discussed in chapter 3). In time-selected migration, birds are under selection pressure to complete the journey as quickly as possible, and incorporate time-saving methods – such as increasing airspeed and extending daily flight periods – in order to reduce overall journey duration, at the expense of greater energy expenditure (Alerstam and Lindström 1990). Ospreys migrating north in the spring might therefore be expected to incorporate time-minimisation strategies, but there is evidence of this in both northward and southward journeys.

TIME MINIMISATION

Seasonal differences in Osprey migration speed have been reported for European birds and also those from North America. Alerstam *et al.* (2006) found that Swedish Ospreys incorporated fewer stop-over days in spring, which is likely indicative of time minimisation – although other factors, such as pre-migratory fuelling and stop-over site quality, may also be important. Meanwhile, Martell *et al.* (2014) describe how Ospreys returning to nest-sites on the East Coast of North America use a shorter route in spring. Autumn migration coincides with the hurricane season, and so Ospreys travelling to South America continue over land from Cuba to Hispaniola to reduce the distance they must fly across the Caribbean Sea. However, in spring, when the threat of hurricanes has passed, they are able to make a longer but more direct flight across the sea, reducing the overall distance of the northward journey and potentially facilitating a more timely arrival on the breeding grounds.

In Europe, weather conditions encountered by Ospreys and other migrant species during spring are likely to be less favourable than autumn, particularly as they approach the breeding grounds (Shamoun-Baranes *et al.* 2003, Mellone *et al.* 2012). Although Ospreys migrate predominantly by soaring-gliding flight, their ability to switch to flapping flight in adverse weather conditions means they are better adapted to a final sprint towards the breeding grounds than pure soaring migrants (Kerlinger 1989, Agostini *et al.* 2015). Alerstam (2006) suggested that some Ospreys migrating to nests in Sweden exhibit such behaviour.

It is noteworthy that our satellite-tracking studies have shown that Ospreys migrating between the UK and West Africa appear to incorporate time-minimisation strategies into both southbound and northbound journeys (Mackrill 2017). Nocturnal flight (Figure 6.7), which can be indicative of time selected migration, was more commonly undertaken by adult birds during autumn, with 61 per cent of 61 night-time flights undertaken by 11 of 14 tracked birds occurring on the southbound journey. This usually involved the birds extending flight times after sunset (mean duration 319 minutes (±203)). Nocturnal flight was most frequently undertaken in Europe during autumn, with 21 flight sectors including elements of nocturnal flight, nine of which related to crossings of the Bay of Biscay. It seems that Ospreys predominantly use nocturnal flight as a time-minimisation technique or to complete flights across ecological barriers (Alerstam 2006). There was a tendency for later-departing males to incorporate more nocturnal flight and, significantly, three of these individuals flew continuously for more than 24 hours on the first day of their southward migration. One bird, Blue XD, a breeding male from Strathspey, flew 2,115km direct to Extremadura in central Spain over the course of 36 hours, including an overnight crossing of the Bay of Biscay between Brittany in northwest France, and eastern Asturias on the north coast of Spain. The power of adult birds to make such flights is testament to their ability to utilise differing flight modes – exploiting thermal updrafts when available and also undertaking long periods of flapping flight at night and when crossing the ocean. Under such circumstances, flying with a tailwind significantly reduces energy expenditure and greatly increases their flight range compared to still air or headwind conditions (Alerstam 2000).

These results are in keeping with a growing body of evidence that raptor migration speed is influenced by selection pressure during autumn as well as spring. Panuccio *et al.* (2014) and Mellone *et al.* (2015) suggest that Black Kite *Milvus migrans* and Booted Eagle *Aquila pennata* migration may be time constrained in autumn because of competition for resources in the

Migration

Figure 6.7. Ospreys readily extend flying times into the hours of darkness, especially when making sea crossings or as a time-minimisation technique (© John Wright).

African wintering grounds. Ospreys have been shown to remain highly faithful to the same wintering site each year (Dennis 2008, Washburn *et al.* 2014), and all of the birds in the UK study returned to the same site in successive winters. The reclamation of these winter territories may therefore act as a strong selective cue, influencing daily travel routines and time-budgets during autumn migration given the potential fitness benefits of returning to a known site.

Weather

The long flights at the beginning of autumn migration by later-departing male Ospreys from the UK illustrates the ability of adult birds to recognise and take advantage of favourable meteorological conditions. Wind is a key factor influencing migration speed and routes by both adult and juvenile Ospreys. At the basic level, tailwinds enable Ospreys to reduce their relative airspeed and, thus fly for longer – and therefore further – each day, or to fly at faster ground-speeds. Greater daily distances achieved by Ospreys on their southward migrations from the UK was predominantly due to the positive effects of tailwinds and also stronger thermals, which increased flight speeds (Mackrill 2017). Both factors are known to enhance both hourly and daily speeds of migrating raptors (Borher *et al.* 2012, Mellone *et al.* 2015, Vansteelant *et al.* 2015). Mellone *et al.* (2012) found that tailwind strength was the most important factor in determining the daily distance flown by Ospreys and three other species of raptor across the Sahara.

STOP-OVERS

As already described, stop-overs – where a bird interrupts its journey at a specific site for one or more days – are a key factor influencing the overall duration of migration. Satellite-tracking studies have revealed that both adult and juvenile Ospreys are highly individual in their use of stop-overs. Some complete migrations without them, while others pause for prolonged periods.

ADULT OSPREY STOP-OVER BEHAVIOUR

The individual variation is well illustrated by the migratory behaviour of 14 satellite-tagged adult Ospreys from the UK, some of which were tracked on multiple return journeys (Mackrill 2017). Five individuals completed a total of nine autumn migrations without stop-overs (26% of all autumn journeys). Furthermore, two of the birds – an English female and a Scottish male – each completed three successive autumn migrations without any stop-over days, both using extremely direct routes. Female 30(05), from the Rutland Water population, migrated to the Senegal coast midway between Dakar and St Louis, a flight of 4,433km (±37km) in a mean 12 days (±1 day), with an average straightness of 0.98 (±0.01); male Blue XD, from Strathspey in northern Scotland, took a mean 16 days (±2 days) to fly 5,253km (±96km) to the Casamance coast in southern Senegal, with an average straightness of 0.97(±0.02) (see Figure 6.4). In each case, favourable meteorological conditions facilitated the fast journeys.

The remaining 25 autumn migrations, completed by 10 birds, included stop-over days, with a mean travelling fraction of 0.67 (±0.17). Adult birds typically return to the same stop-over site each year, which is often a place they discovered on their early migrations (see below). Nimrod,

Figure 6.8. An Osprey fishing at Île d'Oléron on the Atlantic coast of France. This is a favoured stop-over area for Ospreys from the UK (© John Wright).

an adult male from northern Scotland, stopped over on Île d'Oléron, a small island south of La Rochelle on the Atlantic coast of France, in autumn 2008, 2009 and 2010. In each year the bird favoured the same forested area and fished along the coast once or twice a day. In 2010 Nimrod spent a total of 19 days on the island, which constituted almost half of the total days (40) he spent migrating to his wintering site in Guinea-Bissau. I visited Île d'Oléron in September 2010 with John Wright and Paul Waterhouse, and it was easy to see why a bird like Nimrod would return there each year. Separated from the French mainland by a wide estuary, the abundant fish stocks provide a rich hunting ground for migrant Ospreys, and we saw at least 15 different individuals during a two-day visit (Figure 6.8). Nimrod's extended stop-over meant that despite wintering just 250km further south, he spent almost four weeks longer on migration than Blue XD, who reached the Casamance River in southern Senegal in just 13 days on one occasion.

Stop-overs are also a feature of the spring migration of adult Ospreys, but as in autumn there is considerable individual variability. As described above, Ospreys are under selection pressure to make a timely arrival at their nest in the spring. This perhaps explains why Nimrod completed three spring migrations without stop-overs, contrasting greatly with his behaviour in the autumn. Overall, however, there was no seasonal difference in stop-over use when all birds were considered (mean travelling fraction in autumn = 0.77 ±0.21, mean travelling fraction in spring = 0.74 ±0.25). Meteorological conditions are generally less favourable during spring, particularly as returning adults move northwards through Europe, and so a greater proportion of stop-overs may be associated with inclement weather, but this is not always the case. A female known as Beatrice (Figure 6.9), first ringed as a chick in Aberdeenshire in 2000 by Ian Francis and Stuart Rae, and then re-caught and satellite-tagged by Roy Dennis as a breeding adult in Moray in July 2008, wintered beside the River Guadiaro in southern Spain (Dennis 2016a). This is a relatively short migration for an Osprey, but Beatrice still incorporated a three-day stop-over beside the River Ardour in the south of France on the first autumn she was tracked. It turned out that this was a place well known to Beatrice, because she stopped there on each of her subsequent 14 migrations. It was a particularly favoured site in spring, when she would arrive as early as late February and then linger for between 20 and 26 days. Over the years it was possible to record her daily ranging pattern, and this varied significantly between years. In 2012 and 2013 her daily flights were all within 3km^2, suggesting that fishing was easy, but in other springs she ranged much more widely: 25km^2 in 2013, 49km^2 in 2011, 62km^2 in 2010 and 112km^2 in 2014. In spring 2016, she arrived earlier than ever before on 27th February to find the river in flood. This would have made fishing much more difficult than usual and after several days she moved 90km south-west to the north coast of Spain. She was seen and photographed at Txingudi estuary on 3rd March and then spent several days moving along the Basque coast, attempting to find alternative fishing sites. Sadly, persistent heavy rain made for swollen, muddy rivers and extremely difficult fishing conditions, even for an experienced adult Osprey. Eventually she was found dead on the banks of the River Urumea close to Hernani, having starved to death. This was an extremely sad demise but demonstrated that even old, experienced Ospreys make mistakes; had she left her wintering site in southern Spain later she may have missed the worst of the bad weather.

The fact that some individual Ospreys migrate to distant wintering grounds without a break is indicative of the species' ability to use a fly-and-forage migration strategy, where birds pause to feed before, during or after a day's flight (Strandberg and Alerstam 2007, Klaassen et al. 2008). In fact, when Ospreys pass migratory watchpoints they are sometimes seen carrying fish with them. Satellite tracking has shown that when migrating through Europe and North Africa, Ospreys usually choose to roost near a waterbody, thereby enabling them to feed in the evening or early morning. As a result, daily flying times tend to be shorter in Europe than Africa

Figure 6.9. Beatrice, an adult female Osprey satellite-tagged by Roy Dennis in Moray, northern Scotland, had a regular stop-over beside the River Adour in southern France, where she lingered each spring, despite only wintering in southern Spain (© John Wright).

(Mackrill 2017). This, however, does not always negate the need for stop-overs. Monti *et al.* (2018b) highlighted the importance of stop-overs from an energetics perspective. They found that long-distance migrants from Sweden spent 50 per cent of their migration duration on stop-overs, and that these individuals spent 30 days longer on stop-overs than those undertaking short-distance migrations in the Mediterranean region, 80 per cent of which migrated without any stop-overs at all. The Swedish birds incorporated an average of 2.3 stop-overs (±1.04) per migration, with sites predominantly in Central Europe, and southern Spain or North Africa. This corroborates our own findings in the UK, with most stop-overs undertaken in France, Spain or Portugal, and a small number in Morocco, north of the Sahara (Dennis 2008, Mackrill 2017). Stop-overs thus represent a valuable opportunity for migrant Ospreys to refuel as they journey to distant wintering sites, but a combination of pre-migratory fuelling, favourable meteorological conditions and the ability to utilise the fly-and-forage migration strategy enables some birds to complete long-distance migrations of more than 4,500km without any breaks.

JUVENILE OSPREY STOP-OVER BEHAVIOUR

Juvenile Ospreys, like adults, can be surprisingly variable in their use of stop-overs. The most significant difference, of course, is that adult birds are navigating to known stop-over locations that they use every year, whereas juveniles might linger at a site they encounter, by chance, en route. Individual variations were well illustrated by the group of eight juvenile Ospreys satellite-tagged in northern Scotland by Roy Dennis, which I analysed as part of my PhD research. Each migrated to wintering sites in West Africa, but there was considerable variability in journey duration. The fastest was a remarkable eight-day flight to Cape Verde, a total distance of

5,469km, by a juvenile male known as Stan. Meanwhile, two other birds, Fiddich and Spey, arrived at sites in Senegal and Mali after 22 days of continuous migration, flying a mean 240km and 283km per day respectively. The remaining five birds incorporated stop-overs of varying lengths, with a mean travelling fraction of 0.34 (range = 0.28–0.54). Four of the five birds spent twice as many days on stop-overs as they did travelling. What is the cause of such variation? It is clear that stop-overs provide a valuable opportunity for juvenile birds to replenish fuel reserves (Yosef *et al.* 2006) and, perhaps most importantly, an additional chance to develop and improve foraging skills (Mellone *et al.* 2013). Juvenile Ospreys are dependent on their parents for food prior to setting off on migration and, as a result, there is little incentive to catch their own fish. Once they begin migrating, though, it is essential that they learn quickly – otherwise they will almost certainly starve to death. Thus, if a juvenile Osprey encounters a lake, reservoir, river or estuary where fishing is relatively easy, it will often linger for a prolonged period (Figure 6.10). This is exemplified by the fact that the five Scottish juveniles all stopped over for extended periods in Europe: one in southern England, three in Spain and a fifth in northern Portugal. In addition, a juvenile male, known as Ozwald, made an apparent false-start to migration, flying a loop of Scotland before spending a total of 50 days, 100km south-east of his nest-site. This, and a subsequent stop-over to the west of Madrid, meant it took Ozwald 90 days to reach his eventual wintering site on the coast of Mauritania.

It was significant that the three birds that completed migrations without any stop-over days, spent a mean 55 days at their nest-site between fledging and migration, whereas the corresponding figure for birds that incorporated stop-overs was 35 days. This indicates that the longer-staying individuals deposited more fuel before setting off, as suggested by Kjellén *et al.*

Figure 6.10. Juvenile Ospreys often linger at stop-over for prolonged periods during their first migration, particularly if they are catching fish relatively easily. Stop-overs thus offer a valuable opportunity for young birds to refine their foraging skills (© John Wright).

(2001) who found that later-departing Ospreys flying south from Sweden incorporated fewer stop-overs, possibly as a result of additional fuel deposition on the breeding grounds.

It seems that the need to conserve energy is a key determinant of the migration speed of juvenile Ospreys. Given that they migrate according to an endogenous programme of distance and direction, there is little, if any, selective advantage to be gained from migrating faster, whereas the urge to reclaim a known wintering site potentially provides stronger motivation for adult birds. The more cautious approach adopted by juvenile Ospreys likely reduces mortality risk during migration and facilitates individuals arriving at wintering sites in better condition (Newton 2008). Previous research has shown that juvenile Ospreys wander widely after arriving in potential wintering areas (as described in chapter 7), indicating that birds in optimal physical condition may have a competitive advantage when selecting wintering sites. Miller *et al.* (2016) reported similar behaviour during Golden Eagle *Aquila chrysaetos* migration, leading them to suggest that autumn migration is predominantly energy-selected.

Sea crossings and navigation

It was known from ringing studies that Ospreys migrate on a broader front than many other migratory raptors, and satellite tracking has added further fascinating detail in recent decades, particularly in relation to the species' ability to make long ocean crossings.

As mentioned earlier, juvenile Ospreys are known to migrate alone, which potentially puts them at a disadvantage compared to species that follow experienced conspecifics on their first migration. Juvenile geese, cranes and swans migrate with their parents, while raptor and stork species often migrate in mixed-age flocks, enabling them to learn optimum routes from the experienced adults (Newton 2008). Instead, young Ospreys must rely on an endogenous programme of direction and distance, referred to as vector summation, to migrate to distant wintering areas, as described earlier (Alerstam 2000).

The rules of vector summation predict that if strong crosswinds blow a migrating Osprey off course during any one flight step it is unlikely to compensate in subsequent steps because it will simply continue to migrate in the same inherited direction. This has been well illustrated by the migratory flights of juvenile birds from the UK. Easterly crosswinds can be particularly problematic for young Ospreys from the UK because they may necessitate very long ocean crossings. As described earlier, the Osprey's morphology – and specifically its long narrow wings, which reduce drag – means that it is well adapted for long ocean flights, but such journeys can be very hazardous for juveniles with underdeveloped flight skills.

OCEAN CROSSINGS BY JUVENILES

A good example of the influence of crosswinds is a highly unusual migration of a juvenile Osprey satellite-tagged by Roy Dennis in 2012 (Figure 6.11). The juvenile male, known as Stan, left his nest in Moray on 12th September when a brisk northwesterly wind offered good tailwind assistance. He continued flying through the night and at 7am next morning had crossed the Brittany peninsula in northern France and was heading powerfully across the Bay of Biscay. By now the wind was almost due north, and at 4pm that afternoon Stan made landfall on the Spanish cost at Santander and eventually settled to roost a further 120km south-west after a non-stop flight of 1,700km from northern Scotland. Next day Stan continued to make rapid progress and that night roosted near Beja in southern Portugal. Next morning, when Stan resumed his migration, a stiff northeasterly wind resulted in him heading south-west from his overnight roost and by 4pm he had reached the southwest tip of Portugal. Research by Klaassen

Migration

Figure 6.11. Map showing the migration tracks of two satellite-tagged juvenile Ospreys, Rothiemurchus (purple) and Stan (orange), from Scotland. Both birds made long flights over the ocean due to wind drift (Roy Dennis and Tim Mackrill).

et al. (2011) demonstrated that under such a scenario an experienced adult Osprey would have the ability to compensate for the strong northeasterly and adjust its heading in order to fly south-east towards the Moroccan coast. Stan, however, continued on the same south-westerly heading, necessitating a very long flight across the Atlantic. He eventually made landfall in Lanzarote in the Canary Islands 25 hours later, having travelled just under 1,000km across the sea, at altitudes ranging from a minimum of 10 to a maximum of 290 metres.

Next day Stan island-hopped across the Canary Islands, and by evening he was on the southern tip of Gran Canaria. He was only 200km from the African mainland, but getting there would have necessitated a compensatory change of direction against a strengthening northeasterly wind. Instead, he continued south-west across the Atlantic next morning. By 10pm he had flown more than 700km over the sea and, with no land in sight, was forced to continue through the night. Eventually, at around 3:45pm the following afternoon, he reached the island of Ribeira Brava in Cape Verde, having flown more than 1,500km across the Atlantic in 38 hours of non-stop flight. Over the course of the next week Stan remained on Cape Verde, initially on Ribeira Brava, before moving north-west to San Vicente and then Porto Novo. The final transmission from his satellite-tag was during the afternoon of 26th September on the west coast of Porto Novo. It may be that Stan's transmitter failed, but it seems more likely that having had time to regain strength on Cape Verde he set off again on the same south-westerly trajectory as before. With the next land some 2,700km away in eastern Brazil he was highly unlikely to survive such a flight, and the fact that no further transmissions were received from the tag, and that he was never seen again, suggests that he drowned somewhere in the Atlantic. His remarkable migration to Cape Verde – a flight of over 5,000km in just eight days – demonstrated the hazards of the first migration to juvenile Ospreys and the implications of crosswinds. There is in fact a small sedentary population of Ospreys on Cape Verde, and it seems possible that on occasion vagrants, like Stan, may become resident and join the breeding population.

Although Stan did not correct for displacement during his long ocean crossings, there are examples of juveniles compensating for the effects of crosswinds during flights over the sea. Rothiemurchus was another juvenile Osprey satellite-tagged by Roy Dennis in northern Scotland. This young male, who was late to fledge, left his nest-site after a relatively short post-fledging period and three days later, on 10th September, set out across the English Channel from Plymouth. From here the crossing to the Brittany peninsula is just over 150km but Rothiemurchus was pushed further south-west by a strong east-northeasterly wind, causing him to miss the French coast and necessitating a much longer flight across the Bay of Biscay (see Figure 6.11). Flying at an altitude of around 250m, the easterly wind continued to be a problem and caused Rothiemurchus to drift west. He continued flying throughout the night and by dawn had missed the north coast of Spain and was over 150km west of the coast. However, unlike Stan, Rothiemurchus made a clear attempt to return to land once it was light, altering his course to a south-easterly heading. He was now flying against the wind, necessitating a significant increase in airspeed that would have made the flight far more arduous, particularly having been on the wing for more than 24 hours. A series of GPS fixes indicated that he had landed on a boat, or something else floating in the sea, at 10am and remained on board for the next three hours. The opportunity to rest for a few hours proved vital because once he resumed his migration Rothiemurchus maintained a south-easterly heading at an altitude of less than 50m that would have lessened the effect of the strong wind. He eventually made landfall at around 8pm just to the south of Porto. The sudden change of direction after dawn on the morning of 11th September and the opportunity to rest on a boat undoubtedly saved the young Osprey. Had he continued to drift with the wind he would have been blown out into the wilds of the Atlantic and drowned. It is significant, therefore, that the first signs of this compensatory change of direction only occurred after dawn. The GPS fix at 7am was some 150km west of the Spanish coast. It seems possible that the bird could see land from its altitude of 47 metres, indicating that this was the stimulation needed to change course. In contrast, Stan – who was further out to sea – was highly unlikely to have been able to see any land as he crossed the Atlantic, which perhaps explains the differing behaviour of the two birds.

It seems likely that the UK's proximity to the Atlantic means that in years with strong easterly winds there might be increased loss of juvenile Ospreys drowned at sea during their first migration. However, some researchers have suggested that juvenile Ospreys do possess the ability to compensate for crosswinds even during the hours of darkness. In North America, Ospreys breed along most of the Eastern Seaboard, from Florida north to Nova Scotia. Satellite-tracking studies by Rob Bierregaard, Mark Martell and others have revealed that adult birds breeding in northern states, such as Massachusetts and New Hampshire, follow the coast as they head south through Connecticut, New Jersey, Maryland and Virginia, preferring to take a longer overland route than risk a more direct flight across the western Atlantic. Juveniles, in contrast, are less predictable in their initial orientation. Many follow the same coastal route as the adults, but there are some who do attempt long crossings of the Atlantic. Of 24 juvenile Ospreys tagged in New England, 14 used the coastal route, but 10 flew south across the Atlantic, and nine successfully reached the Caribbean Islands (Horton *et al.* 2014). These remarkable flights consisted of a mean 52 hours of continuous flying, during which time the birds covered an average of 2,162km in a near-constant course. The authors propose that the birds were able to fully compensate for wind drift given that, on occasions, they were subjected to perpendicular winds as strong as 40km/h, from both the left and the right of their forward movement direction, and yet rarely strayed more than 10km off course. This ability to compensate for the effect of crosswinds contradicts earlier work of Thorup *et al.* (2003), who proposed that such behaviour is age dependent in Ospreys. Under the classic triangle of velocities theory described by Alerstam (2000), migrating birds alter their heading in order to compensate for the effects of crosswinds; by doing so they counteract the effect of perpendicular winds and maintain a ground track in the intended direction. Horton *et al.* (2014) argue that while crossing the Atlantic the New England juveniles were subjected to such highly variable wind conditions that a change of heading by as much as 90° would have been necessary to maintain the direct-track paths recorded in some instances. Had they conceded to wind drift, their ground tracks would have been highly variable due to the dynamic nature of the winds they experienced. The juvenile Ospreys also adjusted their airspeed according to wind conditions. When flying with tailwinds, the Ospreys reduced their airspeed and thereby made valuable energetic savings, but when flying into a headwind, they significantly increased their airspeed in order to maintain the fast groundspeeds recorded throughout the ocean crossings. Such behaviour conforms to the rules of optimum bird migration first described by Alerstam and Lindström (1990). Horton *et al.* (2014) suggest that their findings indicate that the navigational system utilised by juvenile Ospreys during trans-oceanic migration enables spatially precise and temporally modulated positional orientation at the kilometre to sub-kilometre spatial scale over periods of less than one hour during both day and night. However, such a complex navigation system cannot be explained by vector summation, because drift compensation and airspeed regulation require a system that integrates both positional and temporal information.

Given the lack of visual clues available during trans-oceanic flight, Horton *et al.* (2014) suggested that the earth's magnetic field likely plays a key role and thus proposed a 'chord and clock' system of navigation. Under this scenario they suggest that migrating Ospreys are able to use external information, in particular the earth's magnetic field, as well as celestial cues, to determine the scalar distance between two locations – in their terminology, the chord.

Importantly, they suggest that the bird's response to magnetic cues is defined relative to its natal nest, which acts rather like a magnetic pole for the bird. They can gauge the passage of time calibrated against exogenous time-dependent cues – in other words, the clock. The authors of the research acknowledge that further multidisciplinary work is needed to investigate the existence of such a complex navigation system, and it is important to bear in mind that the method they used to categorise wind drift was different to earlier work of Thorup *et al.* (2003). Horton *et al.* (2014) based their analysis of migration tracks on each individual Osprey's final destination, whereas Thorup *et al.* (2003) compared the movement tracks of individual Ospreys against a group mean, arguing that such a method avoids a bias towards compensation. Indeed, they suggest that the susceptibility of juvenile Ospreys to wind drift may actually contribute towards the maintenance of genetic variation in the inherited migratory orientation programme essential to the evolution of new migration routes, particularly in the face of global environmental change. This is because changing winds favour different inherited directions in different years. For example, as discussed above, in years with a prevalence of easterly winds, it is likely that a greater proportion of young Ospreys setting off on a south-westerly course from the UK will be lost at sea; instead, such conditions would favour those using a more easterly route where long and perilous sea crossings are not required. Conversely, in years with a greater proportion of winds from the west, juveniles using an easterly route could be more prone to being lost in the Sahara, as an easterly route involves a longer flight across the desert in order to reach the preferred wintering areas in West Africa, as described later in this chapter.

OCEAN FLIGHTS BY ADULTS

Further research will help to shed more light on the navigational capabilities of juveniles, but it is clear that adult Ospreys use experience gained during previous migratory journeys to adapt their response to crosswinds according to local conditions, drifting when favourable and compensating to avoid being blown into unfavourable habitats (Klaassen *et al.* 2011). In North America, adult Ospreys keep to the coast as they migrate south along the Eastern Seaboard of the United States, which contrasts with the behaviour of some juveniles, as described above (Horton *et al.* 2014). You might think, therefore, that adult birds would avoid long flights across the Bay of Biscay as they migrate south from the UK, but this is not the case. In an analysis of the migratory flights of 14 adult Ospreys from Scotland and England, I found that 40 per cent of autumn journeys included a flight of 300km or more across the Bay of Biscay. Yet these ocean crossings were only undertaken when they were aided by tailwinds from the north-east (Mackrill 2017). During autumn, winds along the French continental shelf show considerable variability, but northwesterly winds tend to prevail in late summer (Puillat *et al.* 2006). Under such circumstances, it is beneficial for migrating Ospreys to keep to land as they migrate south through France. However, when the wind shifts to the north-east, adult Ospreys appear to intentionally concede to wind drift, and fly across the Bay of Biscay. In doing so, they save time according to the triangle of velocities described by Alerstam (2000), by maintaining a south-westerly heading to maximise tailwind assistance. This enables the birds to reduce airspeed (while maintaining a fast groundspeed) to a greater extent than if they flew south along the coast, because the latter option would involve some degree of compensation for wind drift through an increase in airspeed (Liechti 1995). Nevertheless, the usual energetic savings gained by conceding to wind drift are reduced over the ocean because flapping flight is usually, but not always, required (Agostini *et al.* 2015).

Monti *et al.* (2018b) found that Ospreys making short-distance migrations in the Mediterranean region readily cross the open sea, with flights ranging from 86.48km to

Migration

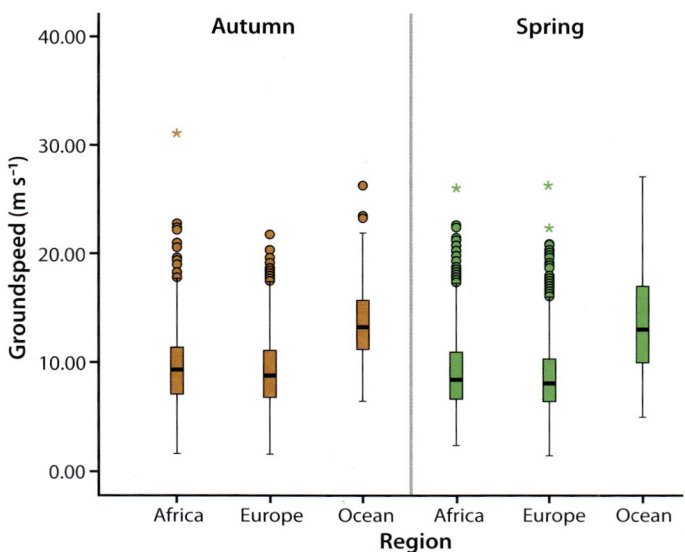

Figure 6.12. Seasonal and regional variation in flight speeds of adult Ospreys from the UK, demonstrating that Ospreys fly significantly faster when migrating over the ocean (Mackrill 2017).

1,023.51km, mean = 430.37, ±269.3km, n = 15 tracks) and flight times varying from 3 to 25 hours (mean = 10.4h, ±7.6h), including some at night. Cross country speed during these ocean flights was considerably faster than over land (sea = 38.2km h^{-1} ±11.3km h^{-1}; land = 29.8km h^{-1} ±11.5km h^{-1}). Ospreys usually made ocean crossings with tailwind support, but even so, airspeeds were still 5km h^{-1} higher during sea crossings than when flying over land. The authors posit that this may have been due to the fact that ocean crossings were undertaken predominantly by flapping flight and so the birds were not delayed by periods spent circling in thermals, or by the zigzag track that can occur when wind direction is different from desired track direction and the soaring bird corrects for wind drift. However, it probably also indicates an increased urgency to minimise time spent over the sea. A similar trend for faster flight speeds over the ocean has been observed in other studies. Klaassen *et al.* (2008) found that Swedish Ospreys migrated faster over the Baltic and Mediterranean than when flying over land, and likewise adult birds from the UK travel significantly faster during sea crossings than over land in both spring and autumn (Figure 6.12). It is noteworthy that most Mediterranean birds performed a pre-migratory movement to a secondary feeding site. This probably allowed them to build fat stores and to wait for favourable tailwinds, which greatly reduces the risks associated with long flights across the sea (Monti *et al.* 2018b).

FLIGHT METHOD OVER THE OCEAN

Recent advances in satellite-tracking technology have allowed very high-resolution data to be collected as Ospreys migrate over the ocean, and this has shed new light on the flight method that birds use during sea crossings. These data have shown that, as expected, the birds usually migrate by flapping flight. On occasions, though, they are also able to exploit weak lift that facilitates soaring-gliding flight, as observed over land. I analysed 12 ocean crossings made by three adult male Scottish Ospreys (Mackrill 2017). The transmitters carried by the birds logged their location every minute during the middle part of the day and this allowed detailed three-dimensional flight paths to be reconstructed for sections of the 12 sea crossings (Figure 6.13). This analysis revealed that soaring-gliding was achieved during three sea crossings, with climbing rates during two of these flights within the range reported for migrating Ospreys tracked by radar over land in southern Sweden (Kjellén *et al.* 2001). Overall, the mean rate of climb when circling over the sea was 1.35ms^{-1} (±0.56ms^{-1}), compared to 1.8 ms^{-1} (±0.46 ms^{-1}) over land. This implied that Ospreys are able to exploit weak thermals that sometimes develop over the sea under specific environmental conditions. It is notable, therefore, that the soaring-gliding flights were all made during autumn, when sea-surface temperatures were warmer. It is known that sea thermals can develop when cool air flows over the warmer water surface Elkins (1995).

Duriez *et al.* (2018) were able to undertake an even more detailed analysis of ocean flights of Ospreys satellite-tagged in Italy by programming Global System for Mobile Communication (GSM) tags to log a 3D location once every second while five juveniles were crossing the

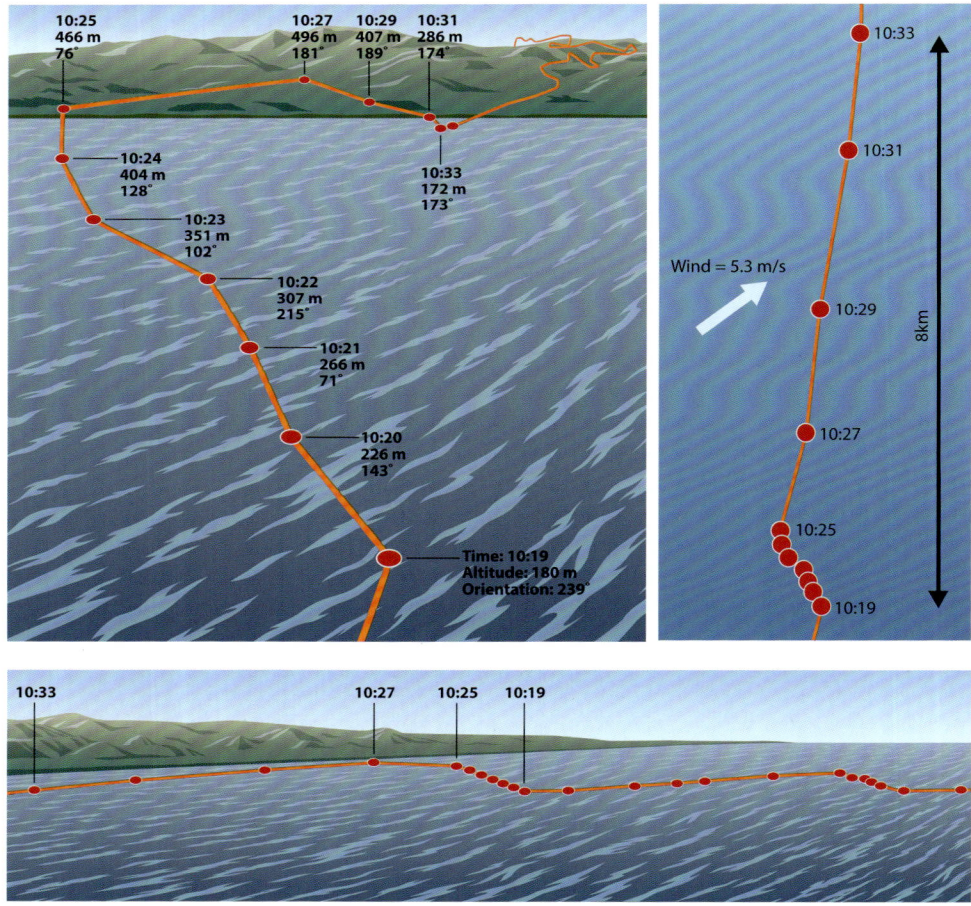

Figure 6.13. Climbing and gliding behaviour of Blue XD during a crossing of the Bay of Biscay. Red circles show the location of GPS fixes, with accompanying time and instantaneous altitude and orientation data. All times GMT. Wind direction indicated by white arrow (Roy Dennis and Tim Mackrill).

Mediterranean Sea. This confirmed that Ospreys are able to achieve soaring-gliding flight over the ocean, albeit with lower performance and greater energy expenditure than over land. All of the tracked birds used thermal soaring over the sea, climbing in thermals on average once every 20.3km over the sea, compared to every 6.4km over land. Overall, they spent 32 per cent of time soaring over the sea, compared to 55 per cent of time over land. The average climb rate was 1.6 times slower, and average flight height 200 metres lower, over the sea. Significantly, the frequency of thermal soaring increased when sea temperature became warmer than the air, and even a bird crossing the sea at night was able to use thermal soaring-gliding flight. This is because atmospheric convection only requires a small air/water temperature gradient, and the different dynamics of air and water temperature changes after sunset favour a positive sea–air temperature difference.

Duriez et al. (2018) suggest that one of the key functions for thermal soaring over the sea might be to gain altitude for safety rather than energy, given that in many cases Ospreys continued flapping even while circling in thermals. The Ospreys continued flapping their wings for 42 per cent of the time while using thermal uplift over the sea, compared to just 24 per cent over land. However, there were certain circumstances, particularly when the birds were flying in crosswinds, that they were able to stop flapping for extended periods, and thus save energy.

Some seabirds are known to use the wind in order to gain altitude, and it is possible that Ospreys exploit the wind in a similar way. In dynamic soaring, albatrosses swoop across an area of wind shear – a sudden change in wind velocity and/or direction over a short distance – just above the ocean surface to increase their airspeed (Sachs 2005). This is then converted to potential energy through a gain in height, similar to the motion of a pendulum, and may account for 80–90 per cent of the total energy required for sustained soaring (Richardson 2011). Lee waves and wind shear – which can occur anywhere in the atmosphere – may well be beneficial to migrant Ospreys as they traverse the sea. These waves and vortices are difficult to predict, but can extend as far as 300–400km downwind of the topographic features generating them, and it could be that migrating Ospreys are able to recognise them by the associated cumulus clouds which form in the rising air (Elkins 1995).

These findings again emphasise the Osprey's ability to take a flexible approach to migration. While it is clear that Ospreys do not require sea thermals in order to complete long ocean crossings, the recent evidence from high-resolution satellite tracking suggests that they do aid migrants as they cross the sea, particularly during autumn.

SEASONAL DIFFERENCES IN OCEAN CROSSINGS

Our satellite-tracking research has shown that Ospreys from the UK are much more likely to attempt sea crossings during autumn. Ten adult birds made a total of 18 flights in excess of 300km across the Bay of Biscay (Mackrill 2017). Fourteen of these flights (mean distance = 519km, ±286km) were undertaken during autumn, as described above, compared to only four (mean distance = 403km, ±111km) in spring. Thus, 40 per cent of all autumn migrations included a significant flight across the Bay of Biscay, compared to just 14 per cent of spring migrations. All of the autumn flights were made when northeasterly winds provided tailwind support (mean tailwind component = 4.2 ms^{-1}, ±1.7 ms^{-1}), as described above, but in spring a prevalence of northwesterly winds appeared to prohibit flights across the ocean. Those flights that were undertaken were all made with a headwind component (mean = 4.1 ms^{-1}, SD = 0.6 ms^{-1}), which would make them significantly more demanding from an energetics point of view. It is also notable that eight of the autumn flights across the Bay of Biscay included elements of nocturnal migration, compared to just a single one in the spring. Indeed, three of the autumn flights included continuous migration through the night. This suggests that more favourable wind conditions encountered during autumn and the potential availability of sea thermals makes flights across the Bay of Biscay less challenging for migrant Ospreys in autumn than in spring.

Seasonal differences in ocean crossings have also been documented for North American Ospreys (Martell *et al.* 2014). Birds that breed in eastern parts of the United States and Canada use a migration route that involves crossing the Caribbean Sea between Hispaniola and South America. During autumn, birds using this flyway cross from Florida to Cuba and then on to Haiti and the Dominican Republic, before making a 600–800km crossing of the Caribbean Sea. In spring, they use a similar route, but make a longer, more direct crossing of the Caribbean making landfall in Haiti, Jamaica or Cuba – a flight across the ocean of 680 to 1,200km that usually takes between 27 and 40 hours to complete. Martell *et al.* (2014) suggest that, as described earlier, this more direct route saves time in spring when Ospreys are eager to reclaim nest-sites. The longer overland route through the Dominican Republic is necessary in autumn because migration coincides with the hurricane season, which considerably increases the risk associated with long flights across the ocean. Previous migratory experience therefore appears to play a central role in shaping this subtle seasonal variation in routes.

Repeatability of routes – goal areas

It has been possible to study the repeated journeys of individual Ospreys in increasing detail as the reliability and longevity of satellite transmitters has improved, and this has provided new insights into the navigational capacity of adult birds. Alerstam *et al.* (2006) analysed the repeated journeys of Ospreys migrating from Sweden. They found that the flight paths of the same individual were often 120–405km apart, with a maximum east–west separation of 1,400km. As they point out, this exceeds the expected normal range of vision, which indicates that Ospreys do not find their way by following familiar landmarks. However, the flight paths did converge in some regions and at the same sites, indicating the existence of up to three intermediary goal areas along the migration flyway of individual birds. Route fidelity was low between the goal regions, presumably because adult Ospreys submit to wind drift when further away from the goal, and then compensate to an increasing degree as they approach it, as predicted by the rules of optimum bird migration (Alerstam and Lindström 1990). Alerstam *et al.* (2006) suggest that Ospreys are able to do this through map-based navigation, where the bird uses multisensory cues, such as visual landmarks, gleaned from previous migratory experience, to navigate to a known location (Mouritsen *et al.* 2016). This may be aided by path integration, which refers to the capacity of a migrating bird to use idiothetic cues (i.e. internal cues), or cues generated by the bird's movements, to calculate its location by monitoring its trajectory in relation to a start location (Gallistel 1990, Whishaw and Wallace 2003).

Since this initial study, further work on the repeated journeys of individual Ospreys has confirmed the existence of these goal areas. Vardanis *et al.* (2016) identified four or five goal areas for four Swedish Ospreys tracked for at least two complete years. All four of these Ospreys visited at least one of the sites on every journey in autumn and spring. The study found that the total recurrence of visits to goal areas was on average almost twice as high for Ospreys as it was for another solitary migrant raptor, the Marsh Harrier, and that Ospreys were much more likely to visit these sites during both autumn and spring. Moreover, the Osprey's habit of revisiting individual-specific goal areas, which were often used for stop-overs, meant that routes converged towards these sites on repeated migrations, as reported by Alerstam *et al.* (2006).

Learning the Way – The Case Study of Rothiemurchus

It is clear from this evidence that these goal areas play a critical role in the migratory journeys of individual adult Ospreys, but how do they establish these sites in the first place? The key to answering this question is analysis of long-term tracking of juveniles through to adulthood. Although such data are comparatively scarce, there are some notable cases that provide a fascinating insight into how experiences during the first migration shape subsequent journeys. In the UK, the best example is Rothiemurchus, a juvenile male Osprey described earlier in this chapter who was satellite-tagged at a nest in Strathspey in northern Scotland by Roy Dennis in August 2010 (Figure 6.14). Further details of this bird's movements can also be found on the Roy Dennis Wildlife Foundation website www.roydennis.org/rothiemurchus.

First Migration – Autumn 2010

After an arduous crossing of the Bay of Biscay, described above, Rothiemurchus spent more than a month fishing along the Rio Paiva in northwest Portugal, just over 40km south-east of Porto. Juveniles often stop over for lengthy periods during their first southward migration at sites where they readily catch fish, and this is likely to have been the case in this instance. It would certainly have been important in helping the young male to recover from the long sea crossing and to deposit fuel prior to recommencing his migration.

The young male resumed his migration in mid-October, following the Portuguese coast before making a 400km crossing of the Atlantic to Casablanca in Morocco. He then skirted around the western end of the Atlas Mountains before setting out across the Sahara. On 22nd October he was at the southern end of Ras Nouadhibou, a 40km-long peninsula on the border of Western Sahara and Mauritania, and that afternoon made a 312km flight across the sea, before eventually making landfall on the Mauritanian coast after dark. Rothiemurchus continued south two days later and arrived at Djoudj National Bird Sanctuary in northern Senegal, close to the Mauritania border on 28th October. He remained at this vast wetland for almost three months before continuing further south to the Sine-Saloum delta in mid-January, where he explored an area of some 1,400km^2.

The young male moved on again at the end of March: this time to a wetland on the north bank of the Gambia River, approximately 50km south-west of his favoured area at the Sine-Saloum. Once there he became far more sedentary, living on a quiet 4km^2 section of river bordered by mangroves. By this stage adult Ospreys would have headed north on spring migration, and this probably gave Rothiemurchus the opportunity to establish this site as his own. He spent the rest of the summer in the same location and by autumn was well established in an area of less than 1km^2. He had now been in West Africa for a year and his satellite transmitter had provided a fascinating insight into how juvenile Ospreys choose a wintering site.

Second Migration – Spring 2011

As expected, Rothiemurchus remained at his newly established wintering site until spring 2011 when he began his first migration north. He set off on 9th May and headed north-east through Senegal and then across the Sahara, some 900km to the east of his route in autumn 2009. By the time he reached Morocco, Rothiemurchus was still considerably further east than his southbound track, necessitating a crossing of the High Atlas. On 16th May he crossed the Mediterranean to the east of Gibraltar and then continued north towards Madrid. He then changed his heading to the north-west – likely influenced by his autumn journey – and arrived at the Sor estuary in Galicia on 20th May. He remained in northern Spain for nine days and then flew direct across the Bay of Biscay to Devon, making landfall at Hope Cove to the east of Plymouth after flying 830km in 28 hours across the ocean. Amazingly, he came ashore at almost

The Osprey

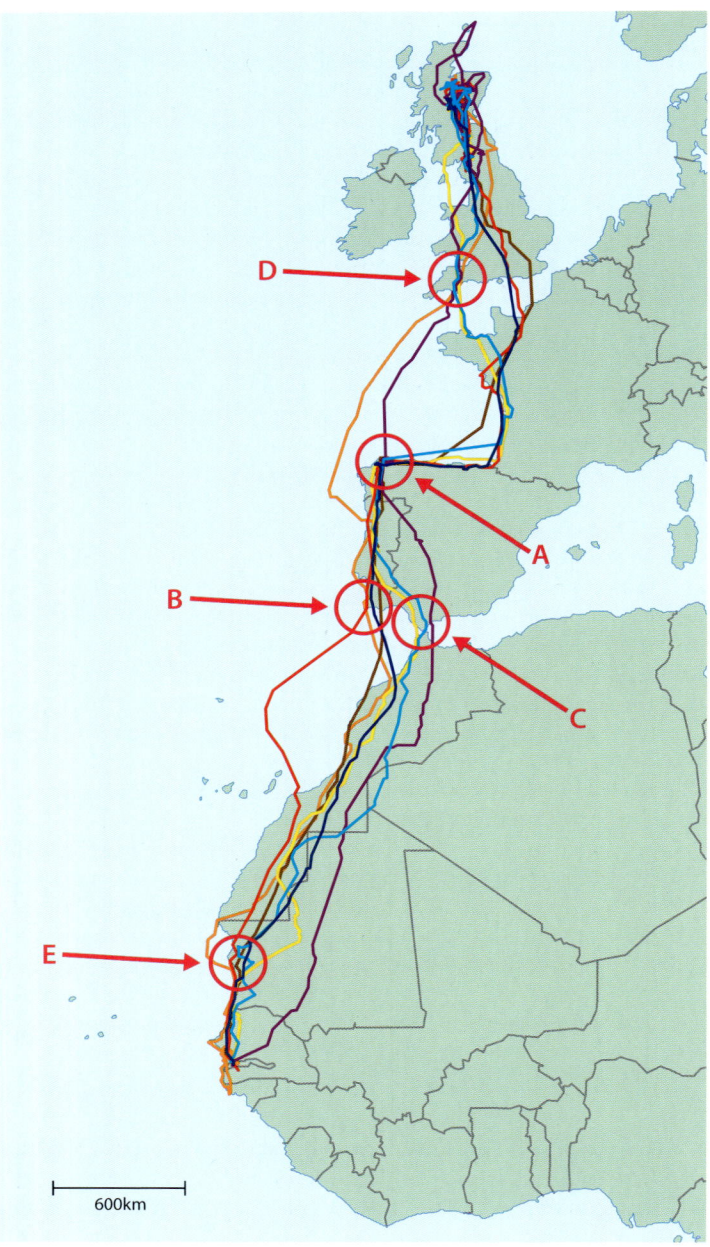

Figure 6.14. Migrations by Rothiemurchus, a male Osprey tagged as a juvenile by Roy Dennis in 2009. Orange = autumn 2009; purple = spring 2011; red = autumn 2011; cyan = spring 2012; blue = autumn 2012; yellow = spring 2013; brown = autumn 2013. A = intermediate goal area in Galicia, Spain; B = autumn migration goal area, SW Portugal; C = spring migration goal area, Strait of Gibraltar; D = spring migration goal area, Prawle Point, Devon; E = autumn migration goal area, Mauritania coast (Roy Dennis and Tim Mackrill).

exactly the location he had departed from in September 2009. Having made it back to the UK for the first time, Rothiemurchus spent the summer exploring Scotland and northern England, including visits to his natal nest in June and July, as described in chapter 3.

THIRD MIGRATION – AUTUMN 2011

Rothiemurchus headed south in early September. He reached Noirmoutier Island just off the Atlantic coast of France on 15th and followed the French coast south, as many adult Ospreys do each autumn. Upon reaching Spain, he made a highly unusual 90° turn to the west and flew purposefully through the Basque Country, Cantabria and Asturias, to the Sor estuary in Galicia. This was the estuary from which he had departed on his spring crossing of the Bay of Biscay, but his detoured route around the Bay of Biscay was 450km further than a direct flight across the sea from Noirmoutier to Galicia. This deviation from the most direct migratory path was a clear indication that he was navigating towards an intermediate goal area in Galicia, but was perhaps wary of making another long flight across the Bay of Biscay. He subsequently spent a week beside the River Eume, 45km to the south, another location he had visited on his flight north.

Rothiemurchus resumed his migration on 2nd October and headed south through Portugal and southwest Spain before setting out across the Atlantic from a location just 12km east of his corresponding location in autumn 2009. On this occasion he flew in a wide arc over the sea before making landfall in southern Morocco after a 24-hour non-stop flight of 1,278km across the Atlantic. He then continued south across the Sahara, eventually arriving on the Mauritanian coast, 11km north of the location where he had made landfall after his flight across the Atlantic two years previously. From there he continued south along the Mauritanian coast before stopping at Djoudj National Bird Sanctuary and the Sine-Saloum delta, prior to returning to his wintering site close to the Gambia/Senegal border on 9th October.

FOURTH MIGRATION – SPRING 2012

As expected, Rothiemurchus remained at his favoured site all winter, living in an area of just 0.5km². He headed north again on 12th April 2012, making stuttering progress across the Sahara and then passing through the Atlas Mountains on 21st April. He flew north-east through Spain, after crossing the Strait of Gibraltar, and returned to the Sor estuary in Galicia where he remained for several days, for another spring stop-over at this fish-rich site.

Once he resumed his migration, Rothiemurchus headed east along the north coast of Spain and then north up the Atlantic coast of France, following the track of his autumn detour. As in previous journeys, he then crossed the English Channel from Brittany to Devon and made landfall at Prawle Point, just 11km east of the corresponding location a year earlier. He then spent another summer ranging extensively in Scotland in search of a nest-site and mate.

FIFTH MIGRATION – AUTUMN 2012

Rothiemurchus began his journey south in autumn 2012 with a long flight from the Lake District to Normandy and then south along the Atlantic coast of France. He then made the usual dogleg along the north coast of Spain before returning to the River Eume in Galicia for his regular autumn stop-over. He remained there almost four weeks before continuing south through Portugal and then heading across the Atlantic to Morocco, 30km east of his position the previous autumn. He made good progress across the Sahara before heading south along the Mauritanian coast and then over Djoudj as he had done in previous journeys. By 6th October he was back at his regular wintering site.

SIXTH MIGRATION – SPRING 2013
Rothiemurchus headed north again on 3rd April and crossed the Strait of Gibraltar into Spain 15 days later. He then took his usual route north through Iberia before arriving for his spring stop-over at the Sor estuary on 24th April. He remained there for a week before heading east along the north coast of Spain on 2nd May, and then flying 300km across the Bay of Biscay from Villaviciosa to Bayonne. He made his usual crossing from Brittany to Devon close to Prawle Point, three days later, before continuing north to Scotland, but again failed to breed.

SEVENTH MIGRATION – AUTUMN 2013
In September 2013, Rothiemurchus avoided the dogleg through France and Spain by flying direct across the Bay of Biscay from just north of La Rochelle to Villaviciosa and then west to his autumn stop-over beside the River Eume, where he remained for 21 days. He then headed south through Portugal and crossed to Morocco from Faro, 30km east of where he did so the previous autumn. Having made landfall south of Casablanca, he headed across the Sahara before continuing south along the Mauritanian coast as in previous autumn journeys before arriving at his usual wintering site on 10th October.

EIGHTH MIGRATION – SPRING 2014
After a winter in his favoured haunt, Rothiemurchus headed north on 5th April 2014, crossing the Sahara in four days and then skirting around the western end of the Atlas Mountains. As usual on the journey north he crossed the Strait of Gibraltar and then continued north-west through Iberia, before arriving at the Sor estuary on 18th April. He remained there for a shorter period than on previous journeys, before heading direct across the Bay of Biscay to the French coast between La Rochelle and Nantes. Two days later he crossed the English Channel from Brittany, but this time made landfall on the Devon–Dorset border, 65km east of previous years. He finally arrived back at his previous summer's haunts in Perthshire on 27th April.

NINTH MIGRATION – AUTUMN 2014
Having failed to breed again, Rothiemurchus headed south on 30th August and crossed from Dorset to Brittany on 4th September. Next day he set off across the Bay of Biscay in a south-south-west direction, but when he was off Noirmoitier he turned back towards mainland France, crossing over Île d'Yeu and then passing down the French coast to roost that night just south of Lac d'Hourtin-Carcans. On 6th September he crossed into Spain and by evening next day he was back in his usual haunts beside the River Eume in Galicia. He was there for 17 days before resuming his migration south into Portugal and then across the Atlantic from Faro to Rabat. After a fast migration across the Sahara – flying 1,203km in two days – he headed south along the Mauritanian coast and arrived back at his winter quarters on 2nd October.

Sadly, Rothiemurchus's transmitter stopped sending data after 7th November 2014, but the five-and-a-half years that it worked offered a fascinating insight into how an Osprey learns from its first migration and how the initial route, which in Rothiemurchus's case was profoundly affected by the weather, influences subsequent journeys and, thus, intermediate goal areas. The most obvious goal area for Rothiemurchus was Galicia in northern Spain, where he favoured the River Eume in autumn and the Sor estuary in spring. Most Ospreys from the UK migrate through central Spain, and it is clear that Rothiemurchus's atypical behaviour can be traced back to his first migration. Having been blown out into the Atlantic, Rothiemurchus spent over a month living along the Rio Paiva to the south-east of Porto. Although he did not go back to this particular location, his north-west track through Iberia each spring was indicative of a desire to return to this region and, ultimately, the establishment of Galicia as a goal area. This, combined with the experience of two arduous sea crossings early in life, resulted in the highly unusual dogleg migration along the French and Spanish coast during autumn which became a feature of subsequent journeys. It was significant, however, that as Rothiemurchus became more experienced he did make more direct flights across the sea again, which reflects the general trend among adult Ospreys to undertake sea crossings when conditions are favourable.

Other possible goal areas appear to differ between spring and autumn migrations. It is notable that on every autumn journey Rothiemurchus migrated south across the Atlantic from southwest Portugal to Morocco, but in spring made the shorter crossing across the Strait of Gibraltar. Likewise, in spring he flew from Brittany to the Devon coast near Prawle Point on three successive migrations, a route learned on his first flight south, but in subsequent autumn migrations he crossed between Sussex and Normandy. Finally, he flew south along the Mauritanian coast between Nouakchott and Djoudj during every autumn migration, but was more variable in his choice of route north. This probably results from the fact that as they near their final destination Ospreys tend to compensate for wind drift, but in the early stages of their journey (i.e. Mauritania on the flight north, and southern England on the journey south) they allow themselves to be drifted by the wind to a much greater extent (Alerstam and Lindström 1990). This saves energy in the early stages of migration when there is no clear advantage of compensating for wind drift. However, as they approach a goal it is necessary to compensate for the effects of crosswinds in order to navigate to a specific location. Analysis of Rothiemurchus's autumn flights through the UK revealed that this was indeed the case, with brisk northwesterly winds aiding his migration from Cumbria to Sussex and then across the English Channel. Once he reached northern Spain, though, it was necessary on occasions to turn away from the most favourable winds in order to reach Galicia.

The example of Rothiemurchus provides an insight into why Ospreys migrate on a broad front compared to other migrant raptors. Satellite tracking has shown that even siblings from the same nest use contrasting routes, through Europe in particular. Vardanis et al. (2016) suggest that this may be possible due to the Osprey's reliance on fish. High-quality sites are widely dispersed throughout Europe, meaning that even if juveniles use an atypical route they are likely to encounter suitable foraging sites regardless. Hunting efficiency may also improve with local knowledge and this probably explains high site fidelity to specific stop-over sites, which are often initially encountered on the first migration. It is notable, however, that when European Ospreys reach Africa for the first time, atypical routes across the Sahara can be more problematic for young birds.

The Sahara

Covering between seven and nine million square kilometres, the Sahara presents a clear ecological barrier to Palearctic-African migrants (Moreau 1972). Even the shortest crossing requires a flight of approximately 1,500km, and this means a period of fasting for adults and juveniles, coupled with a high risk of dehydration (Strandberg et al. 2010). Furthermore, dust storms have increased in frequency since the late 1950s and this further heightens the risk associated with the crossing (Goudie and Middleton 2006, Niang et al. 2008).

The first documentation of flights across the Sahara was made by Kjellén et al. (1997), who tracked two adult females from the same region of Sweden. Unexpectedly, the birds used contrasting routes, with one following a westerly track south through Central Europe and then across the Sahara to Ivory Coast, and the other using a more easterly route through the Middle East and then onwards to Mozambique. The latter bird avoided a long flight across the Sahara, but the other took 10 days to cross the Sahara from Algeria to Burkina Faso before continuing to Ivory Coast. The satellite data collected by the Swedes demonstrated that during its long flight across the Sahara this adult female Osprey rested at night and made the crossing during daylight hours only.

Since the groundbreaking study of Kjellén *et al.* (1997), knowledge of Osprey migration has increased dramatically, most notably with the recent advent of GSM technology. Ospreys migrating south from the UK tend to use a westerly route across the Sahara, setting out across the desert in southern Morocco and then continuing south through Western Sahara and Mauritania before arriving in northern Senegal some four to five days later (Dennis 2008, Mackrill 2017). Birds from other populations in Western Europe, including France and Germany, typically use a similar route, while birds from further east, such as Finland and Estonia, make a more easterly crossing from Libya or Egypt south towards either central or eastern Africa (Väli and Sellis 2016). Nevertheless, there is considerable variability and some Finnish birds have used the more westerly route between Morocco and Senegal (Saurola 2005, 2014).

FLIGHT METHOD IN THE DESERT – THE VALUE OF THERMALS

The flights of experienced adult birds offer valuable information about the strategies used by Ospreys to cross the Sahara. One might expect adults with previous migratory experience to follow the coastline in order to feed, but various satellite-tracking studies have revealed that they chose the most direct path across the desert, preferring to minimise time in this inhospitable terrain. This heightened urgency is manifested in increased daily travel times. Ospreys satellite-tagged in the UK flew for an hour longer per day in Africa than in Europe during both autumn and spring migrations (Mackrill 2017). Klaassen *et al.* (2008) identified the same trend among Swedish Ospreys, with birds flying for a mean 2.7 hours longer in Africa than Europe. This variation occurs because, as described earlier, Ospreys behave in the manner of fly-and-forage migrants in Europe, regularly interrupting migration to feed – whereas in Africa, and particularly during crossings of the Sahara, they maximise daily flight times. Nevertheless, there is a clear trade-off between time and energy minimisation. Despite their status as facultative soaring migrants and an apparent desire to complete the crossing as quickly as possible, Ospreys usually limit flight times to periods when thermal updrafts are available when crossing the Sahara. This is exemplified by the fact that nocturnal flight is rare in the desert, being recorded on only 3 per cent of daily flights through the desert by 14 adult Ospreys from the UK (Mackrill 2017). In addition, a satellite-tagged Osprey translocated from Scotland to the Basque Country in northern Spain, flew 533km through the Sahara after 9pm on its first autumn migration (Galarza *et al.* 2017). Ospreys also initiate daily flights later than another facultative soaring migrant, the Marsh Harrier, preferring to wait until thermals develop before beginning their daily flights (Mellone *et al.* 2012).

My analysis of trans-Saharan flights by three adult Ospreys satellite-tagged by Roy Dennis in Scotland indicated that as much as 90 per cent of time was spent in soaring-gliding flight during periods when high-resolution data were available, normally the middle part of the day when solar-charged battery voltage is highest and thermals most likely to be available (Mackrill 2017). A three-dimensional view of an Osprey's flight across the Sahara using these GSM data illustrates the changes in altitude associated with soaring-gliding flight very well (Figure 6.15). Using thermals in this way enables Ospreys to achieve relatively fast cross-country speeds, while minimising energy expenditure. Boundary layer – the height at which cumulus clouds form, and a proxy for thermal conditions – was the key determiner of daily distance flown during 184 days of trans-Saharan migration undertaken by a total of 12 adult Ospreys. On days when thermal conditions were good, Ospreys migrated greater distances than when thermal

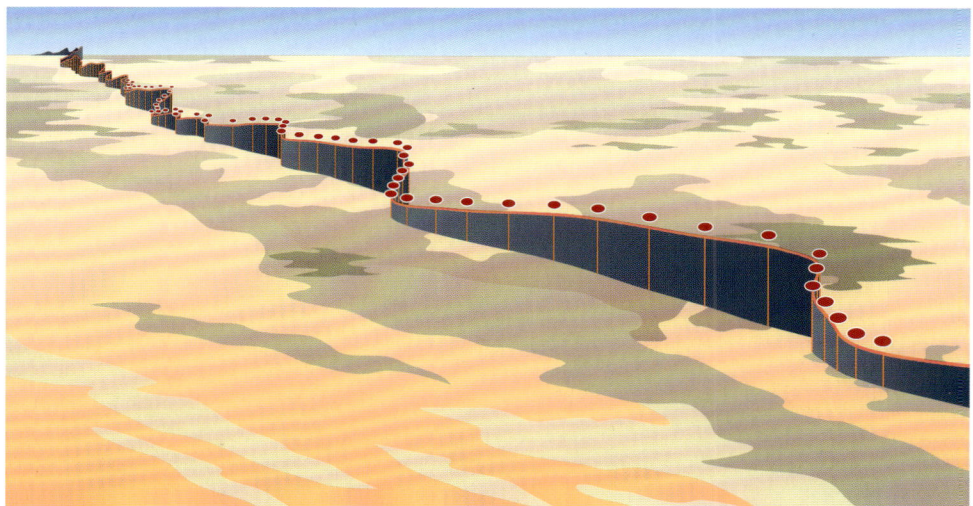

Figure 6.15. Climbing and gliding behaviour of a Scottish adult male Osprey, Blue XD, across the Sahara. When making desert crossings, Ospreys tend to limit their daily flight times to periods when thermals are available. Each red circle shows a single GPS fix, with approximately one minute between each. The shaded area below the GPS fixes shows the changes in altitude along the flight track (Roy Dennis and Tim Mackrill).

updrafts were weaker or more sporadic (Mackrill 2017). Thus, the delicate balance between time and energy minimisation during Saharan crossings appears to be greatly influenced by thermal conditions.

MORTALITY IN THE DESERT

The most comprehensive analysis of the dangers associated with crossings of the Sahara was undertaken by Strandberg *et al.* (2010) who analysed the flights of Ospreys and three other species of migratory raptors across the desert. They found that mortality associated with the Sahara passage made up a substantial fraction of total annual mortality, particularly in juveniles, for which it constituted about half. Adult Ospreys are highly competent navigators, and less likely to die in the Sahara, but dust storms in the Sahara are capable of forcing even the most experienced Osprey off its preferred course, or in extreme circumstances can result in aborted crossings or even death. The associated strong winds also have the potential to break up thermal updrafts, which further inhibits the northward migration of Ospreys and other migrator raptors. Strandberg *et al.* (2010) identified cases of aberrant behaviour, retreat migration or death of Ospreys as they crossed the desert. Aberrations (classified as either course change, slow speed, interruption or retreat) were identified in six of 24 (25 per cent) migration tracks of adult birds in autumn, and in five of 17 (29 per cent) adult tracks in spring. Of the four juvenile tracks, two (50 per cent) included aberrations and one bird died in the desert. Although these aberrations may not result in death of adult birds, they may delay arrival on the breeding grounds. Of birds that exhibited aberrant behaviour, 70 per cent arrived later than the species-specific median arrival date, while only 29 per cent of birds showing no aberrations were late. This has knock-on effects, because Strandberg *et al.* (2010) found that Ospreys and Marsh Harriers that arrived late at the breeding grounds due to aberrant behaviour in the Sahara had a significantly lower breeding success than birds that migrated north without such issues.

A lack of safe roost sites is another risk for migrant Ospreys as they traverse the Sahara. Analysis of satellite imagery indicates that many migrant Ospreys roost on the desert floor, which leaves them vulnerable to predation (Figure 6.16). This was exemplified by the fate of 09(98), an adult male Osprey that we satellite-tagged at Rutland Water in 2011 (Figure 6.17). 09(98) set out on migration on 5th September the following year, having raised two chicks, and made rapid progress south through France and Spain, averaging over 400km per day. By the evening of 10th September, he had reached a ridge on the northern edge of the Sahara, and settled among large rocks and boulders for the night. It looked a relatively safe place to roost, but three days later the next batch of data from the satellite transmitter indicated he was still in the same location – a worrying sign. I posted information about the bird's location online in the hope that someone may be able to go and investigate. Within hours, Farid Lacroix, an ex-search and rescue helicopter pilot living in Agadir, offered to help and next day drove five hours south from his house. Using GPS to guide him, he trekked up to the ridge in searing desert heat. As feared, he found that 09(98) had been predated, likely by a raptor or owl because the feathers appeared to have been plucked, rather than chewed. This demonstrated that even for an experienced adult Osprey – as a 14-year-old, 09 had completed 24 migrations before – the Sahara presents a very real danger.

As the death of 09(98) showed, the Sahara is a hazard for all Ospreys, regardless of age or experience, but it is a particularly demanding challenge for juveniles on their first migration. This is well exemplified by an analysis of the first migrations of 35 juvenile Ospreys satellite-tagged in northern Scotland and central England over a 20-year period between 1999 and 2019 (Table 6.2). During this period satellite-tracking technology advanced significantly, but even the early tags provided valuable data. Six of the transmitters failed during the first migration, but 18 of the remaining 29 birds either stayed in Iberia or reached suitable wintering grounds in West Africa. Eleven of these individuals completed an initial migration to Senegal, three to Mauritania and another spent time in Mali before moving to Guinea-Bissau. The other bird, Stan, migrated to Cape Verde, as described earlier. Two that remained in Iberia both wintered in Portugal. Meanwhile, of the 11 birds that did not survive, the satellite data indicated that four died in the Sahara, three perished in the Atlantic and two in the Atlas Mountains. The other birds died in Morocco and Scotland. Thus, 14 per cent of all tagged birds where the outcome of migration was known died in the Sahara, and 36 per cent of known mortality occurred in the desert. These data indicate that long sea crossings and the Sahara pose the most significant threats to juvenile Ospreys on their first migration from the UK.

Adult Ospreys nearly always use a westerly route across the Sahara, passing through Morocco, Western Sahara and Mauritania. However, those that died followed a more easterly route through Algeria and Mali. This

Figure 6.16. Satellite-tracking data has shown that Ospreys often roost on the desert floor when crossing the Sahara (© John Wright).

Migration

Figure 6.17. Ex-search and rescue helicopter pilot Farid Lacroix trekked up to a remote ridge in the northern Sahara to locate the remains of satellite-tagged male 09(98) from Rutland Water, who was predated while roosting (Clockwise from top left: © John Wright, © Farid Lacroix, © Farid Lacroix).

requires a longer flight and even if birds that use this route do survive the desert crossing, they must then head west towards the West African coast. It is no surprise, therefore, that juveniles that successfully cross the Sahara generally use the westerly route favoured by adult birds. Using this route means that they arrive at the Mauritanian or Senegalese coast soon after completing the flight across the Sahara, where foraging opportunities are rich and diverse.

Table 6.2. Outcome of first migration by 35 juvenile Ospreys from the UK

Outcome	No. of birds	Comments
Completed migration	18	Initial wintering locations: Senegal (11), Mauritania (3), Guinea-Bissau (1), Cape Verde (1), Portugal (2)
Transmitter failed	6	
Died during migration	11	Location of deaths: Sahara (4), Atlantic Ocean (3), Atlas Mountains (2), Morocco (1), Scotland (1)

Conclusions

- The majority of European and North American Ospreys are migratory, although those at lower latitudes tend to remain sedentary or make short-distance movements. European Ospreys migrate to sub-Saharan Africa, although some winter in Southern Europe and the Middle East. Ospreys from eastern Russia migrate south to coastal Arabia, the Indian subcontinent and South-East Asia. Most North American Ospreys winter in Central or South America. Routes used vary between populations.
- Individual Ospreys migrate alone, and on a broad front. Adult birds migrate to a known wintering site, whereas juveniles follow an inherited programme of direction and distance referred to as vector summation to reach potential wintering areas.
- Ospreys typically fly 200–400km per day during migration, although some birds may travel >2,000km in a single flight. Juveniles usually fly similar daily distances to adult birds.
- The total distance flown by Ospreys on migration varies considerably both among and between populations, but some individuals travel >8,500km. Those undertaking very long migrations may spend >60 days on each migratory journey. It is more typical for a flight of about 5,000km to take approximately three weeks to complete.
- Some Ospreys complete migrations without stop-overs, whereas other birds may linger on every journey at favoured stop-over sites that they encountered on their early migrations.
- Ospreys employ the fly-and-forage strategy during migration, where birds pause to feed before, during or after a day's flight.
- Wind can have a profound effect on the migration routes used by juveniles, who usually do not compensate for wind drift, on their first southerly migration. Adults allow themselves to drift with the wind when far from their migratory goal, but compensate to avoid unfavourable habitats or when approaching a goal area. Research has shown that individual Ospreys may have three to five goal areas that they pass through on each journey, with routes diverging significantly in between.
- The Osprey is a facultative soaring-migrant, meaning that it alternates between soaring-gliding and flapping flight according to environmental conditions. Its long narrow wings, which reduce drag, enable it to make long sea crossings, sometimes in excess of 1,000km. Wind drift may result in some juvenile Ospreys making very long flights across the ocean. Adult birds usually only attempt such flights across the sea when they have tailwind support. They may also fly through the night when crossing ecological barriers, or if migration is time-selected.
- Ospreys show elements of time-selected migration on their northward journeys, when it is advantageous for adult birds to make a timely arrival at breeding sites. This can include extending daily flight times or flying in unfavourable conditions. There is evidence that later-departing birds also incorporate time-minimisation techniques during autumn, with the reclamation of winter territories key.
- The highest mortality of juvenile birds tends to occur when crossing ecological barriers such as oceans and deserts.
- The Sahara is a major ecological barrier to Ospreys from Europe, and may account for a significant proportion of annual mortality. Ospreys tend to limit daily flight to when thermal updrafts are available in the desert, but sometimes extend flight times into the night.

CHAPTER 7
Wintering behaviour

It is January 2016 and I am in a small boat in the Sine-Saloum delta with a group of fellow Osprey enthusiasts from the Rutland Osprey Project. We had been in the same location the previous day and had seen an Osprey with a blue ring on its right leg. This indicated that it was a bird from either England or Wales but, frustratingly, we had not been close enough to read the inscription on the colour ring. Knowing it might be a bird from the local Rutland population we returned in the hope of seeing it again. As our boat cut through the calm waters, we could see the characteristic profile of an Osprey perched on the edge of the mangroves, bathed in the gentle evening light. This had to be the bird. As we inched closer it took off. Fortunately, it flew towards the boat, allowing my colleague, John Wright, to fire off a few camera shots as it banked overhead. We were elated when John confirmed that the inscription was white 32: revealing this to be a bird that had fledged from the Manton Bay nest in 2011 and bred for the first time locally in 2015.

When I returned to the Sine-Saloum delta with another group, three years later, the bird was there again – perched in almost exactly the same spot. Travelling to West Africa to study Ospreys on their wintering grounds is a great privilege, and the thrill of seeing a bird you know well makes the experience even more memorable. It is tempting to think of the Ospreys that breed in the UK as 'our' birds, but that is certainly not the case. Adult birds are faithful to the same wintering site each year, meaning that 32(11) would have known 'his' part of Senegal just as well as his English nest-site.

As our sightings of 32(11) proved, you do not necessarily need the latest in GPS tracking technology to find a known Osprey on its wintering grounds, but the recent advances in satellite telemetry have added greatly to knowledge of the behaviour of individual birds during winter, particularly when coupled with field observations.

Wintering destinations

As discussed in the previous chapter, the wintering locations of Palearctic and Nearctic Ospreys are widely dispersed, with the majority of birds undertaking long-distance migrations. Most Ospreys from Northern Europe migrate to sub-Saharan Africa, with individuals from populations in the west, including the UK and Germany, wintering in tropical West Africa – from Mauritania to Ghana – and those from further east, including Finland, Estonia and Russia, generally favouring eastern and central Africa, with some individuals migrating to the southern coast of South Africa (Hake *et al.* 2001, Dennis 2008, Väli and Sellis 2016, Mackrill 2017, Babushkin *et al.* 2019, Østnes *et al.* 2019). In addition, a small but increasing number of birds are also choosing to winter at more northerly latitudes, most notably on the Iberian Peninsula (Martín *et al.* 2019). Those breeding further east in Russia migrate south to coastal Arabia, the Indian subcontinent and South-East Asia (Ferguson-Lees and Christie 2001, M. Babushkin pers. comm. 2019). Meanwhile, the winter distribution of Ospreys that breed in North America ranges in latitude from the southern United States to southern reaches of South America and in longitude from the west coast of Mexico eastward throughout the Caribbean and the eastern coast of South America (Washburn *et al.* 2014).

Some Ospreys breeding at more southerly latitudes do not migrate. Satellite-tracking work in the Mediterranean has shown that Ospreys from Corsica, the Balearic Islands and mainland Italy have heterogenous migratory behaviour. Monti *et al.* (2018a) satellite-tagged a total of 41 Ospreys in the region and found that 73 per cent of the individuals migrated, and the remaining 23 per cent stayed at breeding sites all year round. These latter birds regularly travelled between nesting areas and feeding sites situated within 20km. More than half of the adult birds tracked in the study (9 of 16, i.e. 56.3 per cent) did not migrate, while the remaining seven birds completed short-distance migrations within the Mediterranean region. In contrast, all except two of 25 tagged juveniles migrated (92 per cent). There were also differences in the proportion of resident versus migratory birds between populations: 83.4 per cent of birds from Corsica migrated, compared to 71.4 per cent of birds from the Balearic Islands, and 66.7 per cent of tagged birds in Italy.

Arrival on the wintering grounds

Most females leave the breeding grounds before their mate and usually arrive at established wintering sites earlier. Washburn *et al.* (2014) published an overview of the wintering behaviour of North American Ospreys from satellite-tracking data. They found that Ospreys arrived on their wintering grounds as early as 31st July and as late as 4th December, with females arriving earlier than males (21st September (±2 days) vs 7th October (±3 days)). Observations and satellite data for whole-family groups from North America showed that in six pairs, females arrived 36.8 days (range 20–70 days) before their mates; while in three pairs males arrived on average nine days (range = 2–20 days) before females (Bierregaard *et al.* 2020).

A similar pattern of arrival is evident among Ospreys migrating from the UK to wintering sites in Iberia and West Africa, with arrival dates of satellite-tagged adult birds ranging from 19th August to 25th October (n = 49). Females again arrived earlier, with a median arrival date (12th September) 3.5 weeks earlier than males (7th October) (Dennis 2008, Mackrill 2017). Similarly, three Estonian satellite-tagged adult females arrived at established wintering sites in Congo, Sudan and South Sudan between 11th September and 15th October, and all were earlier than an adult male who reached his wintering site beside the White Nile in Angola on 24th October after a 39-day migration (Väli and Sellis 2015). Likewise, Swedish adult females returned to wintering sites (mean = 27th September, n = 6), 2.5 weeks earlier than males (mean = 13th October, n = 5) (Hake *et al.* 2001).

Juveniles, unlike adults, are not migrating to a known wintering site, and so it can be difficult to determine exactly when migration is completed, particularly as they wander extensively throughout their first winter. Seventeen juveniles from the UK arrived at potential wintering sites in Iberia and West Africa between 8th September and 13th November, with a median arrival date of 6th October (Dennis 2008, Mackrill 2017). Meanwhile, two Swedish juveniles completed initial flights to Cameroon and Guinea on 5th and 26th October respectively (Hake *et al.* 2001).

Habitat requirements in winter

As a strictly piscivorous species, the key requirement for wintering Ospreys is a reliable source of fish, whether they are wintering at temperate or tropical latitudes, or in freshwater, brackish or marine environments. Ospreys tend to be opportunistic and are well adapted to exploiting a range of different wintering localities.

Field observations and satellite-tracking studies have revealed that Ospreys from Western Europe favour coastal parts of West Africa for wintering. Some of the highest densities of wintering Ospreys occur around river deltas and estuaries, where shallow water makes fishing a simple task for experienced adult birds (Prevost 1982, Dupart 2021). Ospreys readily fish along the coast too and so linear stretches of coastline also support large numbers of wintering birds, with individuals flying out to sea to catch fish once or twice a day. Senegal supports some of the highest concentrations of wintering Ospreys in West Africa, and colleagues and I have encountered colour-ringed Ospreys from the UK, France, Germany, Spain, Finland and even Latvia during field visits over the last decade (Figure 7.1).

Detailed surveys undertaken by ornithologist Jean-Marie Dupart and colleagues in Senegal have demonstrated the importance of this part of West Africa for wintering Ospreys (Table 7.1). Monthly counts were carried out between Djoudj National Bird Sanctuary close to the Mauritanian border, and then along a 190km stretch of coastline between St Louis and Dakar during winter 2020–21 and again in 2021–22. Three sites to the east of Dakar were also surveyed. A maximum count of 822 individual Ospreys was made in January 2021, with the highest concentrations recorded at three protected coastal sites: the St Louis Marine Protected Area (MPA), Langue de Barbarie National Park, and the Somone MPA – as well as along the Grand Côte between Potou and Dakar and at the Pointe Sarène lagoon (Dupart 2022). These sites account for 85 per cent of the population of wintering Ospreys in the area surveyed. Dupart (2022) suggests that the preference for these coastal areas is due to the abundance of easily caught marine fish and a lack of human activity. Three of the sites have conservation designations, while the latter two are relatively quiet unpopulated areas with little disturbance. There tends to be more active human presence along rivers and other wetland sites in inland areas, which renders them less suitable for Ospreys, and may lead to an increased rate of persecution, as discussed later in this chapter. A comparison with a 1979 survey undertaken at many of the same locations by Yves Prevost as part of his PhD studies (Prevost 1982) revealed a 195 per cent increase in the numbers of wintering Ospreys in January 2021 and a 194 per cent increase the

Figure 7.1. Colour rings have provided valuable information on the winter destinations of Ospreys from Europe. Rings shown here are (clockwise): German, Scottish, English, Scottish, French, Latvian (© John Wright).

Table 7.1. Results of winter surveys of Ospreys in northern Senegal (sources: Prevost (1982) and Dupart (2022))

Site	January 1979	January 2021	Percentage change vs 1979	January 2022	Percentage change vs 1979
Northern Zone					
Djoudj National Bird Sanctuary	25	32		21	
Nord Gandiol		15		21	
Trois Marigots		12		19	
St Louis MPA	100	64		22	
Langue de Barbarie		186		189	
Grand Côte	76	370		373	
Toubab Dialaw	46	17		18	
Lagune de Yene Tode		3		10	
Popenguine		14		14	
Lagune La Somone		44		66	
Lagune de Nianing Nord	32	4		3	
Lagune de Nianing Sud		10		10	
Lagune de Pointe Sarène		45		49	
Lagune de Mbodiène		6		4	
Subtotal	279	822	+195%	819	+194%
Delta du Saloum					
Joal MPA	55			16	
Reserve de Palmarin				26	
Sangomar MPA				58	
Delta du Saloum 1				90	
Delta du Saloum 2	145			9	
Subtotal	200			199	
Casamance					
Reserve de Kalissaye	186			167	
Elinkine				21	
Subtotal	186			188	+1%
Grand Total (1979 vs 2022 sites)	665			1206	+81%

following year. This likely reflects a significant increase in the European Osprey population over the same period. For example, there were fewer than 30 breeding pairs in the UK in 1979 (Dennis 2008) compared to over 300 by 2021 (R. Dennis pers. comm. 2021). The most startling increase is along the Grand Côte, where the wintering population increased by 386 per cent in 42 years, although observer effort may be a contributing factor in these counts (Table 7.1) (Dupart 2021).

The surveys were extended to include parts of the Sine-Saloum delta and Casamance region further south during the winter of 2021/22. Dupart (2022) acknowledges that the sampling only covered limited parts of these vast regions, but nevertheless made it possible to identify the areas most frequented by Ospreys. Numbers across the two areas were similar to the 1979, but there was a marked change in distribution, with the largest concentrations in 2022 recorded on the coast, whereas birds favoured mangroves further inland in 1979. Dupart (2022) suggests that sandbanks and sandy islands are favoured by Ospreys today because they are close to productive foraging grounds, and there is little or no human disturbance.

Overall, numbers of Ospreys wintering in Senegal have approximately doubled in the four decades between the two surveys, resulting in significantly higher densities in the best habitat compared to 1979 (Figure 7.2). Indeed, Dupart (2022) observed that the increasing numbers mean that the birds have to be more tolerant of each other than in the past.

Other Osprey populations have different wintering habits according to habitat availability. Washburn *et al.* (2014) carried out a detailed analysis of the habitat used by 79 satellite-tagged Ospreys from North America while on their South American wintering grounds. Half of the birds (50.6 per cent) wintered on river systems, 30.4 per cent in coastal areas and 19.0 per cent on lakes – and, as expected, all wintering sites contained at least one major waterbody. They found that female Ospreys wintered on river systems more than coastal areas or lakes, whereas males wintered on all three waterbody types equally. Forest-dominated areas were preferred (50.6 per cent), while agricultural areas (25.3 per cent) and wetland areas (13.9 per cent) were also commonly used. Elsewhere, Saggese *et al.* (2014) published an overview of wintering Ospreys in Argentina. They found that Ospreys use river systems and their major tributaries in northern and northeastern Argentina, and frequent reservoirs in central and northwestern regions.

Monti *et al.* (2018a) analysed the winter habitat choice of satellite-tagged Ospreys from the Mediterranean region, the majority of which wintered in southern Spain, Morocco, Algeria, Sardinia, Corsica, the Balearic Islands and mainland Italy. Coastal brackish-water sites were

Figure 7.2. Ospreys winter at high densities along the coast of Northern Senegal and often perch in close proximity to each other on favoured beaches (© John Wright).

preferred (58.92 per cent), with Ospreys also utilising freshwater (21.35 per cent) and marine (19.73 per cent) environments. They found that 62.75 per cent of birds were opportunistic and frequented different habitats within their winter range. None of the tagged birds had a home range fully associated with the marine environment.

There are indications that the wintering habits of Northern European Ospreys are currently undergoing a shift, with an increasing number of birds wintering at more northerly latitudes, such as the Iberian Peninsula. Martín et al. (2019) estimate that three per cent of the Osprey population of Western Europe now spend the winter on the Iberian Peninsula, with an increasing trend for juveniles to linger and eventually establish a permanent wintering site there. The highest densities tend to occur in coastal areas, such as the Bay of Cadiz, where there was a 6 per cent increase in the number of wintering Ospreys between 2000 and 2016. Reservoirs in Andalucía also provide valuable, and relatively new, overwintering habitat given that the number of reservoirs in the region has doubled since the 1960s (Casado and Ferrer 2005). It is likely that this shift is being driven, at least in part, by climate change and, given that shorter migrations are likely to enhance survival (Klaassen et al. 2014), it seems probable that the number of wintering Ospreys in Iberia will continue to increase.

Adult behaviour in wintering areas

When considering the wintering behaviour of Ospreys, it is important to differentiate between adults and juveniles. By the mid–late 1990s ringing recoveries had provided valuable information on wintering areas of Ospreys, but the advent of satellite tracking has provided new insights into the behaviour of known individuals. These data have revealed that adults are faithful to the same site every winter, but juveniles wander extensively when they first arrive in potential wintering areas.

SITE FIDELITY

As demonstrated by our encounters with the Rutland Water male, 32(11), in the same mangroves three years apart at the Sine-Saloum delta, adult Ospreys show strong winter site fidelity, and this has been borne out by long-term tracking of individual birds. Ten adult Ospreys satellite-tagged in northern Scotland and at Rutland Water between 1999 and 2019, and followed for at least two years, returned to exactly the same site in successive winters (Dennis 2008, Mackrill 2017). One of them, adult female Green J, was the first Osprey to be satellite-tagged in the UK, when Roy Dennis caught her at her nest-site near Carrbridge in the Scottish Highlands in 1999. The initial satellite-tracking data revealed that she wintered at Embalse de Gabriel y Galán, a large reservoir in Extremadura in central Spain. She was re-caught by Roy 14 years later in summer 2013 and fitted with a more sophisticated solar-powered GPS tag, which provided more accurate locations and greater temporal resolution. This showed that she continued to remain faithful to Embalse de Gabriel y Galán, living in an area of up to 26km^2 during the winter. She would typically roost in gorges south of the dam, and then fly 3–5km north to hunt and rest (Dennis 2015).

Similarly, Washburn et al. (2014) demonstrated that North American Ospreys are also extremely faithful to the same wintering site each year. All 18 adult Ospreys tracked for a second wintering period returned to exactly the same site (i.e. within 1km distance) as the previous year.

Wintering behaviour

WINTER RANGE

Most adult Ospreys are highly sedentary during winter, with the majority of daily movements associated with fishing activity. The essential requirements for an Osprey during winter are reliable fishing grounds, relatively undisturbed daytime perches and a safe roost site. It is for this reason that large densities of wintering Ospreys can be encountered along the West African coast. The long sandy coastline between Dakar and St Louis in Senegal is one of the prime wintering locations for Western European Ospreys. As described above, Dupart (2022) recorded 373 wintering Ospreys along this stretch of coastline, which represented a very significant increase since a previous survey undertaken by Yves Prevost in 1979. The density of wintering Ospreys is particularly high on the section between Potou and Lompoul and also in the area just south of Khondio (Dupart 2021). Here, Ospreys spend most of their time on the beach, often perched just metres apart. Satellite tracking of birds that spend the winter on this coast has shown that they require a very small winter home range. Most birds make just one or two short flights out to sea to catch fish and then spend the rest of the day perched on the sand. They then retreat into nearby coastal woodland or isolated trees to roost.

Two satellite-tagged birds from Rutland Water have wintered on this section of coastline. In 2011, with the help of Roy Dennis, we caught and satellite-tagged 09(98), a 13-year-old male who had been translocated to Rutland Water from northern Scotland in 1998. That autumn he completed a 16-day migration to the Senegal coast and settled at a wintering site just to the south of Lompoul. Once there he fell into a predictable daily routine, perching in the same location on the beach each day and roosting 150m inland. It was curious, though, that he would sometimes fly a considerable distance out to sea in order to catch fish. The satellite data, which logged a single GPS fix every hour, indicated that the majority of foraging flights were within

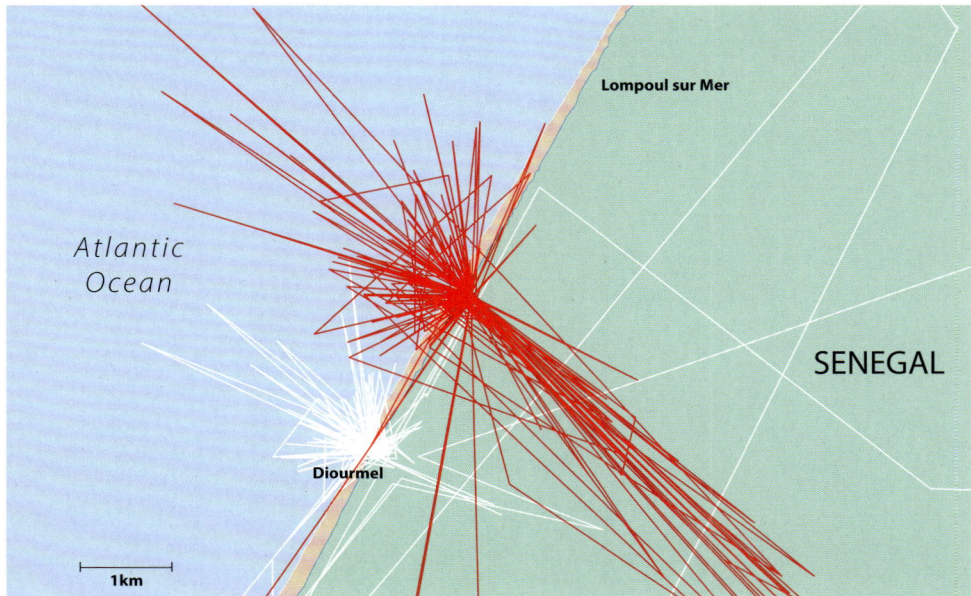

Figure 7.3. The winter range of two satellite-tagged Ospreys from Rutland Water, 09(98) (red) and 30(05) (white), showing how sedentary adult Ospreys tend to be on the wintering grounds (data courtesy of Leicestershire and Rutland Wildlife Trust).

1.5km of the shore, but on occasions he flew farther. 09(98) was logged >1.5km from the coast on 10 occasions during the course of the winter, up to a maximum distance of 7.3km. Such behaviour was first documented by Prevost (1982), who watched Ospreys flying 1–5km out to sea off the coast of West Africa in order to catch flying fish. He noted that they would rise to an estimated height of approximately 300m in order to locate fish, before slowly losing altitude and then diving from approximately 100m. The mean altitude of 10 GPS-fixed points was 43.9m (range 20–86m), and it seems possible that 09(98) was targeting flying fish, as Prevost had observed. Despite these occasional longer hunting fights out to sea, 09(98)'s winter range was very small. An analysis using fixed kernel density contours and least-square cross-validation factors revealed that 09(98)'s 95 per cent utilization distribution (UD) was 10.72km^2, with a core area (50 per cent UD) of just 0.68km^2.

A second bird, adult female 30(05), wintered just 2km along the coast (Figure 7.3). Her winter range and daily behaviour was extremely similar to her compatriot from Rutland Water, spending the majority of each day perched on the beach, and then flying out to sea to catch fish once or twice every day. The majority of these fishing trips were within 500m of the coast, although like 09(98) she occasionally made longer flights out to sea, up to a maximum of 3.5km. Her winter range was even smaller than 09(98)'s, with a 95 per cent UD of 1.59km^2 and a 50 per cent UD of 0.25km^2. Despite wintering in such close proximity to each other, the ranges of the two birds rarely overlapped, as shown in Figure 7.3. Although they were not actually satellite tracked in the same year (both were satellite-tagged as adults) it seems likely that they rarely encountered each other despite wintering at such close proximity. This neatly demonstrates how adult Ospreys like 30(05) are extremely faithful to their favoured perches and fishing areas (Figure 7.4).

North American Ospreys behave very similarly during winter. Washburn *et al.* (2014) analysed the movements of four satellite-tagged adult birds wintering in Brazil, Venezuela and Guyana. The winter home range of the four birds varied from 2.53km^2 to 25.98km^2 (95 per cent UD), and the core range from 0.67km^2 to 2.51km^2 (50 per cent UD). Of the locations provided by the GPS satellite transmitters, 97 per cent were within 5km of the centre of the Ospreys' core-use areas. A male wintering at Lake Valencia in Venezuela was tracked for two successive winters and, as expected, its behaviour was extremely consistent between years. His home range varied from 2.24km^2 to 2.53km^2 and core range from 0.67km^2 to 0.76km^2 over the course of the two winters. This bird had a smaller home range than the other three, which all wintered along rivers. The home ranges of birds living along rivers were more linear in shape than the male at Lake Valencia, which is indicative of the fact that home-range size varies according to local factors. The Ospreys were exclusively diurnal during winter, and the majority of activity occurred between 10am and 4pm local time. This highlights that the key priorities for adult Ospreys in winter are energy conservation, and reducing the risk of mortality.

Figure 7.4. Adult Ospreys, such as female 30(05) from Rutland Water, are faithful to the same wintering site each year (© John Wright).

The populations of Ospreys that breed around the Mediterranean are interesting because, as described earlier, a significant proportion of adult birds do not migrate. Monti *et al.* (2018a) performed a GIS analysis of the movements of satellite-tagged adults in the region, and found that home-range sizes were small, regardless of whether individuals migrated or remained resident. The mean winter home range (95 per cent UD) was 70.0km^2 (±83.7km^2) but the core area (50 per cent UD), used on a day-to-day basis, was much smaller, measuring 8.0km^2 (±6.9km^2). As an example, one Corsican bird lived in a core area of 3.12km^2 on the north coast of Morocco during 2015–16, with average daily movements of 5.2km^2 (±4.3km). The satellite-tracking data showed that resident birds regularly commuted to feeding sites located within 20km of nests. The wetlands of Albufera and Ses Salines represented the main feeding areas in the Balearics, Corsican birds favoured marine coves close to breeding sites, and Italian birds frequented the wetland systems of southern Tuscany. This behaviour is similar to adult Ospreys in other non-migratory populations. In Australia, for instance, adults remain on territory throughout the year and occupy the same site over multiple seasons (Dennis 2007a). Studies on Kangaroo Island in South Australia demonstrated that some young birds dispersed to the mainland once they reached independence (range 80–400km from the natal site), but that philopatric recruitment was high, with 22% of surviving young remaining on the island or returning to join the breeding population (Dennis 2007b).

Most Ospreys occupy a small winter range, but there are sometimes exceptions. A Norwegian male, 81114, was tagged as a juvenile at a nest-site in central Norway and completed a 68-day migration to Nigeria (Østnes *et al.* 2019). This bird went on to establish a wintering site centred on the Kainji reservoir on the River Niger, favouring three core areas, with a 50 per cent UD of 52.3km^2 and a 90 per cent UD of 146.7km^2. As expected, he flew north to Norway in the spring of his third calendar year, before returning to Kainji reservoir the following autumn and for a further four winters after that. Surprisingly, he continued to move significant distances along the River Niger each winter – contrasting markedly with the behaviour of most adult Ospreys on the wintering grounds. As Østnes *et al.* (2019) point out, it would be impossible for a bird behaving in this way to defend a winter territory in the manner observed elsewhere in West Africa. A female in the same study behaved in a more typical manner. She established a wintering site 20km north-west of Mount Cameroon in Cameroon during her second calendar year and lived in a core area (50 per cent UD) of just 2.13km^2. She behaved similarly when she returned to the same location after her first northward migration.

Another example of apparently atypical behaviour during winter concerned an adult Osprey from the UK. We caught and satellite-tagged five-year-old male AW(06) at a nest near Rutland Water in 2011, and he completed an initial 14-day migration to the coast of Guinea, arriving on 9th September. The satellite data showed that he quickly settled into a predictable routine, fishing in the sea once or twice every day and favouring an area close to the shore for daytime perching. AW(06) remained in the same location until 17th December when, surprisingly, he moved south. He flew just over 1,000km in five days, passing through Sierra Leone and Liberia before settling at the River Lobo in a forested region of western Ivory Coast. There he inhabited a linear-shaped home range for over seven weeks, rarely venturing more than 500m from the forest-lined river, and favouring three core areas, 6km apart, along its course. However, on 12th February, he flew 32km north-west to Lac de Buyo, a very large artificial lake that he visited en route to the River Lobo. He roosted near the lake that night, before returning to his favoured stretch of the Lobo next morning. This sudden change of behaviour was unexpected and suggested that fishing in the river may have become difficult, particularly when he returned to Lac de Buyo three days later, on 16th February. He was perched a few hundred metres from a lakeside village that evening, and the following morning the tag was stationary on the outskirts of the village. There were no further movements and transmissions completely ceased next day. Although it was not possible to find anyone to investigate further, it seems highly likely that the bird had died through anthropogenic means – either intentionally or perhaps by becoming tangled in fishing net. Zwarts *et al.* (2012) suggest that persecution in this part of Africa is often more prevalent in inland areas than on the coast, and AW's sudden demise appeared to corroborate this suggestion. Indeed, AW's death soon after moving to a new location inland was very similar to the loss of satellite-tagged juveniles described later.

AW's midwinter movement from coastal Guinea to inland Ivory Coast was extremely unusual for an adult Osprey, but very likely something the bird repeated each winter; it almost certainly represented learned behaviour from his first winter in West Africa. Sadly, the bird's demise the first winter after the tag had been deployed meant it was not possible to prove whether this was the case, but as a five-year-old Osprey, his migratory habits would have been well established. There are few other examples of similar movements by adult Ospreys, but a female that was satellite-tagged in North Carolina in the United States spent three months wintering in the Amazon Rainforest in western Peru before moving 700km north to the Cauca Valley in Colombia (Washburn *et al.* 2014). She remained there for a month before starting spring migration. This suggests that the bird regarded the site in Colombia as a spring stop-over, whereas AW's movement further south during winter indicated that his time in Guinea was more likely an extended autumn stop-over.

SOCIAL BEHAVIOUR IN WINTER

The fact that Ospreys exhibit such strong site fidelity in winter means that they are often very territorial over favoured perches and fishing grounds, particularly towards juveniles searching

Figure 7.5. An adult male Osprey (right) chasing a juvenile male. Adults can be very aggressive to juveniles at some sites during winter (© John Wright).

for their own wintering site. It seems that the type of habitat probably plays a role in the tolerance of individual Ospreys towards others. On fieldtrips to West Africa, we have recorded more aggressive interactions and territorial behaviour along linear habitat features, such as rivers, than in more open areas such as the Sine-Saloum delta. A favourite trip of mine is to take a boat along the Allahein River which forms the southern border between The Gambia and the Casamance region of southern Senegal. Here, Ospreys can be found perched in mangroves and other trees every few hundred metres along the river, and we frequently encounter colour-ringed individuals that are present every winter. The *tioop* warning call is a regular sound on the Allahein in winter and those adult birds that return to the same perch every year can be very aggressive towards juveniles in particular (Figure 7.5).

The recent increases in the number of wintering Ospreys in Senegal has meant that, as a general rule, adult birds now live in higher densities during winter than in the past (Dupart 2021). However, there has always been some degree of social behaviour during winter. Adult Ospreys can be surprisingly tolerant of each other, sometimes congregating in groups of 10–20 birds at the best quality wintering sites. Energy conservation is an adult Osprey's principal concern during the winter, and so if fishing is easy then there is little threat from birds they encounter at the same site every year – it is likely they recognise the same individuals from winter to winter. Juveniles are sometimes tolerated in larger congregations of Ospreys at the best quality sites, such as the Sine-Saloum delta and Langue de Barbarie in Senegal, but they have to be persistent in order to become established there, as described later in this chapter (Figure 7.6).

Figure 7.6. Juveniles are sometimes tolerated by wintering adults, but they have to be persistent if they wish to remain in areas with the highest concentrations of adult birds, like here at Tanji in The Gambia (© John Wright).

There can be benefits to associating with other Ospreys during the winter. Prevost (1982) sometimes recorded high densities of Ospreys at the mouth of the Saloum River in Senegal, logging up to 22 birds perched within 100 metres. He noted that in such circumstances there was often a social component to foraging: Ospreys that started foraging often did so near other individuals that were already hunting; and, after a lull in foraging, one Osprey often stimulated

Figure 7.7. Ospreys congregate in small groups at the best wintering sites, such as Île des Oiseaux at the Sine-Saloum delta (© John Wright).

Figure 7.8. Ruddy Turnstones sometimes peck at fish being held and eaten by Ospreys on the sand (© John Wright).

others to hunt. He recounts in his PhD thesis that the best example of this occurred when he was observing 12 Ospreys perched in close proximity to each other. An Osprey began hunting nearby and caught a fish from a shoal passing in the shallow water. Nine of the other Ospreys subsequently commenced hunting in the same location within the next eight minutes, and six of them caught fish before the shoal dispersed. In these circumstances there was a clear advantage to foraging in a location where the first bird had been successful. Prevost reports that he frequently observed such behaviour at the mouth of the Saloum River where birds hunted in a

Figure 7.9. Ospreys can become tolerant of regular and predictable human activity on their wintering grounds (© John Wright).

well-defined and restricted area, but less so at the mouth of the Senegal River where birds hunted over a wider area and were more dispersed (Prevost 1982). The closest I have come to observing this kind of behaviour is at Île des Oiseaux in the Sine-Saloum delta, south of where Prevost made his observations. Here adult Ospreys congregate in good numbers and catch fish in the shallow water surrounding the island. The fishing is so good that the Ospreys are extremely tolerant of each other (Figure 7.7), and also of avian scavengers looking for an easy meal. We have observed Slender-billed Gulls *Chroicocephalus genei* and Ruddy Turnstones *Arenaria interpres* walking up to Ospreys perched on the ground or low stumps, and pecking at the fish the Osprey is holding in its talons (Figure 7.8). Some particularly brazen individuals will even take fish from an Osprey's bill on occasion.

TOLERANCE OF HUMAN ACTIVITY

The highest densities of wintering Ospreys are usually found in areas with the least disturbance (Dupart 2021), but some birds settle in locations with some human activity, and show some degree of habituation to this. The long sandy coastline of the Grand Côte in Senegal is used by locals to transport fruit and vegetables by horse and cart, and they pass numerous perched Ospreys as they travel along the tideline, with little or no reaction from the birds (Figure 7.9). Ospreys can be equally tolerant of fishermen in small boats or canoes, which are a feature of places such as the Sine-Saloum delta; moreover, Ospreys fishing at the Somone Lagoon in Senegal often dive into the water close to sunbathing tourists. Dupart (2021) comments that, in his experience, first-winter birds can be much more wary and rarely tolerate a close approach in the same way as some experienced adults, who seem to become habituated to certain human activities.

Juveniles

While adult birds migrate to a known site and then remain there throughout the winter, young birds wander extensively after they first arrive in potential wintering areas. The long-term satellite tracking of specific individuals has shown that these explorations play a key role in the eventual establishment of a winter territory. There are comparatively few examples of juveniles being tracked through to adulthood – principally due to high juvenile mortality – but one of the best from the UK is Rothiemurchus, a juvenile male tagged by Roy Dennis at a nest in northern Scotland in 2009 (Figure 7.10). As described in the previous chapter, Rothiemurchus completed a 47-day migration to northern Senegal, comprising 13 travelling days and 34 stop-over days.

ESTABLISHING A WINTERING SITE – THE CASE STUDY OF ROTHIEMURCHUS

After arriving in Senegal on 28th October, Rothiemurchus lingered at Djoudj National Bird Sanctuary. This vast wetland is situated on the southern edge of the Sahara and is often the first major wetland that Palearctic migrants encounter after traversing the desert. A staggering 1.5 million birds occur annually and 120 species of Palearctic migrants winter here, including up to 50,000 Garganey *Spatula querquedula* and very large numbers of Shoveler *Spatula clypeata* and Pintail *Anas acuta*. When I first visited in January 2011 with John Wright, Paul Stammers and Frederic Bacuez, the sheer abundance of birds was breathtaking and it was wonderful to see wintering Ospreys mixing with the vast concentrations of wintering wildfowl (Figure 7.11). The site supports large numbers of wintering Ospreys and Rothiemurchus remained there for almost three months before heading further south on 15th January. The satellite data provides

Wintering behaviour

Figure 7.10. A male Osprey, known as Rothiemurchus, was satellite-tagged as a juvenile by Roy Dennis in northern Scotland in 2009. The transmitter, which logged data for five years, provided a fascinating insight into many elements of the early years of an Osprey's life. We located Rothiemurchus (left) at his wintering site in Senegal in January 2012, where John Wright took this image of him being chased by another adult male Osprey. The satellite transmitter is just visible on the bird's back (© John Wright).

no explanation for such a move, but young Ospreys exhibit considerable wanderlust during their first winter.

Rothiemurchus travelled south along the Senegalese coast after leaving Djoudj, and then passed through The Gambia and the Casamance region of southern Senegal, to Guinea-Bissau. The islands off the coast of Guinea-Bissau are another favoured wintering region for Ospreys from Western Europe, but after reaching Ilha de Jeta, Rothiemurchus turned north again and returned to Senegal. On 22nd January he arrived at the eastern side of the Sine-Saloum delta in Senegal. This huge area of mangroves and intertidal shallow water is another extremely important area for wintering Ospreys, as described earlier, and Rothiemurchus spent two months exploring an area of over 1,400km^2. The delta supports abundant populations of mullet and other fish species, and so it is a good area for juvenile Ospreys to refine their foraging skills (Figure 7.12). Prevost (1982) studied the foraging behaviour of immature Ospreys in Senegal and found that 19 per cent of dives by birds aged approximately six months were successful, compared to 40 per cent of dives by adult birds. Thus, at this early stage in his life, Rothiemurchus would still have been some way off attaining the hunting proficiency of experienced adult birds. This is in keeping with my own observations in West Africa. I remember on one occasion watching a juvenile Osprey make at least 17 unsuccessful dives in shallow water close to Bijoli Island, a small offshore island in The Gambia. It subsequently returned to the island to rest on the sand (Figure 7.13), before trying again sometime later. Adult birds, meanwhile, were catching fish with relative ease in the same location.

One of the major issues for juvenile Ospreys during their first winter is that they are subordinate to adult birds. A frequent sound along the coast of West Africa – or in any location where Ospreys winter for that matter – is a plaintive *tioop* call which is identical to the guard call used

Figure 7.11. Djoudj National Bird Sanctuary in northern Senegal supports vast numbers of wildfowl, including Garganey (above) and White-faced Whistling Ducks (below), as well as wintering Ospreys (© John Wright).

Figure 7.12. The fish-rich waters of West Africa provide ample opportunities for juvenile Ospreys, such as this German colour-ringed male, to refine their foraging skills (© John Wright).

Figure 7.13. It often takes numerous dives for a juvenile Osprey to make a catch. This individual made many unsuccessful dives into the sea around Bijoli Island in The Gambia, before alighting on the sand to rest (© John Wright).

by breeding Ospreys to warn intruders not to approach an active nest. It is used by adults during winter for a similar purpose, usually in response to the sight of another fishing Osprey. The calling bird will often be perched inconspicuously in the mangroves (Figure 7.14), and it is only once it utters the familiar call that you are alerted to its presence. Juveniles are frequently chased away from the best foraging areas by adult birds, which can be fiercely territorial of their favoured perches and fishing sites, particularly in linear habitats, as described earlier.

Adult birds inevitably congregate at the highest densities at the best sites, such as the Sine-Saloum delta, and this may have been why Rothiemurchus moved on again in March – this time to a wetland on the north bank of the Gambia River, approximately 50km south-west of his favoured area at the Sine-Saloum (Figure 7.15). There would have been fewer Ospreys at this site, particularly as adult birds began to head north to their breeding grounds during March and early April, and it was notable that Rothiemurchus became much more sedentary after arrival, living in a quiet 4km² section of river bordered by mangroves. Satellite tracking and colour-ringing studies have shown that the vast majority of immature Ospreys remain on the wintering grounds for the whole of their second calendar year, and the spring departure of adult birds is likely a key milestone for them, easing pressure at the best sites and helping young birds to become more settled.

Rothiemurchus spent the rest of the summer in the same location and by autumn was well established in a favoured area of less than 1km². Even the return of adult birds appeared to have no discernible impact on his daily movements; indicating that after a year in West Africa he had established a wintering site. This was borne out by his behaviour in future years. After heading north to Scotland for the first time in the spring of his third calendar year, Rothiemurchus returned to the same wintering site the following autumn, returning there on 9th October and

Figure 7.14. Adult birds often perch inconspicuously in mangroves during winter (© John Wright).

Wintering behaviour

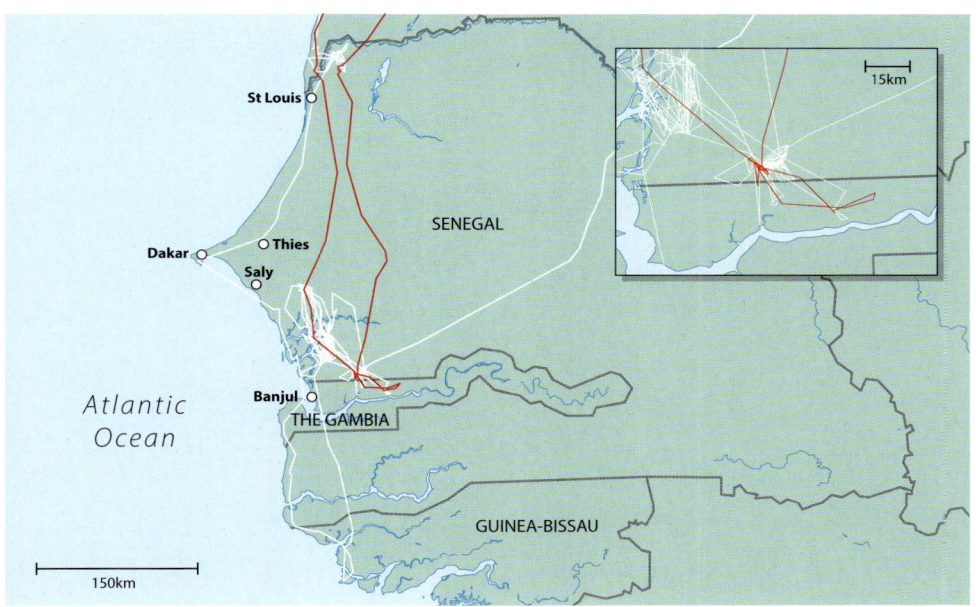

Figure 7.15. Rothiemurchus wandered extensively during his first 18 months in West Africa (white track), before eventually establishing a wintering site just north of the River Gambia (see insert). He subsequently returned to this location after his first northward migration and was extremely sedentary (red track) (Roy Dennis and Tim Mackrill).

remaining for the next six months, living in an area of just $0.5km^2$. When myself and a group from the Rutland Osprey Project visited the site on 30th January 2012, we saw Rothiemurchus being chased by another adult Osprey – indicating that such disputes are probably part of the day-to-day life of Ospreys on their wintering grounds. The satellite-tracking data showed that Rothiemurchus remained faithful to the same wintering site for the next two winters, and even after the tag stopped transmitting, a field visit by Joanna Dailey, Chris Wood, Junkung Jadama and Fansu Bojang revealed that he was still at the same site in December 2017.

MORTALITY OF JUVENILES

Rothiemurchus's movements prior to establishing the winter territory are typical of young Ospreys when they first arrive in West Africa. It is clear that young birds have a desire to explore when they arrive in potential wintering areas, but they are also greatly influenced the actions of experienced adult Ospreys who can make it difficult for juveniles to become established in optimum wintering localities, such as the Sine-Saloum Delta (figure 7.16). It is perhaps no surprise, therefore, that a significant proportion of young birds do not survive the initial 18 months on the wintering grounds. Only approximately 30 per cent of young Ospreys survive to two years (Dennis 2008) and you might think that the highest rates of mortality would occur on the first migration. However, an analysis of satellite tracking of 35 juvenile Ospreys tagged in northern Scotland and at Rutland Water over a 20-year period between 1999 and 2019, indicates that significant mortality also occurs on the wintering grounds (Dennis 2008, Mackrill 2017) (Table 7.2). Discounting six birds, whose satellite-tags failed during the first migration, 18

Figure 7.16. The vast Sine-Saloum delta is prime wintering habitat for Ospreys, but it can be difficult for juveniles to become established at such sites, due to the presence of large numbers of wintering adults. Here, a juvenile male is perched on the sand at the Sine-Saloum with Slender-billed Gulls flying overhead (© John Wright).

birds (i.e. 62 per cent of birds with a known outcome) migrated successfully to Iberia or West Africa, and 11 (38 per cent) died en route, as described in the previous chapter. However, of the 18 birds that completed their first migration, only three are known to have survived to two years, although tag failure was an issue in some cases. Many of the transmitters used in the early years, from the late 1990s onwards, were much less reliable and did not have inbuilt solar panels, which limited longevity. It is possible, therefore, that some of the birds whose tags failed during migration or on the wintering grounds survived and later returned to the UK. The figure for survival to two years should thus be taken as an absolute minimum.

What is clear from these data, however, is that completing the first migration does not guarantee that young Ospreys survive. The satellite data indicated that at least six of the remaining birds died in West Africa. Four of these individuals died in inland areas away from the best overwintering habitat on the coast, where they may be more vulnerable to human persecution (Zwarts *et al.* 2012). It seems likely that young birds are pushed into these areas by the actions of dominant adults. A growing network of fieldworkers means it is sometimes possible to investigate the cause of death, which can be both natural and anthropogenic in nature. Two such cases are described below.

JOE (WWW.ROYDENNIS.ORG/JOE)

Joe, a juvenile male, was satellite-tagged by Roy Dennis at a nest in Caithness in 2011 and completed a 37-day migration to northern Senegal, arriving near the capital Dakar on 29th September. He subsequently flew south to the Sine-Saloum delta, before returning north again and visiting Lac de Guiers and then a small wetland near Rao, south-east of St Louis. The

Table 7.2. Survival of satellite-tagged juvenile Ospreys from the UK (n = 35)

Outcome of first migration	Total birds	Percentage
Completed first migration	18	51.5%
Died on first migration	11	31.5%
Tag failure on first migration	6	17%
Survival on wintering grounds		
Died on wintering grounds	6	17%
Tag failure on wintering grounds	5	14%
Unknown	3	8.5%
Died on return migration	1	3%
Survival to two years (return to UK)	3	8.5%

All percentages relate to total number of birds tagged.

satellite data indicated that the transmitter did not move after 11th October and I was able to arrange for a friend, Frederic Bacuez, to go and check out the last known location. Frederic located the remains of the bird and the satellite-tag on his second visit in early December. The small wetland was almost dry, with just 15–20cm of water in some places, indicating that the bird may have got into difficulty when diving into the shallow water, in pursuit of catfish. One possibility is that he was injured and then predated by a mammalian predator, such as a Side-striped Jackal *Canis adustus*. Naïve young Ospreys which, like adults, frequently perch on the ground are susceptible to mammalian predators such as feral dogs or jackals, and even more so if they are injured. The fact that Joe died at a suboptimal, ephemeral wetland demonstrates the difficulties young Ospreys face if they are kept away from the best wintering sites by established adults. Eriksson and Wallin (1994) detected a significant relationship between rainfall in West Africa and survival of third-year Swedish Ospreys, indicating survival of young birds was enhanced by higher rainfall. In drier years when ephemeral waterbodies dry up quickly, it is likely that mortality rates of young Ospreys are higher than in years when there are more extensive inland waterbodies capable of sustaining them when they first arrive on the wintering grounds. This subject undoubtedly warrants further attention, particularly with regard to a changing climate.

FIDDICH (WWW.ROYDENNIS.ORG/FIDDICH)
Intriguingly, satellite-tracking studies indicate that mortality of young Ospreys in West Africa often occurs when a bird moves to a new site. Juvenile Ospreys frequently move from site to site during their first winter, partly due to the behaviour of aggressive adult birds. Fiddich, a juvenile female satellite-tagged by Roy Dennis in Moray in 2012, migrated to Senegal in just 22 days, without stop-overs that autumn. She initially settled on a remote section of the River Bounoum, south-east of Lac de Guiers and remained there until late November. During this period she made one excursion to the coast near Dakar, but otherwise stayed fairly sedentary throughout, presumably in the absence of other Ospreys, given the remote nature of the site. The river flows into Lac de Guiers during the rainy season and it is possible that by late November it was beginning to dry up. This perhaps explains why the young female began exploring elsewhere. She

completed a two-week return trip to Guinea – a return flight of over 1,500km – and then, in early January, moved to Djoudj National Bird Sanctuary. She remained there for the next six weeks, living in an area of 24km². On 14th February she returned briefly to the River Bounoum, but by this stage in the year it had probably become unsuitable for an Osprey. She subsequently headed south-east and eventually settled on the River Gambia, in eastern Senegal. She lingered there for two weeks, favouring an area of just 200 × 70m on the banks of the river. However, on 15th March the transmitter started sending signals from a village, situated 2km away. It was clear something had happened and so I organised for Junkung Jadama and Colin Cross to make a special overland drive from coastal Gambia to the village. The villagers were highly suspicious and denied knowledge of the transmitter, even though we knew it was in one of the compounds in the village. This meant we were unable to prove what had happened to the bird, but it seems likely that it was caught in fishing nets or perhaps intentionally captured. Some Ospreys can be extremely unguarded and would be relatively easy to catch. Clearly, if Fiddich had remained at Djoudj then she would have not died in this manner, but it is likely that pressure from adult birds, coupled with the drying up of the ephemeral wetland she had favoured earlier in the winter forced her to search elsewhere.

MOVEMENTS DURING SECOND CALENDAR YEAR

Satellite tracking has shown that the vast majority of young Ospreys remain on the wintering grounds for the whole of their second calendar year, but there have been some notable exceptions. Two birds from Rutland Water are known to have returned to the UK as one-year-olds, and both were individuals that wintered in Iberia rather than crossing the Sahara to West Africa. The first was 06(01), a female translocated to Rutland Water in 2001 who spent her first winter at Vila Franca de Xira on the River Tagus near Lisbon – a site that has attracted increasing numbers of wintering Ospreys from Western Europe in recent years. She flew north to Yorkshire and southern Scotland the next spring, whereupon her tag ceased transmitting. However, to our great surprise she returned to Rutland Water, without the satellite-tag, the following April, and successfully reared two chicks. This was the first known breeding of a two-year-old Osprey in the UK (Mackrill *et al.* 2013). The second bird, T3(16), a male colour-ringed at a nest in Rutland in 2016, was seen at Fishlake Meadows in Hampshire in July 2017. A subsequent colour-ring sighting, received in January 2019, revealed that this bird winters near Rianxo on the northwest coast of Spain – the northernmost wintering location recorded for a bird from Rutland Water, which undoubtedly influenced the early return.

A female satellite-tagged by Roy Dennis at a nest in Moray in the Scottish Highlands in 2012 also flew north during her second calendar year. This individual, known as Fearna, had spent her first winter on the Senegal River in northern Senegal and by late April she was living in a fairly small area of 13km². However, on 1st May she set off north across the Sahara and then crossed the Strait of Gibraltar into Spain on 10th May. She continued north through Spain, but then was found dead in the Rioja region in the north of the country on 26th. Had she survived it would have been fascinating to see if she continued north to Scotland, or summered in Iberia.

Monti *et al.* (2018a) reported that five of 25 tracked juveniles from the Mediterranean region survived more than two years and, as expected, four remained on the wintering grounds for the whole of their second calendar year, before returning to their natal area for the first time the following spring/summer. However, one individual from Corsica spent its first winter in Algeria and then returned to its nest and a secondary site in Corsica during June of its second calendar year. It remained there until November, before returning to Algeria for the winter. Next spring it migrated back to Corsica much earlier, during March, and remained until August. Like the two Rutland birds, the relatively close proximity of its wintering site to its natal nest was

Wintering behaviour

Table 7.3. Annual spring departure dates of satellite-tagged Ospreys from the UK tracked for more than one year

Bird	Sex	Wintering location	Spring migrations tracked	Median	Range
Green 5B (Beatrice)	Female	Spain	8	1st March	23rd February– 6th March
Green J	Female	Spain	4*	28th March	20th March– 5th April
30(05)	Female	Senegal	4	10th March	10th–11th March
White PE (Morven)	Female	Mauritania	4	14th March	9th–17th March
Blue XD	Male	Senegal	2	17th March	16th–18th March
Red 7J (Nimrod)	Male	Guinea-Bissau	3	22nd March	21st–27th March
Red 8T	Male	Senegal	4	18th March	14th–22nd March
Blue AE (Rothiemurchus)	Male (non-breeding)	Senegal	3	7th April	5th–12th April

*satellite-tagged twice, 14 years apart

probably a factor. Indeed, the straight-line distance (652km), was considerably less than the corresponding figure for the two Rutland Ospreys (1,650km and 1,270km respectively).

Departure from the wintering grounds

As described in chapter 6, adult Ospreys tend to show a predictable pattern of departure from the wintering grounds, likely prompted by endogenous cues, such as hormonal changes (Sharma *et al.* 2018). Breeding birds tend to be the first to depart in order to make a timely arrival back at their nest-site, with no clear differences between the sexes (Martell *et al.* 2014, Mackrill 2017). Younger non-breeders leave later, particularly third-calendar-year birds flying north for the first time, which may not depart wintering areas until late April or May. Most adult birds are fairly consistent in their timing of departure. For example, female 30(05) from Rutland Water departed on 10th or 11th March for four successive springs, while Green 5B (Beatrice), who was tracked for eight years, always departed early, in late February or early March, despite wintering in southern Spain, as described in the previous chapter. The annual spring departure dates of satellite-tagged Ospreys returning to the UK is shown in Table 7.3.

Conclusions

- Most Ospreys from Northern Europe migrate to sub-Saharan Africa, with individuals from populations in the west wintering in tropical West Africa, and those from further east migrating to eastern and central Africa, or in some cases, the southern coast of South Africa. Birds from eastern Russia winter in coastal Arabia, the Indian subcontinent and South-East Asia. The winter distribution of North American Ospreys ranges in latitude from the

southern United States to southern South America, and in longitude from the west coast of Mexico eastward throughout the Caribbean and the eastern coast of South America.
- There is a growing trend for Ospreys from Northern Europe to winter in Iberia.
- Ospreys tend to be opportunistic and are well adapted to exploiting a range of different wintering localities, including freshwater, brackish and marine environments. Adult birds show strong site fidelity to a known wintering locality and most have a small home range after arriving.
- Adult Ospreys are often very territorial of favoured perches and fishing grounds, particularly towards juveniles searching for their own wintering site. However, they can be more tolerant at the best quality sites, where small numbers may perch together during diurnal hours. Adult birds probably recognise each other from winter to winter because of strong site fidelity.
- Juvenile Ospreys are highly mobile during their first six months in potential wintering areas. They are often kept away from the best quality wintering sites by established adult birds, and may be forced into suboptimal habitat such as ephemeral waterbodies. This can be a particular problem in dry years when these waterbodies dry up more rapidly.
- Significant mortality of juveniles occurs on the wintering grounds, particularly when young birds are forced into poorer quality areas. Satellite tracking of Ospreys from the UK has indicated that mortality in West Africa often seems to occur when juveniles move to a new site, particularly in inland areas.
- Most juveniles remain on the wintering grounds for the whole of their second calendar year. The absence of adult Ospreys for six months allows them to become established at a site that will become their regular wintering locality in future years. On rare occasions some individuals move north during their second calendar year.
- Established breeders are the first to depart the wintering grounds. This usually occurs during March, although some leave earlier. Third-calendar-year birds may not fly north until late April or May.

CHAPTER 8
Our relationship with the Osprey

The Osprey is a bird that has been admired by humans for centuries, and yet our relationship with this iconic species is a complicated one. From the writings of Aristotle in the fourth century BCE, to Shakespeare, who uses the fishing prowess of the Osprey as a metaphor in *The Tragedy of Coriolanus*, the Osprey has been referenced and indeed revered in many historical texts. And yet, perversely, the activities of humans have had longlasting and serious implications for the species across much of its range. The Osprey's penchant for raiding medieval fishponds led to widespread persecution in Europe (Dennis 2008), while harmful insecticides resulted in severe declines across much of the species' range during the mid-twentieth century (Poole 2019). These anthropogenic factors mean that, despite its cosmopolitan distribution, the Osprey is absent from many former breeding areas (Dennis 2016). Fortunately, it is a species that responds well to conservation interventions, and proactive conservation measures such as artificial nest building and translocations have facilitated a welcome comeback in recent decades (Dennis 2008, Poole 2019).

Historical writings

There are numerous references to Ospreys in historical literature. The Greek philosopher Aristotle (Figure 8.1) wrote his *History of Animals* in the fourth century BCE, and included an observation of *haliaetos* (the Osprey) in Book IV: 'It has the neck both big and thick, bowed wings and a broad tail. It bides upon the sea-coast and shores. It often happens, when it cannot lift what it has taken that is submerged beneath the tide' (Creswell 1902). Five hundred years later, in the first century CE, the Roman writer Pliny the Elder added his own observations in Book 10 of his *Natural History*: 'The Haliaetus remains, with eyesight of the keenest, poising itself aloft when it spies fishes in the sea below, then dashing headlong on them, the waters being parted by its breast' (Whalley 1982). Later in his account, Pliny goes on to describe the early belief that Ospreys forced their young to look directly at the sun, and then dispatched any that blinked, preferring instead to 'bring up and cherish that whose eye will abide the light of the sun, as she looks directly upon him' (Whalley 1982). It is also pertinent that even 2,000 years ago there was some

Figure 8.1. Aristotle referred to Ospreys in his *History of Animals* in the fourth century BCE (© Hulton Archive/Stringer/Getty).

confusion over the Osprey's classification: 'Moreover these Ospreys are not thought to be a several kind of Eagles by themselves, but to be mongrels, and engendered of diverse sorts.'

Like Pliny's musings, many of the early observations and ideas about the Osprey were incorrect. There was a widely held belief in medieval England that fish were so mesmerised by the sight of a hunting Osprey that they would turn belly up in an act of surrender. This was described by William Turner in his *Principal Birds Noticed by Aristotle and Pliny* (1544): 'When the Osprey hovers in the air, whatever fishes be below turn up and show their whitish bellies (as it is believed, the nature of the Aquila compelling them to this), that it may choose that one which it prefers' (Evans and Turner 1903). The dramatist George Peele refers to the same behaviour in his 1594 play *The Battle of Alcazar*: 'I will provide thee of a princely Osprey, that, as he flieth over fish in pools, the fish shall turn their glistering bellies up, and thou shalt take thy liberal choice of all.' Perhaps the most famous reference is in the *The Tragedy of Coriolanus*, written by William Shakespeare between 1605 and 1608. In Act IV, Scene vii, Aufidius uses the Osprey as a metaphor to describe Coriolanus' military skills and hold over Rome: 'He'll be to Rome as is the Osprey to the fish, who takes it by sovereignty of nature.' This wonderful line captures the Osprey's fishing prowess perfectly, and also indicates that the birds were a familiar part of the English landscape in the early seventeenth century, even though persecution would have been limiting numbers by that stage, as described below.

Figure 8.2. An early painting of Ospreys, taken from *The Birds of Great Britain*, published in 1795, at which time the species was in steep decline due to human persecution (© Pete Murray).

Figure 8.3. The accuracy of Osprey artwork improved over the centuries, as this hand-coloured lithograph of an Osprey, taken from John Gould's *The Birds of Great Britain Vol. 1* (1873), demonstrates (© The Natural History Museum/Alamy).

Many writers were fascinated by the Osprey's feet, and another commonly held misconception in medieval times was that the birds had one webbed foot for swimming and one foot with sharp talons to grasp hold of fish. The clergyman William Harrison, in his 1586 *A Description of England*, explains: 'The Osprey hath one foot like a hawk's to catch hold withal and another resembling a goose, wherewith to swim.' It was not until the late eighteenth century that this error was identified by John Walcott in Volume 1 of his 1789 *Synopsis of British Birds*: 'Old authors, and Linnaeus after them, erroneously describe the left foot of this bird as semi-palmated; the right, say they, being designed to hold the prey; the left to swim with: the same is related of the Sea Eagle, a bird which, from similarity of manners, has often been confounded with it.'

Walcott's descriptions, as well as those of William Lewin in *The Birds of Great Britain* (1795), were far more accurate than those of medieval writers, but both authors incorrectly suggest that Ospreys would also take waterfowl in addition to fish. Lewin writes: 'This bird frequents the sea-shores, and large rivers, in various parts of Great Britain, and feeds chiefly on fish; plunging with great violence into the water after them; they will also feed on water fowl.' Lewin's book was also notable for the fact that it included a very early coloured illustration of an Osprey, finished by hand (Figure 8.2). Up until this point, all other illustrations had been in black and white, and of varying degrees of competence. A later hand-coloured lithograph published in John Gould's *The Birds of Great Britain* Vol. 1 (1873) was also much more accurate (Figure 8.3). Unfortunately, by the time of its publication the Osprey had been eradicated from most of Britain, its penchant for taking fish within sight of those charged with protecting fishponds and lakes making it a target for widespread and relentless persecution.

Historical persecution and population declines in the UK and Europe

The Osprey's fishing prowess has been well known for centuries, and evidence suggests that it was its piscivorous diet that led to extensive persecution across Europe. Yalden and Albarella (2009) estimated that there may have been as many as 3,800 breeding pairs in Mesolithic Britain and historical references indicate the Osprey was still widespread during the tenth and eleventh centuries. Dennis (2008), however, suggests that the early selective-killing of fish-eating raptors occurred in England and elsewhere in Western Europe during the Middle Ages due to the strict religious requirement to eat fish on Fridays, and that this led to dramatic declines in the numbers of breeding Ospreys. Fishponds, or stew ponds as they were known, were found in the grounds of monasteries, castles and large country houses and would have provided an easy source of food for Ospreys in many areas – particularly given what we now know about the Osprey's hunting habits and its ability to catch fish even in small waterbodies. The naturalist William Turner sums this up well in *The Principal Birds Noticed by Aristotle and Pliny* (1544): 'The Osprey is much better known today to Englishmen than many who keep fish in stews would wish; for within a short time it bears off every fish.' As Dennis (2008) points out, it seems highly likely that keepers of stew ponds would have ranked Ospreys as one of their worst enemies as a result. The conspicuous nature of the nests, coupled with the species' hunting behaviour, would have made them relatively easy targets for anyone intent on eradicating them.

Persecution of several species, including the Osprey, was actively encouraged by the 1566 Act for the Preservation of Grayne, which permitted church wardens to make payment for their heads. As Yalden and Albarella (2009) explain, the rates of payment were specified, and probably give some indication of the rarity or otherwise of the species listed. The price stated for Osprey, four pence, was the same as for 'Iron', which may refer to male White-tailed Eagles, corroborating the hypothesis that fish-eating raptors were highly sought after. Four pence was also the maximum sum offered and perhaps indicates that by this stage in English history the Osprey was becoming rarer as a direct result of targeted persecution.

Dennis (2008) suggests that by the early seventeenth century the Osprey would have been pushed back into more remote and less populated parts of the country, where rates of persecution were almost certainly lower. The decline of the Osprey in England was replicated in many parts of Europe during the seventeenth and eighteenth centuries, with persecution often actively encouraged by individual states through the provision of bounties (Saurola 2005, Dennis 2008). The 1734 State Law of Sweden included Osprey as a pest species. A Royal Decree on Avian Pests, introduced in 1741 for the 'extermination of raptors and pest birds' then actively encouraged persecution by offering bounty payments for the first time. All wild mammals and birds in Finland were later classified under the Imperial Hunting Decree of 1868 as (1) useful species, the populations of which were to be maintained or increased by protection, (2) harmful or pest species, which should be persecuted, and (3) other species, on whose protection or persecution there were no rules (Pohja-Mykrä et al. 2012). Ospreys were included in the second category and municipalities were obliged to pay bounties for their killing. The Osprey was again listed as a pest species on the revised Hunting Decree of 1898.

The targeted killing of Ospreys peaked during the nineteenth and early twentieth centuries across much of their European range (Bijleveld 1974, Saurola 2005, Dennis 2008). It led to severe population declines in some countries, such as Norway (Bijleveld 1974), and extinctions in others, including Denmark, the Czech Republic, Slovakia, Austria and Switzerland (Zachos and Schmölcke 2006).

DECLINES IN THE UNITED KINGDOM AND IRELAND

The Osprey was already in terminal decline in England during the eighteenth century, but persisted in some southern areas including Cornwall, Devon and Hampshire, as well as Cumbria, in the north. There were cliff-nesting Ospreys on Lundy until 1838 and on the North Devon coast until 1842, and the last English pair attempted to breed near Monksilver in Somerset in 1847 (Dennis 2008). It was perhaps no surprise that one of the birds was shot by a gamekeeper, meaning that the final English breeding attempt failed at the hands of man. There are no documented breeding records in Wales, but it seems likely that this is because the species was exterminated there before ornithological records were kept (Dennis 2008). Evans (2014) reported that a Flemish engineer working on wetland drainage systems on the Dyfi estuary in Mid Wales in 1604 mentioned several 'fishery hawks' breeding close together along the banks of the River Dyfi. Meanwhile, the wildlife historian Twm Elias recorded at least 10 different names for the Osprey in Welsh, dating back to 1604 (Evans 2014). Furthermore, the coat-of-arms of the city of Swansea, granted in 1316, and of West Glamorgan both feature an Osprey. This evidence, coupled with the species' ecological requirements, imply that the Osprey was probably widespread in Wales, particularly around the coasts and estuaries, before suffering the same fate as those in England and Scotland. It is also likely that Ospreys bred across Ireland, where there are a range of names for the species, and frequent references in historical texts (D'Arcy 1999). The last evidence of breeding dates from 1779 when Connaught described a nest with calling young on a ruined castle on an island in Loch Key, Country Roscommon. The disappearance of Ospreys in Ireland thus seems to have occurred during the period when they were in steep decline elsewhere.

In Scotland, where there were fewer people and less persecution, Ospreys persisted for longer than they did south of the border. Nevertheless, at the time of the loss of the last English breeding Ospreys, it is thought that there were only 40 or 50 pairs left in Scotland, including a nest on the abandoned castle at Loch an Eilein in Strathspey. Many of the remaining nests were in the remote western half of Scotland, away from more populated areas in the east (Dennis 2008). Data show that 18 Ospreys were killed by gamekeepers at Glen Garry in the Scottish Highlands between 1837 and 1840. This was comparable to the rate of killing of Golden Eagles (15), but considerably less than Red Kite (275) and Common Buzzard (385) during the same period – probably reflective of the relative abundance of the various species (Pearsall 1950).

It was not only gamekeepers who contributed to the decline of Ospreys. As the species became rarer a new threat emerged: there was now a lucrative market for both Osprey eggs and taxidermy specimens. As Dennis (2008) points out, travelling naturalists went to great lengths to kill the remaining birds and to take their eggs, giving no thought to the implications of their actions. The famous Loch an Eilein nest in Strathspey was robbed on 13 occasions between 1846 and 1900, with other failures attributed to shooting and twice to disturbance (Cash 1914, Waterston 1971). One particular individual, Lewis Dunbar, raided the nest on five occasions from 1848 (Poole 1989).

By the end of the nineteenth century, there was just a handful of pairs remaining in Scotland, but some enlightened individuals recognised the severity of the situation. The Grants of Rothiemurchus at Loch an Eilein, and the Camerons of Lochiel at Loch Arkaig (Figure 8.4), made valiant attempts to protect the birds on their land (Lambert 2001). Sadly, successful breeding was very rare, and the last known pair bred near Loch Loyne in 1916 (Dennis 2008).

Figure 8.4. Ospreys attempted to breed at Loch Arkaig in 1908 and a lone male returned each year until 1913. The Camerons made valiant attempts to protect the Ospreys on their land (© Antiqua Print Gallery/Alamy).

Signs of recovery

For many years it was accepted that, as in England, Ospreys had become extinct in Scotland, but extensive research by Dennis (2008) indicates that sporadic breeding occurred between 1916 and 1954. Ospreys bred in Aberdeenshire in 1925 and 1926, in Galloway and Loch Luichart between the wars, and at Loch Garten from the early 1930s. In some years, lone birds would have been present at nests, before eventually being joined by a Scandinavian immigrant or a returning Scottish youngster, thereby enabling breeding to continue.

Reduced persecution during and after the two World Wars provided some respite for Ospreys and other beleaguered birds of prey in the UK and across Europe. Tapper (1992) showed that the number of gamekeepers in the UK declined from a peak of 23,056 in 1911 to 4,391 in 1951 and this, coupled with more enlightened thinking, permitted some recovery in raptor populations in the UK and across Europe (Saurola 2005, Zachos and Schmölcke 2006, Yalden and Albarella 2009).

Rates of killing increased again after the Second World War. For instance, 93 Ospreys were killed at three fish farms in Lower Saxony in Germany in autumn 1953 (Bijleveld 1974). The difference now was that the species was beginning to be protected by law. The Osprey had been given legal protection in Finland in 1926 (Saurola and Koivu 1987) and in Sweden in the late 1920s (Osterlöf 1973), but it was not until the 1950s or later before other countries followed suit: Denmark in the 1950s, Poland in 1952, United Kingdom in 1954, former East Germany in 1954, Norway in 1962, France in 1964, former USSR in 1964, Spain in 1966, former West Germany in 1968 and Italy in 1971 (Bijleveld 1974). This led to a subsequent reduction in killing, although persecution has never been entirely eradicated (Saurola 1997), as discussed later.

Persecution in North America

Persecution was never as prevalent in North America, but some local extinctions that occurred during the first half of the twentieth century can be attributed to intentional killing (Bierregaard *et al.* 2020). Ospreys were eradicated from southern and central California, as well as northwest Baja Peninsula in Mexico, prior to 1920 (Henny 1983), while shooting eliminated Ospreys from southern Minnesota (Roberts 1932). As in Europe, egg-collecting became an issue in North America during the late nineteenth century, with New Jersey the most badly affected state. Bierregaard *et al.* (2020) report that 248 clutches were stolen from nests in New Jersey between 1887 and 1938, and that a combination of this and shooting resulted in a 75 per cent decline in a population of approximately 100 pairs at Seven Mile Beach between 1884 and 1890 (Stone 1937, Bierregaard *et al.* 2020). Overall, however, persecution in North America did not have the drastic effect that it did in many European countries (Poole 1989).

Egg-collecting

In the UK, nature conservation became a government responsibility in 1949 with the creation of the Nature Conservancy. Then, in 1954, the Protection of Birds Act gave formal legal protection to virtually all birds and their nests for the first time. The rarest species, and those that were routinely persecuted, were given added protection in the form of higher fines (Yalden and Albarella 2009). This brought an end to the childhood hobby of bird-nesting, but the nests of rarer species in particular continued to be routinely targeted by egg-collectors.

The threat of egg-collecting to Britain's rarest breeding birds was exemplified by events at Loch Garten in the Scottish Highlands during the 1950s. In 1954, the year that the Protection of Birds Act was passed by Parliament, ornithologist Desmond Nethersole-Thompson found a pair of Ospreys with two young at the south end of Loch Garten (Dennis 2008). With egg-collecting prevalent, the nest was kept a strict secret in an effort to protect it. Sadly, when the pair returned to the same nest in 1955, the eggs were stolen, and the next spring the nest of a pair of Ospreys on the nearby Rothiemurchus estate was also robbed (Dennis 2008). It was clear that egg-collecting posed a severe threat to the recovery of Ospreys in Scotland and so, in 1958, George Waterston, the director of the Royal Society for the Protection of Birds (RSPB) in Scotland, organised a team of wardens to protect the Loch Garten nest during the incubation period (Figure 8.5). Despite their best efforts, an egg-collector still managed to raid it one wet night in early June (Brown and Waterston 1962). Unperturbed, the group was reassembled the following spring under the guise of Operation Osprey, and this time the area around Loch Garten was designated a protected bird sanctuary, making it an offence to approach the nest without permission. The nest was watched day and night, and finally, on 8th June, the behaviour of the birds changed, indicating that the first egg had hatched. Sure enough, a few days later it was possible to make out the heads of three chicks for the first time (Brown and Waterston 1962, Dennis 2008).

It was now proving impossible to keep the Loch Garten nest a secret, and Waterston made a controversial but ultimately visionary decision to publicise the good news and to encourage people to come and view the Ospreys from a special observation point. A remarkable 14,000 people took up the invitation over the next two months, helping to ensure that public opinion was now firmly in support of the Ospreys and the conservation measures the RSPB had put in place to protect them (Brown and Waterston 1962). It became the first example of ecotourism in the United Kingdom and a source of inspiration for much of the work that we see today.

A 24-hour guard was reinstated while the Ospreys incubated eggs in 1960 and 1961, and each breeding attempt was successful; two chicks were reared in 1960 followed by three the following year. More than 40,000 people visited the public watchpoint over the course of the two summers to see the now famous birds (Brown and Waterston 1962). They were met by a 20-year-old Roy Dennis who had been employed by Waterston as one of the wardens to protect the nest and welcome visitors to the observation point (Figure 8.6) (Dennis 2008). It began a life of work with Ospreys for Roy, who without doubt has gone on to do more to help and conserve the species than anyone else in the UK and across Europe.

Figure 8.5. The return of Ospreys to Loch Garten in the 1950s was the start of the recovery of the species in the UK. George Waterston (left) organised a group of volunteers to monitor and protect the nest (© Charles Eric Palmer/photo-scot.co.uk).

With egg-collecting an ongoing threat, it was essential that any new nests were protected to deter egg thieves. Barbed wire was installed around the base of nest trees and electronic warning systems offered additional protection. In 1963, when the Loch Garten nest failed in storms, Operation Osprey was moved to nearby Rothiemurchus in order to guard a second nest where a Swedish-ringed male had taken up residence (Dennis 2008). Although the eggs failed to hatch, the early signs of a population recovery were evident. In 1967, Ospreys bred successfully in Moray for the first time, rearing two chicks, and

Figure 8.6. The recovery of Ospreys in Scotland and the rest of the UK has been greatly aided by the inspirational work of two conservation visionaries, George Waterston and Roy Dennis (back), seen here together in a boat on Fair Isle in 1967 (© Roy Dennis).

then, in 1970, a pair laid eggs at Loch of the Lowes in Perthshire. The Scottish population had now reached seven pairs, and a total of 44 young had been reared since 1954 (Dennis 2008).

Egg-collecting continued to be a factor in the recovery of Ospreys in Scotland during the 1970s, 1980s and 1990s. During this period Roy Dennis and others closely monitored and protected the growing population, which increased to 92 pairs with eggs in 1995. Roy Dennis's annual monitoring showed that in the 41 years between 1954 and 1995 there were a total of 61 confirmed or suspected egg robberies, which accounted for just over a quarter of observed nest failures (26.2 per cent) and 6.7 per cent of all breeding attempts, where eggs were laid. In the worst years during the 1980s, when the population increased from 20 to 52 active nests, 17–19 per cent were robbed. Fortunately, the overall productivity of active nests between 1954 and 1995 (1.54 young) more than compensated for annual losses and ensured that the population was able to recover. However, it is clear that without the efforts of Waterston, Dennis, the RSPB and others, the story may have been very different. The introduction of custodial sentences for egg-collecting under the Countryside and Rights of Way Act 2000 became a serious deterrent and significantly reduced egg-collecting activity in the UK. In 2021, there were just nine reported incidents of egg-collecting and egg thefts of all species (RSPB 2022), a marked difference to the early years of the Osprey recovery in Scotland.

It is not just in the UK that egg-collecting has threatened Osprey populations. Numbers of breeding Ospreys in Cape Verde were suppressed by the overharvesting of Osprey eggs and also nestlings by islanders for food until the mid-twentieth century (Naurois 1969). As Palma *et al.* (2020) point out, this practice decreased thereafter, allowing numbers to grow from 46–71 breeding pairs in the 1960s (Naurois 1987) to 72–81 pairs by the end of the century. Some breeding sites are still raided occasionally, particularly in areas where nests are more accessible, but the population has continued to recover and now stands at an estimated 103–125 pairs. It seems likely that the practice of nest-robbing selected for high cliff nests in this population and others in the Mediterranean region. Recently, however, there has been a trend for the redistribution of nests away from the coastline in some parts of Cape Verde in response to pressure from tourism (Palma *et al.* 2020).

Impact of Dichlorodiphenyltrichloroethane (DDT)

Ironically, at a time when the Osprey was being given legal protection across much of its range, a new anthropogenic threat emerged. The introduction of organochlorine insecticides, most notably Dichlorodiphenyltrichloroethane (DDT), to agriculture following the Second World

War led to severe ecological damage and had significant impacts on Ospreys in both North America and Europe (Carson 1962, Poole 1989, Henny *et al.* 2010). Organochlorines are readily trapped in fatty tissues and thus become increasingly concentrated in successive levels of the food chain. Raptors are particularly susceptible because they can accumulate levels up a hundred times greater than the food they are consuming (Newton 2010). Furthermore, fish absorb organochlorines directly through their gills, in addition to ingesting them with food, and this further increased accumulations in Ospreys and other fish-eating birds higher up the food chain (Poole 1989).

The extent of the damage that DDT did to Ospreys was not immediately apparent, but the pioneering study of Ratcliffe (1967) on Peregrines, Golden Eagles and Sparrowhawks *Accipiter nisus* demonstrated that eggshell thinning, resulting in subsequent breeding failures and significant reductions in productivity, was attributable to the accumulation of DDT and other organochlorines in the environment. It transpired that egg viability was reduced even at very low concentrations, of a few parts per million.

Some of the first evidence of the impact of DDT on Ospreys was recorded by Ames (1966). He found that levels of DDT and Dichlorodiphenyldichloroethylene (DDE) – one of the most common breakdown products of DDT, formed by the loss of hydrogen chloride – were 5.1 parts per million (ppm) per egg in a population in coastal Connecticut, where Ospreys had been declining by 30 per cent annually for the previous nine years, compared to 3ppm/egg in an apparently stable colony at Chesapeake Bay in Maryland. It later transpired that Osprey eggshell thickness was declining in both North America and Europe and that this began to have a profound effect on breeding success. Even small increases in DDE resulted in major shell thinning; indeed, the most significant changes occurred at lower levels of contamination, as described by Poole (1989). An increase in DDE levels from 0 to 12ppm resulted in a 14 per cent decrease in shell thickness, whereas all further contamination resulted in only a 6 per cent decline. Studies demonstrated that DDE levels greater than or equal to 12ppm were sufficient to result in complete reproductive failure in Ospreys due to the associated eggshell thinning (Johnson *et al.* 1975, Wiemeyer *et al.* 1975). Thin shells increase the chance that an egg will break before hatching, and breeding Ospreys incubating clutches with shells 16 per cent or more thinner than pre-DDT levels were likely to lose at least one egg (Poole 1989). DDT also reduced the viability of developing embryos by affecting the pore structure of the eggshell and thereby reducing gas exchange with the atmosphere (Bierregaard *et al.* 2020). As a consequence, breeding Ospreys rarely produced more than one young once DDE concentrations exceeded 5–10ppm (Poole 1989).

The impact of DDT in North America was significant, with the worst-affected populations declining by over 90 per cent, as summarised by Henny *et al.* (2010) based on data presented at the First World Conference on Birds of Prey in Vienna, Austria, in 1975 (Henny 1977). Productivity rates of Osprey populations on the North Atlantic Coast, where the impacts were most pronounced, often because saltmarshes were sprayed directly with DDT, began decreasing during the 1950s and were lowest between the late 1950s and early 1970s. The associated population decline was stark. Reduced productivity meant that recruitment to the breeding population was badly affected and, as a consequence, established breeding adults were not replaced if they failed to return in the spring. As a result, the breeding population on Gardiner's Island, New York declined from 300 breeding pairs pre-1945 to just 38 pairs by 1970. Similarly, Ospreys breeding on the Connecticut River fell from a peak of 200 pairs pre-1945 to just four in 1970; the Rhode Island population declined from 130 pairs to seven pairs; while the southern New Jersey population fell from 253 pairs to 45 pairs. An analysis of changes in productivity in areas where these data were collected demonstrate why the declines were so drastic. The number of young fledged per active nest in the Gardiner's Island, Connecticut River and Rhode Island populations fell from 1.19 in 1950–52 to a low of just 0.23 in 1963–67, well below the figure required to maintain a stable population, which is 0.95–1.30 (Henny et al. 2010). It was telling that eggshells of Ospreys breeding in Connecticut in 1967–69 were 15–18 per cent thinner than pre-1947 shells (Wiemeyer et al. 1975), indicating that DDT was the principal reason for the decline. Similarly, Spitzer et al. (1978) found that mean DDT concentrations in unhatched eggs collected from failed nests in Connecticut and Long Island, New York, were 22.6ppm wet weight, while the corresponding figures for nests with one young and two young were 11.9ppm and 5.82ppm. These very high levels of DDT were attributable to practices such as those of Suffolk County Mosquito Control Commission on Long Island, which routinely sprayed DDT every spring and summer from the late 1940s to 1966 (Puleston 1975).

Elsewhere in North America, the impact of DDT was more variable, but some populations experienced very low productivity akin to that of the North Atlantic Coast. The number of young per occupied nest dropped below 0.5 at Eastern Bay, Chesapeake Bay in 1966–68, in northern Michigan for most of the 1960s and in northern Wisconsin in 1960–65 (Berger and Mueller 1969, Postupalsky 1977, Henny et al. 2010). Population-level effects were generally less severe, however, likely because these populations lost fewer breeders (Poole 1989). Studies of Ospreys in western parts of North America were limited during years when the impacts of DDT were most prevalent, but there was evidence of low productivity at Yellowstone National Park in Wyoming in 1972–74 (Henny et al. 2010).

The impacts of DDT in Europe were generally less severe than the worst-affected parts of North America, but still resulted in low reproductive rates in many areas. A long-term study of eggshell thickness in Sweden by Odsjö and Sondell (2014) revealed that the thinnest shell fragments occurred in 1973 (mean = 0.438mm, n = 71 clutches), a 15 per cent decline compared to thickness of eggs pre-DDT (0.515mm). Furthermore, the average thickness of shell fragments with no intact eggs was 0.393mm, a 29 per cent decline compared to pre-1946 values, and 0.393mm in nests with a single intact egg, representing a 24 per cent decline. This indicates that significant loss of eggs due to breakages occurred once mean shell thickness in the clutch fell below 0.40 mm. This had a severe impact on breeding success, with productivity declining to just 0.25 young per occupied nest during 1971–73 (Odsjö and Sondell 2014). A similar scenario was evident elsewhere in Europe. Mean brood size of Ospreys in Mecklenburg in Germany declined from 2.2 in 1959 to 0.9 in 1966, probably as a result of DDT and associated eggshell thinning (Moll 1967). Data on DDT levels are limited, but 19 unhatched eggs recovered from nests in the former East Germany between 1979 and 1981 contained mean DDT residues of 4.5ppm and a maximum of 10.8ppm, indicating that it was responsible for the observed declines in breeding rates (Meyburg et al. 1996). Meanwhile, the geometric mean of DDT concentrations in addled Osprey eggs collected from nests in Finland peaked in 1971–75 (mean = 63.6ppm/lipid weight) (Saurola 2005).

Post-DDT recovery

DDT was finally banned in the United States in 1972 and the extent to which it supressed Osprey numbers in previous decades was evident as populations began to recover. Breeding productivity in the best-studied and worst-affected populations in Long Island, Connecticut and Rhode Island gradually increased through the 1970s, although mean productivity (0.73) was still below normal rates (Henny et al. 2010). In other areas the recovery was faster, with productivity in populations in Westport River, Massachusetts, increasing from 0.60 in 1963–67 to 1.24 in 1973–75 and in Maine from 0.38 in 1963–67 to 1.19 in 1973–75 (Henny et al. 2010). By the 1980s and 1990s, levels of DDT residue in eggs were greatly reduced and no eggs collected along the North Atlantic Coast contained DDE at levels greater than 5.2ppm; indeed, most were below 4.2ppm (Steidl et al. 1991, Clark et al. 2001, Henny et al. 2010). Nevertheless, localised high concentrations did persist in areas of extensive past use. For example, it is thought that high levels of DDT in Osprey eggs in western New Jersey were due to dredging of shipping channels in Delaware Bay that caused DDE residues in adjacent agricultural land to be re-suspended when sediments were disturbed (Steidl et al. 1991). Similarly, Henny et al. (2010) reported that 24 per cent of Osprey eggs sampled in 1997/98 from nests on the Lower Columbia River, on the border between Washington and Oregon in the northwest of the United States, contained ≥8.0ppm DDE. These were the highest concentrations recorded in North America during the late 1980s and 1990s. Correspondingly high DDE concentrations were also encountered in a key prey species, the Largescale Sucker *Catostomus macrocheilus*. However, by 2004 no eggs contained more than 2.29ppm DDE and the breeding population had increased from 103 occupied nests in 1998 to 225 in 2004 (Henny et al. 2008a).

While initial recovery from DDT was fairly slow, rates of population growth increased more rapidly during the 1980s and 1990s. The first nationwide survey of Ospreys in the United States in 1981 estimated a total of population of 8,000 pairs (Henny 1983). By the time of the next survey, in 1994, the population had grown to 14,200 breeding pairs, an increase of 77.5 per cent (Houghton and Rymon 1997), and in 2001 it was estimated to have reached 16,000–19,000 pairs (Poole et al. 2002). By 2013 the southern New England and Long Island, New York population – one of the worst affected during the DDT era – was found to exceed pre-DDT levels, at over 1,200 pairs (Bierregaard et al. 2014). Factors such as use of artificial structures for nesting, and changes in food availability, contributed to the population increases observed across North America, but the banning of DDT in 1972 was crucial (Henny et al. 2010).

DDT was phased out in most European countries around the same time as the United States, and well before a worldwide ban (except for restricted disease vector control) following the UN Stockholm Convention on Persistent Organic Pollutants in 2001 (UNEP 2009). This resulted in declines in DDT residues in Osprey eggs, similar to those observed in North America. For example, concentrations of sDDT (= total DDT) in Finnish Ospreys decreased by 73.6 per cent between 1971–75 and 1992–93 (Saurola 2005). The long-lasting effects of DDT, though, are exemplified by the fact that it took 30 years for the eggshells of Swedish Ospreys to return to normal thickness (0.515mm) from a DDT-induced low of 0.438mm in 1973, despite the fact that DDT use was banned in the country in 1970 (Odsjö and Sondell 2014). Similarly, despite the fact that DDT was outlawed in France in 1973, Lemarchand et al. (2014) detected DDE (mean = 0.92 mg/kg) in 12 of 27 samples (dead nestlings and unhatched eggs collected during ringing operations, as well as dead adults) from the Loire River catchment in central France. Although levels detected were generally low, there was one very high value of 8.2 mg/kg, and an unhatched egg with a concentration of 4.6 mg/kg, which is above the threshold of 4.5 mg/kg at which an eggshell can break under the weight of an incubating female. These findings indicate possible recent use of DDT in the region.

The gradual reduction in the exposure of Ospreys to DDT has facilitated recent increases in the European population, with other factors such as a reduction in persecution and conservation interventions including artificial nest building and translocations also important (Saurola 2005, Schmidt-Rothmund et al. 2014). The population in Europe, northern Africa and the Middle East stood at approximately 5,500 pairs in the 1980s, and more than doubled to an estimated 9,500–11,500 pairs by the early twenty-first century (Schmidt-Rothmund et al. 2014).

Curiously, DDT had little discernible effect on Ospreys in Australia. Although DDT was detected in Australian Ospreys, it occurred only at very low levels (mean = 0.11ppm)

(Falkenburg et al. 1994). As a result, no significant difference was found in Osprey eggshell thickness associated with DDT, even though other Australian raptors, such as the Peregrine, experienced reductions of up to 10 per cent (Olsen et al. 1993). It seems that the location of Osprey nests meant they were only exposed to very low levels, compared to other species.

Other pollutants

Various other toxic pollutants that accumulate in the environment have been detected in Ospreys, though none have had the population-level effects of DDT.

Dieldrin

Dieldrin was developed in the 1940s as an alternative to DDT and was used widely in the same era (Scott et al. 1959). Wiemeyer et al. (1975) suspected that dieldrin may have increased the mortality rate of adult Ospreys in the Connecticut population which, as described above, was also badly affected by DDT. A lethal concentration of dieldrin was found in the brain of a dead Osprey in Connecticut in 1967, while another victim of suspected dieldrin poisoning was identified in South Carolina the following year (Wiemeyer et al. 1980). However, as Henny et al. (2010) point out, dieldrin and the closely related aldrin, were predominantly used in the corn belt of the midwestern United States and the cotton belt of the southern United States, which lay outside the breeding range of most Ospreys. This meant there was a much lower risk of dieldrin or aldrin accumulating in fish taken by Ospreys, unlike DDT which was directly sprayed into areas where breeding Ospreys were foraging. Instead, bird-eating raptors, such as White-tailed Eagles and Peregrines, were more likely to be impacted (Henny et al. 2010).

Dioxin

Dioxin (TCDD; 2,3,7,8-tetrachloro-p-dioxin) is a by-product of industrial processes and has been detected in Osprey eggs in the United States. Woodford et al. (1998) found that dioxin levels in fish downstream of pulp mills on the Wisconsin River in the Upper Midwest were 30–100 times higher than control areas upstream, and Osprey eggs collected from areas 7km downstream of the mills contained much higher concentrations of dioxin (29–162 picogram(pg)/g wet weight, n = 18) than those upstream (maximum 23.8pg/g, n = 15) as a result. Although there was no discernible impact on overall breeding success, nestlings that were exposed to dioxin in the ova, or through ingestion of contaminated fish, grew more slowly than those with much lower exposure.

Polychlorinated biphenyls

Polychlorinated biphenyls (PCBs), widely used as the dielectric fluids in transformers, have contaminated areas occupied by Ospreys in both North America and Europe (Poole 1989, Elliott et al. 2000). PCBs have had detrimental impacts on the breeding performance of other species (Harris and Elliott 2011), but their effect on Ospreys has been less significant (Poole 1989). Ospreys breeding in Massachusetts were exposed to very high levels of PCBs (25ppm wet weight) but productivity remained high, with a mean 1.91 chicks per nest (Poole 1989). Similarly, breeding rates of Osprey populations in Connecticut and Long Island, New York, improved following the ban on DDT even though high levels of PCBs continued to be recorded in eggs (Spitzer et al. 1978). Likewise, there was no effect of PCBs on brood size in Swedish Ospreys during the 1970s, although levels were lower than those reported in parts of North America (Odsjö 1982). Grove et al. (2009) observed that, other than in the locations with heavy contamination, there has been a gradual decline in PCBs in aquatic environments inhabited by breeding Ospreys in recent decades, with a 50 per cent decrease in tissues recorded every five to ten years.

Mercury

Mercury is soluble in fat and blood, bioaccumulating in fish and birds as a result (Bryan 1984, Poole 1989). It was routinely discharged in untreated effluents until 1970, when control measures were implemented (NRCC 1979, Hughes et al. 1997), and also leaked into aquatic habitats from mining operations (Langer et al. 2012). Today, low-level atmospheric release from combustion sources, and subsequent deposition elsewhere, is the primary source (Grove et al. 2009, Guigueno et al. 2012). Mercury levels may also be higher in reservoirs than naturally formed lakes and rivers, because it is thought that flooding of terrestrial soils adds inorganic mercury and organic nutrients to the aquatic environment and results in elevated microbial methyl production, the process by which bacteria turn inorganic mercury into an organic mercury compound that can bioaccumulate in the food chain (Hughes et al. 1997).

Studies have shown that mercury appears to have only minor impacts on Ospreys because it is bound to keratin during feather formation, and sequestered into newly grown feathers as a result. This limits accumulations in body tissues, and the feathers themselves are later moulted (Furness et al. 1986, Hughes et al. 1997). Up to 95 per cent of the body burden of mercury in adult Ospreys, and 85 per cent in chicks, may be lost through feathers in this way (Hughes et al. 1997, DesGranges et al. 1998). Some mercury is also transferred to eggs, and Hughes et al. (1997), who studied mercury levels in Ospreys in Ontario and New Jersey between 1991 and 1994, concluded that observed levels are indicative of the exposure of adult birds over the course of the year. They found that feathers sampled from chicks aged 28–35 days old were better indicators of local contamination because spatial differences in observed mercury levels (3.26–7.40ppm) were very similar to those of a key Osprey prey species, the Yellow Perch *Perca flavescens*.

Laboratory-based studies have revealed that Osprey embryos have high sensitivity to mercury. Heinz et al. (2009) injected doses of methylmercury into the air cells of eggs of 26 species of birds and examined the dose–response curves of embryo survival. Ospreys were one of five species classified in the high sensitivity category, whereby the median lethal concentration was less than 0.25µg/g mercury. Fortunately, this apparent high sensitivity to mercury poisoning has not necessarily translated into population-level effects (Poole 2019).

Anderson et al. (2008) studied the breeding performance of Ospreys at Clear Lake, California from 1992 to 2006, in relation to remediation works associated with the Sulphur Bank Mercury Mine, initiated in 1992. They found that mercury levels in feathers declined from 20mg/kg in 1992 to a low of 2mg/kg in 1998. However, levels then increased to 23mg/kg in 2003 and to 12mg/kg in 2006. The total number of active Osprey nests increased from seven in 1991 to 31 in 2006, indicating that the observed fluctuations in mercury levels had no effect. DesGranges et al. (1998), meanwhile, found that the number of fledged young (1.6±0.7) where mercury exposure was highest (>40mg/kg of mercury in nestling feathers) at reservoirs in Quebec, was not significantly lower than nests in either natural habitats (2.0±0.7) or urban environments (1.9±0.7) (H = 4.39, p = 0.11). Sequestration into newly grown feathers thus appears to protect chicks from any toxic effects. Nonetheless, they suggest that there is potential for toxicology problems to arise after fledging.

Proactive conservation measures

Our relationship with the Osprey has not always been a positive one, but the emergence of more enlightened thinking by some in the late nineteenth century signalled the beginnings of a change in attitudes. It has subsequently become clear that the Osprey responds well to proactive conservation measures, and these interventions have helped the species to recover strongly from the effects of centuries of historical persecution and the impact of DDT and other organochlorine pesticides.

Artificial nests

The construction of artificial nests has proved highly effective conservation technique across the Osprey's range, helping to increase breeding productivity and facilitate geographical expansion.

Building artificial nests helps in a number of different ways. First, young and inexperienced birds are prone to building poorly constructed nests in precarious locations that may collapse during the breeding season, leading to the loss of eggs or chicks (Dennis 2008). In the early years of the Osprey's recovery in Scotland, Roy Dennis and colleagues became skilled at identifying such nests, and replacing them with more secure structures during the winter. They also repaired nests that had been damaged by storms. Even experienced adult Ospreys may waste valuable time and energy repairing nests when they return in the spring, and if the damage to the tree is particularly severe, they may be forced to move to a completely new location.

Poole (1984) found that as many as 50–70 per cent of natural nests in Florida, United States, were lost per year, usually blown out during storms, while 30–40 per cent of nests in New York failed in similar circumstances. As Poole (1989) pointed out, Osprey nests in dead trees or in short, flexible live trees such as mangroves, often used by Ospreys in Florida, are particularly vulnerable. Trees used by Ospreys in northern Scotland, principally Scots Pine and other coniferous species, are more robust, but even so a well-built artificial nest is less likely to be blown out of a tree and result in complete breeding failure. Artificial nests have been key to the recovery of Ospreys in Scotland (Dennis 2008). Roy and colleagues have shown that Ospreys readily accept rebuilt nests, and have rescued young after nests collapsed on a number of occasions – placing them in temporary nests either in the same tree or nearby, which the breeding Ospreys accept immediately.

Artificial nests have also proved an effective way to facilitate range expansion. Satellite tracking and colour-ringing studies have demonstrated that young non-breeding Ospreys explore extensively when they return to breeding grounds for the first time, usually during their third calendar year, as described in chapter 3. The long-term colour-ringing of Ospreys in the UK has shown that, given the opportunity, first-time breeders will preferentially take over an established nest, rather than build their own (Dennis 2008, Mackrill et al. 2013, Evans 2014), and so these early explorations help young birds to identify future breeding sites. This knowledge is particularly valuable for females, whose natal dispersal is usually greater than males (Martell et al. 2002, Dennis 2008). Conspecific attraction further influences settlement patterns, because young Ospreys preferentially breed in relatively close proximity to other pairs (Poole 1989, Dennis 2008, Morandini et al. 2019). As such, artificial nests in existing Osprey colonies are readily used (Spitzer et al. 1983). However, the provision of nests on the edge of the current breeding range, and even in more distant locations, encourages Ospreys to expand to new areas. This is particularly important if there is no recent history of breeding Ospreys, and therefore no old vacant nests to occupy (Dennis 2008).

ARTIFICIAL NEST CONSTRUCTION

Before considering the positive impact that artificial nests have had on Osprey populations around the world, it is important to consider what constitutes a good nest.

Ospreys prefer a prominent site with a clear uninterrupted view, and this can be achieved by constructing a nest in the tallest of a group of trees, or a single tree isolated from others. In other locations, a telegraph pole located away from taller trees is suitable. If there are no surrounding trees, such as on saltmarshes, then a very low pole will suffice. The site should be in relatively close proximity to suitable fishing grounds, but not necessarily adjacent to water. Ospreys can travel considerable distance to hunt, as described in chapter 2, but siting the nest within a few kilometres of water will increase chances of occupancy. In Britain and Europe, it is best to site the nest in quiet locations away from roads and regularly used footpaths. Ospreys can, however, become habituated to some human activity, particularly agricultural vehicles and machinery, and so nests can be situated on the edge of arable fields assuming the land-owner is supportive.

If a nest is to be built in the top of a tree, then there needs to be a substantial fork to support it. Large flat-topped pines are ideal, but some deciduous trees are suitable as well. In central England we have built artificial nests in stag-headed Oak trees with good success. If building in a pine tree, it is usually necessary to cut off the top of the tree down to the first ring of strong side branches (Figure 8.7). Likewise, the upper branches of a deciduous tree generally need to be removed to ensure that the nest is prominent and positioned in a strong fork. An exception might be if tree contains a lot of dead wood, in which case dead branches higher than the nest can be left as perches for the birds, as long as they do not impede access to the nest.

A range of methods can be used to build a tree nest. Some people instal wooden platforms (see below), and then build the nest on top of that to provide a strong base. However, if materials are selected carefully then a more natural-looking nest can be constructed. The key requirement is that three or four strong branches, approximately 12cm thick, and 1.5–1.8 metres in length, are arranged in a triangle or square in the fork, in order to form a sturdy base. Once in position, they are nailed, screwed or bolted into the tree to give it rigidity. Secondary supports are then secured across this base. The nest is then constructed on top using smaller sticks, approximately 5cm thick and 0.5–1.5m in length. These are arranged in a circle, at least 1.25m in diameter, and the bottom sticks secured to the base using cable ties or wire. Some are also laid across the base. The nest should be built up to a height of 0.5–0.75m and then filled with clumps of turf, which are pushed firmly down into the base of the nest – replicating the manner in which Ospreys build up and line the centre of the nest with clumps of grass, moss, dried algae and other aquatic vegetation, or even seaweed, as described in chapter 4. It is important to make

Figure 8.7. Artificial nests built in pine trees have proved highly successful for Ospreys in Scotland (© Tim Mackrill).

the structure resemble a used Osprey nest. In the early part of the year, there is a distinct nest-cup that helps to protect the eggs and small chicks, but as the days and weeks progress and the chicks near fledging, the adults bring in clumps of nest lining in order to create a flat platform from which the youngsters can launch themselves into the air. This means that by the time the family departs in autumn, the nest lining is almost level with the top of the sticks, and that should be the aim when constructing an artificial nest. Wood chip, moss and leaf litter can be used to complete the lining.

If suitable trees are not available, Osprey nests can also be constructed on telegraph poles. As a general rule, poles measuring 9–10m in height are ideally suited for Osprey nests, although shorter poles can be used in more open habitats with few or no trees. Electricity companies have been generous in donating poles to our Osprey conservation work in many areas. We have used various designs of platform in the past but the best compromise in terms of longevity and cost is to use 47 × 100mm graded treated timber to construct a wooden frame of approximately 1m². One side of the frame extends approximately 30cm outwards to provide a fixing point for a perch. The inner spurs of the frame form a 15 × 15cm opening where the frame is bolted to the pole. Welded wire mesh (12-gauge 2.5mm wire with 1.5 × 1.5cm holes, hot-dip galvanized) is then be fixed to the upper side of the frame using fencing staples or similar. Coach screws are used to fix the frame to the pole, the top of which is squared off using a chainsaw (Figure 8.8). Once the frame is secured, a perch that will extend approximately 1m above the nest, at an angle of 45°, is coach screwed to the pole and then fixed to the wooden frame to provide additional support. An Oak branch, or similar hardwood, or a fencepost with a T perch screwed on top are best. The pole is then ready to be erected, a task that electricity companies are often willing to assist with (Figure 8.9). An alternative option is to use a prefabricated specially designed metal frame or basket in place of the wooden platform. This offers greater longevity but is a more expensive option. Once the pole has been installed, at least 1.5m into the ground, a triple-extending ladder is used to access the platform and to build a nest using the method outlined above, with initial sticks cable-tied to the weld mesh to secure them. It is also helpful to cable-tie narrow-gauge chicken wire to the weld mesh to prevent the nest lining falling through when it is added.

Once erected it is important to maintain nests in good condition, even if they are not occupied by Ospreys immediately. It can often take several years for Ospreys to occupy a new nest, even in areas that already hold other breeding pairs, and more so in more distant localities remote from an existing colony. We usually inspect nests in late February and early March in order to repair any winter damage and ensure that they are in the best possible condition when birds return from spring migration (Figure 8.10).

In some areas, Osprey nests built close to

Figure 8.8. A timber frame and weldmesh base, fixed to a telegraph pole, has proved to be the most effective design for artificial nests in the Rutland Water area (© Tim Mackrill).

Figure 8.9. Once the pole has been erected, the complete nest is built on the wooden platform. Triple-extending ladders are usually needed for access. Electricity companies are often willing to assist with the erection of nests (© Tim Mackrill).

Figure 8.10. Artificial nests are lined with soil, turf and moss, almost level with the top of the sticks. This replicates how an Osprey nest looks at the end of the breeding season (© Tim Mackrill).

water may be used for nesting by Canada, Greylag or Egyptian Geese. As described in chapter 5, these species are capable of deterring Ospreys from nesting, especially if they occupy nests early in the spring. At Rutland Water we have sometimes covered nests with builders' sacks to prevent geese getting established before Ospreys are present (Mackrill *et al.* 2013), while the convex lid of a dustbin has been used for the same purpose in Canada (Dennis 2008). Terrestrial predators, such as the Pine Marten *Martes martes* in Scotland, are a threat in some areas, and metal sheeting wrapped around the base of nest trees can provide useful protection in this instance in order to prevent terrestrial predators climbing up.

UNITED KINGDOM

The recovery of Ospreys in the United Kingdom has been a great conservation success. The protection of breeding sites, liaison with local people, provision of artificial nests, and the translocation of young have been key factors in the re-establishment and expansion of the population since the mid-twentieth century.

Scotland now holds an estimated 300 breeding pairs, with the population having increased from the single pair at Loch Garten in the 1950s discussed earlier in this chapter (R. Dennis pers. comm. 2022). As Dennis (2008) describes, 44 young Ospreys were reared in Scotland between 1954 and 1970 – and by that year early signs of geographical expansion were evident, with five pairs in Moray and a pair at Loch of the Lowes in Perthshire. The population continued to increase both numerically and geographically through the 1970s, aided by the erection of artificial nests, and by the end of the decade there were breeding pairs in Aberdeenshire, Easter Ross and Angus, in addition to those in Strathspey, Moray and Perthshire.

The Scottish population reached a significant milestone in 1996, when for the first time in several centuries there were more than 100 breeding pairs. The geographical expansion continued to be relatively slow, with most new pairs becoming established in areas already holding Ospreys, but in 1996 the first pair breed in the Scottish Borders. A total of 153 pairs reared more than 200 chicks for the first time in 2001, and by 2005 breeding pairs were established in Lochaber in the western Highlands and Galloway in southwest Scotland. Today, the 300 breeding pairs are widely distributed across most areas of the country. The value of artificial nests is exemplified by the fact that Dennis (2008) reported that 23 per cent of nests in his study area in northern Scotland were artificial, with a further 14 per cent rebuilt after the loss of a natural nest due to the weather or tree failure.

The construction of artificial nests has also encouraged Ospreys to spread south across the border into England. Ospreys returned to Cumbria in 1999 and Kielder Forest in Northumberland in 2009, and in each case artificial nests have proved key to the establishment of breeding populations (Table 8.1). A pair of Ospreys raised a single chick in a natural nest in Cumbria in 2000 and three at the same site the following year. A second pair bred successfully on an artificial nest at Bassenthwaite Lake in 2001 and since then the majority of breeding attempts in the region have been on artificial nests. By 2021 a total of 122 young had fledged from 59 breeding attempts at 10 artificial nest-sites, while 32 chicks fledged from 18 breeding attempts at 10 natural nests (P. Davies pers. comm. 2022). As such, the productivity of artificial nests (2.07) was higher than natural sites (1.78). Seven of the nests constructed by the Ospreys themselves were only used for a single year before they moved to artificial nests erected after the previous nest was lost in winter storms, or because a nest had been constructed on an electricity pylon and needed to be moved (P. Davies pers. comm. 2022).

At Kielder, the number of breeding pairs increased from one in 2009 to eight in 2021. During this period, 93 per cent of breeding attempts were on artificial nests, with a productivity of 2.07 (J. Dailey pers. comm. 2022) (Table 8.1, Figure 8.11).

At Rutland Water, the provision of artificial nests has also proved highly successful. The first successful nest, at Site B, was a natural one in the top of a stag-headed Oak but, overall, 69.6 per cent of 102 breeding attempts over the course of 20 years between 2001 and 2021 were on artificial nests built on telegraph poles (Table 8.1). Many of the sites we have chosen for artificial

Figure 8.11. The erection of artificial nests in Kielder Forest has greatly facilitated the expansion of Ospreys south across the English border from Scotland (© Forestry England).

nests are adjacent to arable fields or those that are cut for silage, and experience has shown that Ospreys quickly become habituated to the presence of agricultural machinery and vehicles, and are rarely disturbed by farming operations (Mackrill *et al.* 2013) (Figure 8.12). All of the landowners we have worked with are conscientious when working near 'their' Ospreys and proud to have them on their land. Indeed, most eagerly join us to ring chicks during the summer. An added benefit of artificial nests on poles is that they can be easily accessed from the ground with a triple-extending ladder, making monitoring the population very straightforward.

The remaining breeding attempts in the Rutland area between 2001 and 2021 were at three sites where Ospreys constructed a natural nest in an Oak, which was later replaced with a stronger artificial nest in the same tree (Figure 8.13). Forks in the top of Oaks provide an ideal location for Ospreys to build their nests, but are often at risk of collapsing because they are invariably built on dead and decaying wood. When we ringed the chicks at Site B in 2006, we noticed that the main limb supporting the nest was rotten. In view of this we removed the nest once the Ospreys had departed that autumn, cut away the dead wood at the top of the tree and installed a platform at the top of the main trunk. The nest was a few metres lower as a result, but much more secure. The Ospreys immediately accepted the new nest when they returned the next spring. The importance of this work was demonstrated when another nest in the top of an Oak collapsed in 2019. Fortunately, it occurred shortly after the two chicks had fledged, but it could easily have happened while there were small, dependent young. That winter we installed an artificial nest in a secure part of the tree and it was occupied by the Ospreys when they returned the following spring.

The success of artificial nests in the Rutland Water area has encouraged others to erect nests, and this will undoubtedly facilitate the geographical expansion of the population in years to come. Indeed, it has already been a key factor in the recovery of Ospreys in Wales. Two natural nests were found in 2004, but since then all subsequent known breeding pairs, or at least those that are in the public domain, have used artificial nests. This includes the five

Table 8.1. Number of breeding attempts and young raised at different nest-sites in England and Wales 1999–2021

Location	Natural nests				Rebuilt nests				Artificial nests			
	No. of sites	Breeding attempts	Young	Prod.	No. of sites	Breeding attempts	Young	Prod.	No. of sites	Breeding attempts	Young	Prod.
Cumbria	10	18	32	1.78	5	15	31	2.06	5	44	91	2.07
Kielder	3	3	3	1.0	1	4	9	2.25	8	39	80	2.05
Rutland Water	3	9	16	1.78	3	22	43	1.95	7	71	151	2.13
Wales	2	2	1	0.5	1	17	42	2.47	5	33	71	2.15
TOTAL	**18**	**32**	**52**	**1.63**	**10**	**58**	**125**	**2.16**	**25**	**187**	**393**	**2.10**

'Re-built nests' refers to an artificial nest built to replace a natural nest; prod = productivity, i.e. number of young per breeding attempt.

Figure 8.12. Artificial nests on telegraph poles have been a great success in Rutland. Breeding Ospreys become habituated to farming operations near nests (© John Wright).

Rutland females that reared young in Wales between 2011 and 2021. One of two natural nests discovered in Wales in 2004, collapsed after heavy rain and strong winds, killing the two chicks at the site in the Glaslyn valley. An artificial nest was erected in its place and was used from 2005 onwards.

The overall expansion of the English and Welsh population of Ospreys between 1999 and 2022 is shown in Figure 8.14.

GERMANY

The recovery of Ospreys in Germany has been greatly aided by the provision of artificial nests in trees and on electricity pylons. Dr Daniel Schmidt-Rothmund has led much of this work and has, himself, installed more than 250 nests in trees across the country. The German breeding population more than tripled between 1989 and 2009 (Schmidt 2010) and continued to increase thereafter, to an estimated 750 pairs in 2022 (D. Schmidt-Rothmund pers. comm. 2022). The stronghold of the population is situated in Brandenburg, in the north-east, where Ryslavy (2011) reported 314 territorial pairs nesting at an average density of 1.06 pairs/km². The first artificial nests were constructed in the 1980s and this expediated range expansion from the early 1990s, leading to the recolonisation of Lower Saxony in northwestern Germany and Bavaria in the south (Schmidt-Rothmund *et al.* 2014). The construction of nests on the top of pylons has been particularly valuable in some areas, specifically in eastern Germany, where it has compensated for a scarcity of natural nest-sites, and facilitated recent population increases (Canal *et al.* 2018). An estimated 75 per cent of German Ospreys now nest on pylons, with most of them on artificial nests installed by German power companies together with conservationists (Figure 8.15). The value of electricity pylons to the German population was demonstrated by Canal *et al.* (2018), who categorised a total of 396 different nest-sites in Brandenburg according to substrate type. They found that 74 per cent of nests were built on artificial structures (98 per cent of which

Figure 8.13. The natural nest at Site B, Rutland Water, was replaced with an artificial platform in 2006 because the branches supporting the nest were rotten, leaving it vulnerable to collapse (© John Wright).

were electricity pylons) and 26 per cent were built in trees, mainly pine (89 per cent). Pairs using natural nests showed a higher variance in productivity than pairs nesting on artificial supports. Furthermore, although mean productivity (young per occupied nest) was similar in natural (1.87 ±0.026) and artificial (1.98 ±0.051) nesting sites, the productivity recorded in natural nests decreased over the study period, whereas it did not vary for Ospreys nesting on artificial structures.

FINLAND

As experiences in eastern Germany show, the installation of artificial nests is particularly important if natural sites are lacking. Scandinavia is a stronghold of Ospreys in Europe and in Finland there are an estimated 1,300 breeding pairs (Saurola 2011). Artificial nests have been used to great effect since 1965 to counter the negative impact of commercial forestry in the country, and account for approximately half of the nests nationally (Saurola 2005). Osprey nests are large structures that require substantial support, but the fast-growing species favoured in commercial forests are rarely strong enough. Even if Ospreys are successful in constructing a nest in the flimsy tops, the structure increases in size with repeated use and this often results in the tops of trees breaking under the weight, particularly in winter when nests are covered in snow. Modern forestry has therefore drastically reduced the number of high-quality natural nest-sites for Ospreys. In Pertti Saurola's own study area, 160 artificial nests have been constructed and account for more than 90 per cent of active nests (Saurola 2005). Saurola (1997) estimated that without this intervention the population would have been reduced by 50 per cent in this area. Instead, the

Our relationship with the Osprey

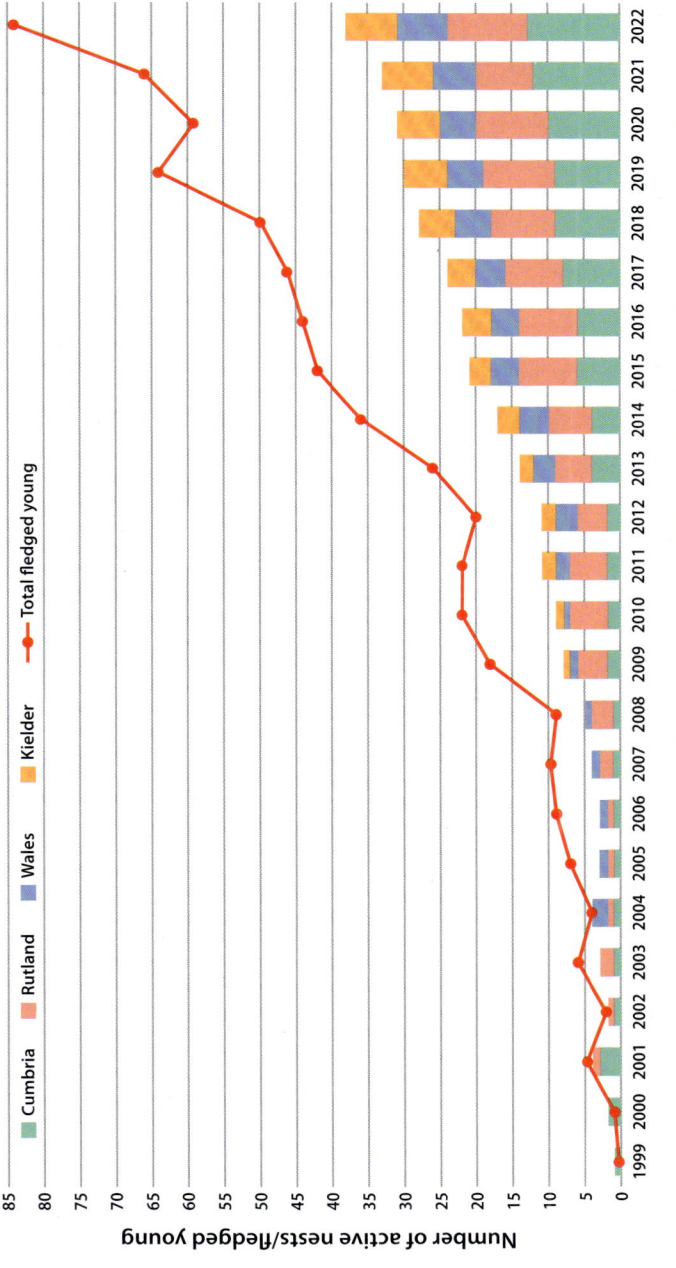

Figure 8.14. The population expansion of Ospreys in England and Wales, 1999–2022. Cumbria and Rutland totals include neighbouring counties (Tim Mackrill).

Finnish population has either remained stable or, in some cases, increased in recent decades (Schmidt-Rothmund *et al.* 2014, Saurola 2021).

UNITED STATES AND CANADA
The Osprey is well known for its ability to utilise a range of anthropogenic structures for nesting in North America, and the recovery of the species during the post-DDT era is closely linked to its ability to adapt to human-dominated landscapes. Artificial nests have been used with great success in many areas as a result (Figure 8.16).

The Osprey population on the Willamette River in western Oregon in the Pacific Northwest of the United States increased from an estimated 13 pairs in 1976 to 78 pairs in 1993 (Henny and Kaiser 1996). Historically, the removal of suitable trees due to intensive farming appeared to limit the number of Ospreys nesting along the river, but once breeding pairs began utilising electricity poles – primarily those in fields relating to farmers' irrigation pumps – the carrying capacity of the area increased significantly. The number of pairs nesting in trees remained constant during the period (13 pairs in 1976 and 12 pairs in 1993), but numbers using poles or transmission towers, first observed in 1977, increased dramatically to 66 pairs in 1993. By 1993 over half (32) of 58 nests on poles had been modified by the power companies to accommodate Ospreys – either by building a platform above the crossarms or through the provision of new poles with an artificial nest nearby. Henny and Kaiser (1996) suggest that the population may have undergone nest-type imprinting, whereby the use of electricity poles by a few pioneering individuals spread through the population. The construction of artificial nests or modification of existing structures to accommodate Ospreys has been similarly successful in other parts of North America (Washburn *et al.* 2014).

Figure 8.15. An estimated 75 per cent of the German Osprey population now nest on pylons, many using artificial nests installed by German power companies together with conservationists. Here, Dr Daniel Schmidt-Rothmund, who has led efforts to conserve and monitor Ospreys in Germany, is visiting a nest to ring a brood of young (© Marie Christine Rothmund).

In southern New England and Long Island the Osprey population increased from 107 breeding pairs in 1975 to 1,274 in 2010, with artificial nests widely used, as described by Bierregaard *et al.* (2014). The provision of artificial nests has been particularly successful in Connecticut (population increase: 10 to 235 pairs, with 87.5 per cent on artificial nests in 2010), at Westport River (14 to 66 pairs, 100 per cent on artificial nests), Nantucket (0 to 14 pairs, 100 per cent on artificial nests) and Martha's Vineyard (3 to 74 pairs, 81 per cent on artificial nests). Platforms accounted for a lower proportion of nests at Rhode Island (47.6 per cent) where the population increased from 9 to 115 pairs during the period. Here a range of artificial structures were utilised as well, with 15.2 per cent of nests on utility poles, 15.2 per cent on light towers and 17.2 per cent on mobile phone/communications masts. A larger proportion of tree nests were

Figure 8.16. Artificial nests, such as this one at Cape Cod, Massachusetts, have facilitated the recovery of Ospreys in the United States during the post-DDT era (© Steve Egerton-Read).

used in Long Island (29 per cent of 189 breeding pairs in 1990), but even here, over 80 per cent of nests were on platforms by 2000, by which time the breeding population had increased to 303 pairs. Overall, 72 per cent of breeding pairs across the region used artificial nests, with the remainder on utility poles (8.1 per cent), light towers (3.5 per cent), mobile phone masts (6.5 per cent), trees (4.2 per cent), the ground (0.2 per cent) and other structures (5.4 per cent). Clearly, these other substrates would have been utilised to a higher degree were it not for the provision of the platforms, but the fact that such a significant proportion of the population utilises them shows the effectiveness of the technique. It also means that populations in some areas far exceed historical levels. In Martha's Vineyard, for example, the 2016 population of 90 pairs exceeded historical population levels tenfold (Bierregaard et al. 2020).

The effectiveness of artificial nests has also been demonstrated in the Canadian Great Lakes region. A detailed study by Ewins (1996) revealed that 35.7 per cent of nests monitored between 1988 and 1993 were on artificial platforms, but the most striking feature was the rate at which newly installed nests were occupied by Ospreys. Ewins reported that 18 of 22 platforms used by Ospreys at Georgian Bay, Lake Huron, were occupied within a year of installation. In one case, a platform erected in May 1992 attracted Ospreys within three days, and the birds were incubating soon afterwards. Similarly, on the St Lawrence River five platforms were installed on rocky islets in autumn 1992, and Ospreys reared chicks in three of them the following year. Ewins (1996) suggests that the fact that nests were occupied so quickly after installation indicates a lack of natural sites in some parts of the Great Lakes, due to timber extraction and urban development.

AUSTRALIA

The increasing use of artificial structures, including artificial platforms, has also been observed in Australia, particularly in urban areas (Dennis and Clancy 2014). In southeastern Queensland, for instance, three-quarters of 20 Osprey nests in Moreton Bay Marine Park were located on artificial structures, three of which were on artificial platforms (Thomson et al. 2019).

Translocations

The banning of harmful organochlorine pesticides, coupled with reduced human persecution led to the recovery of Osprey populations in many areas during the second half of the twentieth century. However, the species' own biology, and specifically a high degree of natal philopatry, particularly among males, means that it is very slow to return to areas where it has been completely eradicated by anthropogenic actions. Artificial nests help to a certain extent, but they are most successful when sited near or within existing colonies. A more proactive approach has been used to restore Ospreys to more isolated parts of their historical range.

Translocation, or hacking, involves moving chicks from well-established, self-sustaining populations to areas, within the historical range, where they are absent. Evidence suggests that juvenile Ospreys imprint on their natal area during the post-fledging phase, and so translocating chicks at five–six weeks, just before they fledge, ensures that when they return – usually in their third calendar year – they are drawn back to the translocation site, rather than their natal nest. This makes it possible to speed up the recovery of the species across its former range.

UNITED STATES

The translocation, or hacking, of juvenile Ospreys was pioneered in the United States, and has since become a widely used conservation management tool on both sides of the Atlantic Ocean. The first translocation projects were initiated in the Tennessee Valley in 1979, northeastern Pennsylvania and New York the following year, and in central Minnesota in 1984 (Poole 1989). Over the years we have learnt that Osprey translocation projects require patience, but the first signs of success were evident by the mid-1980s.

TENNESSEE VALLEY

The first translocation project to get underway was in the Tennessee Valley, where the construction of reservoirs in the 1930s and 1940s greatly increased the suitability of the area for Ospreys. As Beddow (1990) recounts, the first published report of nesting at one of the reservoirs came from Watts Bar Reservoir in 1961 and there were subsequent breeding attempts at several other reservoirs over the next decade, but they were rarely successful during the DDT era. Active management of the population began in 1976, when a long-established but rarely successful nest on a navigation marker at Euchee, Watts Bar Reservoir, was removed and a platform installed above the light on the marker. Ospreys successfully reared three chicks in the nest the following summer, and encouraged by the results, additional artificial nest building was initiated by the Tennessee Wildlife Resources Agency (TWRA) and Tennessee Valley Authority (TVA). Furthermore, when the nest at Euchee was found to have two infertile eggs in 1978, two nestlings from Chesapeake Bay were flown to Tennessee and placed in the nest. Both were accepted by the breeding birds and went on to fledge successfully.

The successful fostering of the two nestlings at Euchee led to an even more proactive approach to the conservation of Ospreys in the area. In 1979, TVA and TWRA gained permission to translocate more young Ospreys from Chesapeake Bay, but this time as part of a longer-term hacking project. A total of 165 Osprey chicks were subsequently translocated and released at a total of 16 different sites (Beddow 1990). This was a success, and the Tennessee Osprey population increased from three nests in 1980 to 131 nests in 1999, the last year that an annual survey was carried out (TWRA 2022).

PENNSYLVANIA

As the first young Ospreys were being translocated to Tennessee, a second project was also getting underway in the Pocono Mountains in northeast Pennsylvania, where several large rivers, glacial lakes and reservoirs provide excellent foraging habitat for Ospreys. The thriving population around Chesapeake Bay again provided donor chicks, with a total of 111 nestlings translocated between 1980 and 1986, as described by Rymon (1989). A key objective of the early translocation projects was to keep the hacking process as natural as possible, and to keep human contact to a minimum. In Pennsylvania 10 hacking towers, each consisting of an artificial nest built on a 2 × 2m platform, 7m above the ground, were built at four locations over a 25km radius. The nest platforms were enclosed with preassembled sides and roof panels to ensure the young birds could not leave before they were ready, and to prevent predator attacks. The front panel was fitted with strap hinges and rope pulleys so that it could be lowered to a horizontal position when the young birds were ready to be released.

Nestlings were translocated at approximately six weeks of age, although this varied from four–seven weeks. They were initially fed finger-sized pieces of fish placed on the end of a stick and offered to the young through a hole in a blind that prevented the young Ospreys seeing the person providing the food, with all observations made through a two-way mirror to reduce the risk of imprinting. At seven weeks, the young Ospreys were able to stand on pieces of fish in order to grip them and tear off morsels to feed themselves. They were then released at approximately eight weeks. Most made short flights to trees or artificial structures nearby. Over 100 of the translocated birds were successfully released in this way. Food continued to be placed on the hacking towers through the post-fledging period, to replicate the natural provision of juvenile Ospreys, until they departed on migration.

The project reached a significant milestone in 1984 when three two-year-old birds, identified by colour rings, were observed back at their respective hacking areas for the first time. By 1986 a total of 23 translocated birds had returned, confirming the notion that the young Ospreys regarded the hacking sites as their natal site. The first signs of breeding activity were recorded in 1985, when four three-year-old males built nests. One of them, which occupied a nest pole near one of the hacking sites, fostered up to nine of the translocated young that visited his nest after release. He readily provisioned fish for the food-begging juveniles, delivering one fish every 10 minutes at the peak of foraging activity. Next year the same male paired with an unringed female on another nest pole, situated 750m away, and successfully reared a single chick. Meanwhile, a second pair of translocated birds fledged three chicks on a nest pole situated just 200m from the hacking site they had been released from four years previously. A further four pairs of translocated birds occupied nests the same year (Rymon 1989). These early results were encouraging because they demonstrated that translocation could be used to restore populations of Ospreys.

MINNESOTA

The early results from Pennsylvania, and also Tennessee, indicated that the Osprey responded well to this innovative management technique and this encouraged others to follow suit. As Martell *et al.* (2002) explain, Ospreys historically nested in east-central Minnesota, which now includes the Minneapolis-St Paul (Twin Cities) urban area. This population was eradicated by persecution and loss of nesting sites by 1900 and Ospreys remained confined to the northern part of the state throughout the twentieth century, despite an abundance of suitable habitat further south. In view of this, and encouraged by work elsewhere, 143 nestlings were translocated from northern Minnesota and released at eight sites in the Twin Cities area between 1984 and 1995, with more than 30 nest platforms also erected (Martell *et al.* 2002).

The first nesting attempt was recorded as early as 1986, when a two-year-old translocated male attracted two unringed females to a nest 3km from its release site. A total of six eggs were laid but unsurprisingly, given the age of the male and the fact that two females were involved in the breeding attempt, none hatched (Englund and Greene 2008). The following year, when eggs again failed to hatch, a translocated chick was placed in the nest and was accepted by the adult birds, who successfully reared it to fledging (Martell *et al.* 2002). From this single pair in 1987, the population slowly increased as more translocated birds returned to breed. Initial population growth was relatively slow, but increased significantly as the number of breeding pairs increased. A total of 194 chicks had fledged from nests in the area by 2000, with a mean productivity of 1.57 young per occupied nest – and this figure jumped to 472 fledged young from 306 occupied nests by 2005 (Englund and Greene 2008). The importance of conspecific attraction in the establishment of Osprey populations was demonstrated by the fact that the breeding population was swelled by immigrants from other areas, with 25 per cent of nesting attempts by birds not from the local population (Martell *et al.* 2002). It is highly unlikely that these individuals would have settled to breed in the absence of other Ospreys. There were also clear sex-based difference in natal dispersal of both translocated and wild-fledged birds. The mean natal dispersal of males was 27km (SE = 5.5, n = 20, range = 1–65km), whereas the corresponding figure for females was 384km (SE = 146, n = 7, range = 8–1,075km). Five females – four translocated and one wild-fledged – dispersed to other populations in Iowa, Michigan, Illinois, Ohio and northern Minnesota (Martell *et al.* 2002).

Artificial nests were readily accepted by young Ospreys returning to the Twin Cities area (Figure 8.17), and all but three sites where breeding was attempted between 1986 and 2000 were on platforms erected specifically for the Ospreys. The only successful nest elsewhere was on the top of a water tower. It is also notable that five sites accounted for 48 per cent of fledged young during the period of initial population establishment between 1986 and 2000. Four of these

Figure 8.17. Ospreys returned to the Twin Cities area in the United States due to a highly successful translocation project. By 2019 there were 90 occupied nests, with many on artificial platforms erected for Ospreys (© Amber Burnette).

sites had been occupied since at least 1992, with only limited turnover of known breeding birds (Martell *et al.* 2002). This demonstrates the importance of a small number of highly productive individuals in the initial stages of population establishment (Poole 1989).

In recent years, as the population has increased, a range of other artificial structures have been used by breeding pairs. In 2015, 39 per cent of active nests were on structures not intended for Ospreys, such as water towers, cellphone towers and light structures (Petersen *et al.* 2020). By 2019, 35 years after the first translocated chicks were released, there were a total of 148 nest-sites in the Twin Cities area, of which 90 were occupied. It is also notable that the efforts in the Twin Cities area inspired releases in the neighbouring states of Iowa and South Dakota, as well as Missouri, using birds from the same source population in northern Minnesota (M. Martell pers. comm. 2022).

BUILDING ON SUCCESS

The early projects in the United States established the principle that translocated Ospreys regarded the release site, rather than the nest where they hatched, as their natal site. It was encouraging that nestlings appeared to cope well with the rigours of translocation, readily fed on fish provided at the hacking site, and were able to be released effectively. The propensity of males, in particular, to return to release sites was in keeping with previous research on the natal philopatry of wild-fledged birds, and indicated that translocation had the potential to dramatically speed up the recovery of Ospreys in the post-DDT era. Experiences in Minnesota and elsewhere suggested that biasing translocation in favour of males was advisable given that they were more likely to return, and then attract wandering females from other populations. This would also enhance the genetic variability of establishing populations.

Translocation has proved to be a highly effective conservation tool in the United States and by 2014 a total of 1,851 Osprey chicks had been hacked in 20 states, greatly increasing the species' range in the post-DDT era (Simnor 2015).

United Kingdom

The pioneering translocation work undertaken in the United States began to generate interest in Europe and in April 1996 Roy Dennis and Tim Appleton travelled to Minnesota to meet Mark Martell and the team behind the translocation of Ospreys to Minneapolis and St Paul, and to learn about the hacking process. They spent a week learning about the work in Minnesota, and how it might be replicated to restore Ospreys in Europe, and specifically at Rutland Water in the English East Midlands.

RUTLAND WATER

By 1995 the Scottish population had reached 92 breeding pairs, and an increasing number of migrant Ospreys were being seen further south. Rutland Water was constructed in the 1970s and, with a surface area of over 1,000 hectares, it immediately attracted passage Ospreys in spring and autumn. Fifteen birds were recorded in 1990 and then 20 in 1992, but most individuals did not linger for more than a few days (Mackrill *et al.* 2013). However, in 1994 a pair were present between mid-May and early September, raising hopes that they may return to breed. Five artificial nests were constructed during the winter and, although the female did reappear the following spring, the male was not seen again. Neither bird was ringed, but it is highly likely that they were both two-year-olds, summering south of their natal area. The urge to continue further north in spring 1995 would have been stronger, and likely explains why the male did not return.

Having pioneered the translocation of Red Kites and White-tailed Eagles in the UK, Roy Dennis was adamant that it was the best means of restoring Ospreys to their former range in the UK and mainland Europe. Rutland Water, being centrally located and with abundant fish supplies, was considered a highly suitable location, and with the support of the reservoir's owners Anglian Water, and the Leicestershire and Rutland Wildlife Trust who managed the nature reserve under Tim Appleton's guidance, an application was made to translocate Scottish Ospreys to the site using the techniques implemented successfully in North America. It was estimated that the reservoir could support a population of 8–10 pairs and that, once established, this population would expand to other waterbodies in the region (Mackrill *et al.* 2013).

A licence was granted by Scottish Natural Heritage (now NatureScot) in June 1996 and just over a month later the first cohort of young Ospreys were driven south from the Scottish Highlands by Roy Dennis and the newly appointed project officer, Helen McIntyre. Over the course of six summers between 1996 and 2001, a total of 64 juvenile ospreys, aged five–six weeks, were collected by Roy and colleagues from nests in Highland and Moray and transported by road to Rutland Water. The sex ratio was biased heavily in favour of males, following advice from the United States, with 49 males and 15 females translocated. The initial licence permitted the collection of the smallest bird from a brood of two or three only, but it was soon apparent that these individuals were susceptible to disease and infection, and some required hand-feeding. Four of the eight birds translocated during the first year died either pre- or post-release, with a *Salmonella* infection thought to have contributed to the death of each individual. In later years the licence was amended so that larger chicks, with a wing length in excess of 320mm, could be collected. This was far more successful and of the 56 birds translocated between 1997 and 2001, 55 set off on migration. A further 11 birds were translocated and released in 2005, with 10 departing on migration a few weeks later (Mackrill *et al.* 2013).

Upon arrival at Rutland Water, the birds were kept in release pens based on the design used in Minnesota. They were initially sited adjacent to two lagoons on the nature reserve, but after the first year they were moved to an elevated position on the edge of a wooded hillside, Lax Hill, with an open field in front, overlooking the wetlands. The birds were held in groups of two or three to replicate conditions in a natural nest, and fed locally sourced Rainbow Trout three times a day (Figure 8.18). The larger chicks, translocated after the first year, did not require hand-feeding and so fish was placed in the pens through a flap in a solid rear wall to keep human contact to a minimum. Each compartment measured 1.8 × 1.2 × 1.2m, with an artificial nest at the rear and a T perch at the front. Solid side walls ensured that birds could not see others in adjoining pens.

The development of the nestlings was monitored from a distance using telescopes and CCTV images, and this enabled the project team to determine when individual birds were ready for release. The birds were kept in the pens for approximately a week later than usual fledging age and during this time they favoured the perch at the front of the pens, making short hops and flights to and from the nest. Prior to release, they were fitted with small tail-mounted radio transmitters to track local movements, and 14 of the 64 birds were also equipped with satellite transmitters.

Release day is one of the major milestones in any translocation project and the optimum methodology was refined over the years at Rutland Water. A key requirement was to allow each bird to leave the pen at their own volition. To achieve this, the front door was lowered on ropes or thin cord, from the rear, so that birds did not have any human contact (Figure 8.19). This was usually done soon after dawn to give the young Ospreys the maximum amount of daylight in which to depart the pens. Some left within minutes, while others were more hesitant and on some rare occasions it was necessary to close the front door for the night and reopen it next

Figure 8.18. Translocated Ospreys in the release pens at Rutland Water. The juveniles were fed locally sourced trout 2–3 times per day (© Leicestershire and Rutland Wildlife Trust).

morning. Over the years it became apparent that a light wind was useful in encouraging the birds to take to the air for the first time. Some made short, simple flights to a nearby perch, while others flew much further. One individual even landed on the spire of the local church, 3.8km from the release pens.

The post-release feeding regime replicated the North American projects, with food placed on top of the pens, and on nearby artificial nests two or three times each day. A predictable pattern of dispersal occurred during the period when the birds were first flying. Most remained within close proximity (<1km) of the release pens for the first seven to ten days, before dispersing further. Some birds made longer exploratory flights thereafter, up to a maximum distance of 21.6km from the release pens, but the satellite data indicated that they predominantly remained much closer, with 85 per cent of accurate satellite fixes (n = 101) within 4km of the release site. The juveniles were frequently observed making dives into the water, but very rarely caught their own fish, and remained dependent on fish provided by the project team until they departed on migration.

In the days prior to departure, the juveniles were often absent from the release site during the day, but were present either at first light or in the evening to feed. They typically departed during late August and early September, four–six weeks after release, as shown in Table 8.2. In some years the birds departed over a relatively short period, but there were 31 days before the earliest and last departure in 1999.

The first signs of success at Rutland Water occurred in late spring 1999 when two translocated males returned to the area. 08(97) was identified in Manton Bay on 29th May and then, on 14th June, a second bird, 03(97), was seen nearby. The next summer, two further translocated males, released in 1998, made it back to Rutland Water.

From 2000 onwards, 08(97) held territory in Manton Bay, 1km from the release site. This

Figure 8.19. On release day, the front of the release pens was lowered using baling twine so that the birds could leave at their own volition. Some took to the air immediately, while others waited much longer (© Leicestershire and Rutland Wildlife Trust).

Table 8.2. Migration dates of translocated Ospreys at Rutland Water 1996–2001

Year	No. of birds	Migration dates (first–last)	Mean days since release
1996	4	28/08–09/09	30
1997	8	30/08–07/09	38
1998	12	26/08–12/09	33
1999	11	22/08–21/09	34
2000	12	22/08–12/09	30
2001	12	23/08–03/09	28

bird often exhibited very aggressive behaviour towards the translocated juveniles, repeatedly divebombing birds perched on an artificial nest in Lax Hill wood adjacent to the release pens, while also preventing them from using perches in Manton Bay, which had been favoured in previous years. In contrast, an unringed female present in the Rutland Water area during 2005 – when a further 11 juveniles were released – delivered fish to food-begging juveniles at the release site (Figure 8.20). This behaviour was first observed on 31st July that year, and 18 'food drops' were recorded, although more were suspected. In each case, the female would land in an

Figure 8.20. An unringed female regularly delivered fish to food-begging translocated juveniles at Rutland Water in 2005 (© John Wright).

artificial nest with a fish, drop it, and then fly off almost immediately. This was similar to behaviour observed in Pennsylvania (Rymon 1989).

Population establishment

The project reached a major milestone in 2001, when the first pair bred successfully. That spring, 03(97) returned to a nest he had built in the top of an Oak tree near Rutland Water the previous summer. He arrived on 28th March, some 19 days earlier than the previous year, clearly intent on breeding. On 3rd April he was joined by an unringed female – possibly an old bird ousted from the Scottish population given a deformity in one of her eyes. Incubation began later that month and the first signs of hatching were evident on 6th June. This was a significant moment for the project and for wider Osprey conservation in the UK and, like the early successes at Loch Garten, was marked with great excitement. I vividly remember the moment I observed the downy head of a chick for the first time as the female delicately offered tiny morsels of fish into the nest-cup; it was a thrilling sight. The single chick went on to fledge successfully and that summer two pairs of Ospreys also bred successfully in the Lake District, while 153 pairs reared a total of 200 chicks in Scotland (Dennis 2008). It was clear that, after the drastic declines of the nineteenth and early twentieth centuries, the species was recovering well, aided by proactive conservation measures.

Since the success of 2001, the Osprey population in Rutland has become well established. The growth of the colony was hampered in the early years by the fact that only a single immigrant female joined the population between 2001 and 2008, despite the presence of unpaired males on nests. It soon became apparent that some females, many established breeders further north, would join a male at one of these nests for a few days, readily accept fish, but then depart again – using the presence of the unpaired males as a tactic to get an easy meal. In hindsight, releasing a more equal sex ratio may have resulted in the population becoming established quicker, particularly as two translocated females bred successfully for the first time in 2003. One of them, 06(01), became the first known two-year-old to rear chicks at a nest in the UK. In view of this, an additional cohort of nine females and two males were released in 2005 in an effort to try and address the initial gender imbalance in the population.

Overall, a total of 13 translocated birds (i.e. 17.3 per cent of total translocated, and 18.6 per cent known to have migrated) returned to the UK and six bred successfully in the Rutland Water area. Unexpectedly, two translocated males dispersed to Wales and re-established a breeding population of Ospreys there, as described in chapter 3 (also see below). Three females are known to have survived and bred successfully, including one that bred in Aberdeenshire, Scotland (Table 8.3).

All of the initial nests in the Rutland Water area were established within 15km of the release site, and this is reflected in the limited natal dispersal of translocated birds that returned to breed locally (Table 8.4). The bias in favour of translocating males, and reluctance of passing migrant females to join the Rutland population, meant that competition for mates among returning males was fierce, resulting in a high age at first breeding among some translocated males. 08(97) did not breed until he was 10 years old and 09(98) had to wait even longer, finally breeding for the first time at 14 years of age. By this age 03(97) had already reared a total of 23 young.

The return of the first wild-fledged birds gave the population a much-needed demographic boost. In 2007, a wild-fledged Rutland female, 5N(04), bred successfully for the first time (Figure 8.21), and then in 2013 one of her offspring, 11(10), reared three chicks with another locally fledged bird, 25(10) – they were the first pair of second-generation Rutland Ospreys to breed. It was originally predicted that the reservoir and surrounding area could support 8–10 pairs of breeding Ospreys, and this milestone was reached in 2015 when eight breeding pairs reared a total of 15 young. The year 2015 was also notable for the fact that it was the last time that a translocated bird bred; all subsequent breeding pairs were comprised of wild-fledged Rutland birds, or immigrants from other populations, thereby demonstrating that the population had

Table 8.3. Known survival of Ospreys from Rutland Water to >2 years (chicks fledged up to and including 2019)

					Breeding locations of returning birds			
Origin	Sex	Total	Total returned	Percentage return	Rutland area	Wales	Scotland	Other
Trans.	M	51	10	19.6	6	2	0	
Trans.	F	24	3	12.5	2	0	1	
Wild-fledged*	M	82	30	36.6	19			1
Wild-fledged*	F	89	16	18.0	8*	5*		3

*3J(12) attempted to breed in Rutland before later moving to Wales, and so is counted in the Welsh column, where she bred successfully

Our relationship with the Osprey

Figure 8.21. Female 5N(04) was the first wild-fledged chick to breed, in 2007. By 2021 she had raised a total of 33 chicks, making her the most prolific breeder in the population (© John Wright).

reached a self-sustaining level (see Appendix 1). By 2022, 21 years after 03(97) and the unringed female had raised a single chick for the first time, a total of 231 chicks had fledged from 113 active nests, giving an overall productivity figure of 2.04 young per active nest, which exceeds many other populations across the species' range.

As shown in Table 8.3, the known survival of wild-fledged males to two years during the initial stage of population establishment was 36.6 per cent, and this was significantly higher than the corresponding figure of 19.6 per cent for translocated males. Significantly, all except of one of these individuals returned to the Rutland Water area, with the exception being S2(15) who bred in the Netherlands for the first time in 2020. The known survival of wild-fledged females to two years was 18.0 per cent which, again, was higher than the figure for translocated females (12.5 per cent). Eight females that fledged between 2001 and 2019 had returned to breed successfully in the Rutland Water area by 2022, five dispersed to Wales, a single female, CJ7(15),

Table 8.4. Age at first breeding and natal dispersal of translocated and wild-fledged Rutland Ospreys (2001–2022)

			Age at first breeding (years)			Natal dispersal (km)		
Origin	Sex	N	Min	Max	Mean	Min	Max	Mean
Translocated	Male	8	4	14	8	1	232	54 (±84)
Translocated	Female	3	2	3	3	4	456	155 (±213)
Wild-fledged	Male	20	3	9	5.3	3	387	31 (±94)
Wild-fledged	Female	16	3	6	3.8	3	349	117 (±122)

became the first to breed on the south coast of England for 200 years (see below), while the remaining two birds bred in northern England and the Scottish Borders respectively. The fact that females readily join other populations makes it possible that other birds have returned, and gone undetected; hence, the lower (apparent) survival figure compared to males.

The age at first breeding among wild-fledged Rutland males is lower than the figure for translocated males (see Table 8.4), but the high return rate means that there is stiff competition for mates and nest-sites. Females have readily nested at three years of age when the opportunity has arisen, and this is reflected in a lower average age at first breeding. Perhaps the most interesting individual is 3J(12), who attempted to breed at Rutland Water for the first time in 2015 as an inexperienced three-year-old. However, when the nest was taken over by Egyptian Geese she subsequently moved to Wales, as described in chapter 3.

Highly productive individuals
A small number of highly productive individuals played a key role in establishing Ospreys in the Rutland Water area, as previously observed in Minnesota. Translocated male 03(97) was particularly important in the early years, rearing a total of 32 chicks at Site B between 2001 and 2015 (Figure 8.22), giving an annual productivity figure of 2.13 young. The importance of this one bird is exemplified by the fact that only six of 49 translocated males bred successfully, rearing a total of 53 young. Thus, 03(97) contributed 60% of chicks reared by translocated males. In addition, 40.6 per cent of his offspring returned to the UK in future years, a figure considerably higher than for young fledged from other nests during the same period (27.1 per cent).

The reluctance of Scottish females to join the breeding population in the early years meant that translocated female 05(00) also played a key role in population establishment. 05(00) raised a total of 17 chicks with 03(97) at Site B between 2003 and 2008, giving a very high annual productivity of 2.83 young. Of these offspring, 41.2 per cent subsequently returned in future years, and this was crucial because only four chicks fledged from other nests during this period, and none returned. Furthermore, only one other translocated female bred at Rutland Water: 06(01), for a single year in 2002.

One of the chicks that 03(97) and 05(00) reared at Site B later superseded her father as the most successful individual in the first 21 years of population establishment. 5N(04) fledged from Site B in 2004, and bred for the first time in Manton Bay in 2007 (see Figure 8.21). She subsequently moved to an artificial nest on private land, known as Site N. She reared a total of 36 chicks between 2007 and 2022, with an annual productivity of 2.25 young. Of her offspring from 2007 to 2019, 22.2 per cent are known to have returned. Moreover, her annual productivity showed no signs of decreasing with age. She successfully reared a brood of four in 2019, and three chicks in 2018, 2020, 2021 and 2022. Sadly, she failed to return in spring 2023.

Although immigrant females were initially reluctant to join the breeding population, there was a notable shift from 2009 onwards. That year two new females joined: a metal-ringed female from Argyll and an unringed bird, most likely of Scottish origin. Then, in 2010, another unringed female, who became known as Maya, settled to breed in Manton Bay. The Argyll female had reared a total of 29 chicks by 2022, with an annual productivity of 2.07. The return rate of her young fledged up to 2019 was 20.0 per cent. Maya's total stood at 34 chicks by the end of 2022, a figure that included two broods of four, and with an annual productivity of 2.62 young. The return rates of Maya's young (fledged up to 2019) were exceptionally high, at 44.0%. These data are shown in Table 8.5.

The clear sex-based differences in natal philopatry are evident when the number of young produced by translocated, wild-fledged and Scottish birds between 2001 and 2022 are analysed. A total of 95 young were reared by Scottish/unringed females (assumed to be from other

Our relationship with the Osprey

Figure 8.22. Translocated male 03(97) reared a total of 32 chicks at Site B between 2001 and 2015 with three different females. Here he can be seen flying back to the nest with a clump of nest-lining material (© John Wright).

populations, as all offspring were ringed). This accounts for 41.1 per cent of fledged young during the period, and far exceeds the 19 young reared by translocated females (8.2 per cent). It also compares favourably with the 117 young produced by returning wild-fledged Rutland females (50.7 per cent of all young) (Figure 8.23a). In contrast, 178 chicks, thus 75 per cent of the total, were fathered by wild-fledged Rutland males, with the remaining birds the offspring of translocated males (Figure 8.23b). The first immigrant male to appear at a nest was in 2022 when a breeding male, 8F(12), disappeared due to an unknown cause, and his two chicks were fostered by an unringed male.

Dispersal to Wales

The success of the Rutland Water translocation has not been restricted to central England. In 2004, two pairs of Ospreys were found breeding in Wales and, unexpectedly, the males at both sites were identified as translocated individuals, 07(97) and 11(98) (Mackrill *et al.* 2013, Evans 2014). Neither had been seen back at Rutland Water since their first migration, but it was significant that they were nesting at almost the same latitude, suggesting that the translocation was a key factor in them settling in Wales.

Table 8.5. Breeding performance of the most productive birds during the first 21 years of population establishment in the Rutland Water area (2001–2022)

Bird	Sex	Origin	Years breeding	Young produced	Annual productivity	Percentage returned
5N(04)	F	Rutland wild-fledged	2007–2022 (MB*, Site N)	36	2.25	22.2%
Maya (unringed)	F	Unringed (Scotland?)	2010–2022 (MB)	34	2.62	44.0%
03(97)	M	Translocated	2001–2015 (Site B)	32	2.13	40.6%
Metal-ringed female	F	Argyll, Scotland	2009–2022 (Site O)	27	2.07	20.0%
33(11)	M	Rutland wild-fledged	2015–2022 (MB)	23	2.88	42.9%
30(05)	F	Rutland wild-fledged	2009-2022 (Site K and Site S)	22	1.69	12.5%
05(00)	F	Translocated	2003–2008 (Site B)	17	2.83	41.2%

*5N(04) bred at Site N for two years (2007–2008) before moving to Site N.
Breeding sites shown in brackets; MB = Manton Bay; 'Percentage returned' relates to chicks fledged up to and including 2019 only.

The Osprey

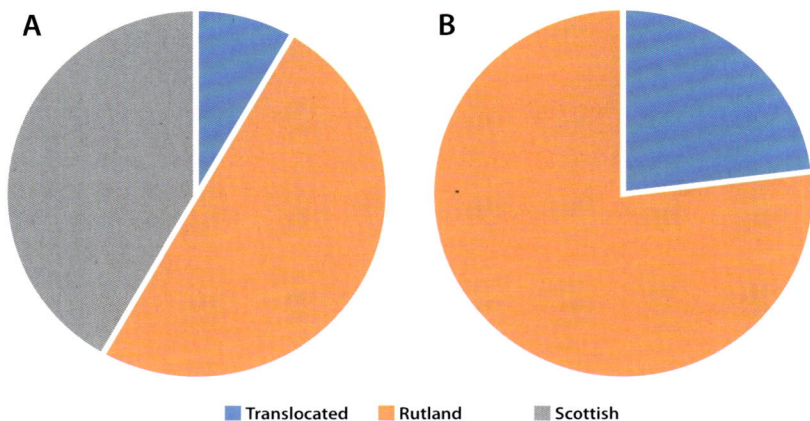

Figure 8.23. The number of offspring produced by translocated, wild-fledged (in Rutland area) and Scottish (including unringed) females (A) and males (B) in the Rutland population between 2001 and 2022 (Tim Mackrill).

11(98) was breeding in the Glaslyn Valley in Snowdonia with an unringed female but, as described earlier, the nest collapsed during heavy rain and high winds, killing the two chicks. Further south, 07(97) had established a nest near Welshpool in Montgomeryshire and partnered with a colour-ringed female from the Black Isle, north of Inverness in Scotland. They raised a single chick, the first to fledge from a nest in Wales for several centuries.

Unfortunately, the Montgomeryshire pair failed to appear in 2005, but 11(98) and his unringed mate returned to their nest, which had been rebuilt and made secure by conservationists during the winter, and raised two chicks (Evans 2014). The two birds continued to breed together until 2015, when the male failed to return from migration. During this period they reared a total of 26 young to fledging, with an annual productivity of 2.17. The male was replaced by an unringed male (thus of unknown origin) and together they raised a total of 18 chicks between 2015 and 2022, giving a high annual productivity of 2.25.

The Welsh Osprey population has slowly increased in number since 2004, with a minimum of six active nests by 2022. Although no other Rutland Water translocated birds joined the population, five wild-fledged Rutland females bred in Wales between 2011 and 2022. During this period they produced a total of 39 young, as described in chapter 3. Overall Rutland Water birds, either translocated or wild-fledged, reared a total of 66 chicks, which accounts for 50.4 per cent of the 131 young that fledged from known nests between 2004 and 2022, as shown in Table 8.6, thereby demonstrating the value of the Rutland Water translocation in the re-establishment of Ospreys in Wales. Given the usual natal philopatry shown by males, the number of young produced by unringed males indicates that there may have been other nests, out of the public domain, that have produced young during the period of population re-establishment.

POOLE HARBOUR
Breeding Ospreys were once widespread along the south coast of England, where the old English name for the species is mullet hawk (Dennis 2008). Following the species' recovery in Scotland and re-establishment in central England and Wales, passage Ospreys became increasingly frequent visitors during spring and autumn migration, but despite the provision of artificial nests in many areas, there had been no breeding attempts by 2017. With the population centred on Rutland Water well established and considered self-sustaining, a licence was granted for a

Table 8.6. Number of young produced by origin of breeding birds in the Welsh population 2004–2022

Origin of breeding birds	Number of breeding males	Number of young	Number of breeding females	Number of young
Translocated (Rutland Water)	2	27	0	0
Unringed	6	95	3	85
Scottish (colour-ringed)	2	5	2	3
Welsh (colour-ringed)	2	4	0	0
Rutland wild-fledged	0	0	5	39
Translocated (Poole Harbour)	0	0	2	4

second English translocation at Poole Harbour in Dorset. This large natural harbour, with abundant populations of Grey Mullet, European Bass *Dicentrarchus labrax* and other estuarine fish, is ideally located to act as the nucleus of a new south-coast colony, and will provide a link between expanding populations in central England, Wales and central France. The first eight juveniles were translocated to Dorset from nests in northern Scotland during July 2017.

The importance of conspecific attraction in Ospreys has been exemplified at Poole Harbour by a female from Rutland Water. On 8th August 2017, I was monitoring translocated juveniles soon after they had been released, when an adult female joined them. She alighted on one of the artificial T perches favoured by the young birds, and I was able to read the colour ring – CJ7 – which showed she was a two-year-old female that had fledged from a nest near Rutland Water in 2015. This was the first sighting of her since her first autumn migration, and she remained in the vicinity for more than a month, often interacting with the young birds. Associating with the translocated juveniles in this way appeared to encourage CJ7(15) to return the following spring. She was first observed at Poole Harbour during May, and then remained in the area for the rest of the summer. She again associated with translocated juveniles after their release in in early August – although, unlike females observed at other sites, did not attempt to feed them.

The following spring, she arrived earlier and paired with LS7(17), the first translocated Osprey to return to the area. They occupied an artificial nest close to the harbour, raising hopes that they would attempt to breed for the first time in 2020. Unfortunately, the male failed to return from migration, and was not seen again, but CJ7(15) remained faithful to the same nest and laid three unfertilised eggs, as described in chapter 3. In 2021, CJ7(15) again returned to the nest and this time laid five unfertilised eggs over an 18-day period. CJ7(15) was finally joined by a male on 18th May, a two-year-old translocated bird, 022(19), five days after laying the fifth egg. She did not lay any further eggs, but the two birds remained together for the rest of the summer. They spent the majority of August and early September with a cohort of 10 translocated juveniles, which they frequently perched with on an artificial nest at the release site.

CJ7(15) and 022(19) subsequently returned in spring 2022 and bred for the first time at the site they had favoured during the latter part of 2021, rearing two chicks (Figure 8.24). At the same time two translocated females, 014(18) and 019(19), were breeding in close proximity to each other in the Glaslyn Valley in North Wales. In each case they had paired with males from the Cors Dyfi nest, supporting the notion that a population on the south coast of England will facilitate links between other Osprey populations in southern parts of the UK and, potentially, continental Europe.

Spain

The Rutland Water translocation was the first of its kind in Europe and, like the pioneering projects in North America, demonstrated that the Osprey responds well to this proactive form of conservation management. The translocation, husbandry and release techniques were refined over the years at Rutland Water; this created a model that could be replicated elsewhere in Europe. Osprey populations were recovering well across much of Europe by the early twenty-first century, but the species' range remained restricted compared to historical times – particularly in southern parts of the continent, where it was absent from many former breeding areas, including in Spain (Dennis 2008). The work at Rutland Water demonstrated that translocation could play a key role in restoring these lost populations.

Andalucía

In 2003, an ambitious project to re-establish Ospreys in mainland Spain, after an absence of more than two decades, began in Andalucía. A total of 191 birds were translocated from Germany, Scotland and Finland over the course of 10 summers and released at two sites, 125km apart: Barbate Reservoir in the south, and Odiel Marshes, close to the city of Huelva, in the west (Muriel *et al.* 2010b, Morandini *et al.* 2019).

The importance of conspecific attraction in the settlement patterns of Ospreys was evident almost as soon as the translocation began. Frequent interactions between released young and adult Ospreys were recorded immediately and three pairs of non-reintroduced Ospreys were recorded nest-building close to the Barbate Reservoir hacking site during spring 2005. One of

Figure 8.24. Translocated male 022 reared two chicks (shown here when they were ringed) with Rutland-fledged female CJ7 at Poole Harbour in 2022; they were the first breeding Ospreys on the south coast of England for approximately 200 years (© Brittany Maxted).

these pairs constructed a nest on top of an inactive power pole at a reservoir 30km from the nearest release point, as described by Muriel *et al.* (2006). However, the nest was dislodged by strong winds two days after incubation began. An artificial nest was installed on another pylon 150 metres away two days later, and it was occupied by the Ospreys the following day. The female subsequently laid a single egg but after 60 days of continuous incubation it became clear that it was addled and would not hatch. Given the significance of this breeding attempt in relation to the restoration of Ospreys in southern Spain, and to encourage site fidelity of the adult birds, permission was sought to translocate two German nestlings earlier than planned, so that they could be fostered by the Ospreys before they abandoned the breeding attempt. This was a technique that had been used effectively in the United States, as described earlier in this chapter. While the necessary arrangements were made – and to ensure that the breeding birds did not lose interest in the nest – a local 12-day-old Black Kite chick was placed in the nest. This species was considered suitable because it is abundant in the local area, similar in size to an Osprey, and can feed on fish. The Ospreys accepted the Black Kite nestling within two hours, and the male returned with a fish soon afterwards, which the female fed to the chick. They continued to care for the kite nestling until two 10-day-old Osprey chicks arrived from Germany. At this point, the kite, which was in good condition, was returned to its natal nest and replaced by the two Osprey chicks. As before, they were immediately accepted by the adult Ospreys, and the female fed them within 33 minutes. Both went on to fledge successfully at 53 and 55 days of age. The process was repeated successfully in 2006 when the same pair again attempted to breed, but eggs failed to hatch. The cross-fostering experiment is another excellent example of how a proactive approach can be of great benefit to Osprey conservation.

The first translocated Ospreys to breed successfully reared three chicks at Odiel Marshes in 2009. The birds – a four-year-old German male and Scottish female of the same age – had occupied an artificial nest on an out of service power pole 800 metres from the hacking cages the previous summer, and returned to the same site again in spring 2009. Non-breeding White Storks were present on the nest when the Ospreys arrived, and so the Osprey project team made efforts to discourage the storks until the Ospreys had definitively occupied the nest a week later. The female laid eggs and started incubating on 3rd April, and three chicks went on to fledge successfully (Muriel *et al.* 2010b).

The first successful breeding attempt by translocated birds occurred a year later than at Rutland Water (i.e. six years after the first releases, compared to five), but the larger number of birds translocated overall (191 compared to 75 at Rutland Water) meant that the population subsequently increased at a faster rate. By 2018, 15 years after the first birds were released, the population had increased to 15 breeding pairs (nine pairs in Cadiz and six in Huelva) (Siverio *et al.* 2018). That year there were nine breeding pairs in the Rutland Water area, even though the project began in 1996, seven years earlier than in Spain. This demonstrates the clear advantages of releasing a larger number of birds even though the Rutland project has, ultimately, been successful.

BASQUE COUNTRY

Migrant Ospreys pass through northern Spain in large numbers each spring and autumn, but there were only isolated breeding attempts during the twentieth century (Galarza and Zuberogoitia 2012). A pair nested on cliffs near Ribadesella in Asturias until approximately 1960 (Bijleveld 1974), while another bird was observed nest-building at Urrúnaga reservoir in the Basque Country in 1973 (Ferrer and Casado 2004). There were later breeding attempts at El Grado reservoir in Huesca during the 1990s, and at nearby Barasona reservoir in 2005 (Lorente 2005). In view of this, and with the nearest breeding population located in Orléans Forest in central France, some 600km north, translocation was considered the best means of restoring a viable breeding population in the short–medium term (Galarza and Zuberogoitia 2012).

The estuary and marshes of the Urdaibai Biosphere Reserve on the Basque coast regularly attract migrant Ospreys (Galarza and Dennis 2009), and the site was considered highly suitable for a translocation project (Galarza and Zuberogoitia 2012). Sixty Ospreys were collected from nests in the Highland and Moray regions of northern Scotland between 2013 and 2017, and translocated to Urdaibai. As at Rutland Water, the collection was biased in favour of males – albeit to a lesser degree – with a total of 38 males and 22 females translocated over the course of five summers, in the expectation that returning males would encourage females from other populations to stay and breed.

The first translocated birds returned in 2015, and by 2017 seven translocated males had been identified in the local area. That summer, one of the 2013 males, P1, was seen with an unringed female on an artificial nest on the River Cubas near Santander, just two days after it had been erected. The pair stayed together for the rest of the summer and then returned to breed for the first time in 2018. Meanwhile, another male bird, N4, also translocated in 2013, set up territory on an artificial nest at Courant d'Huchet Nature Reserve in the Les Landes region of southern France. He attracted a female from Corsica in May and, like the Santander birds, they remained together until September and then returned to breed in 2018. In 2021 a pair attempted to breed at Urdaibai for the first time when male N3, translocated in 2013, paired with a female that had been translocated to Orx Marshes, in Les Landes on the Atlantic coast of France (see below). Three chicks hatched, including one albino individual – the first known case in the species (Figure 8.25). The albino chick died a few days after hatching and then the two surviving young were killed by heavy rain when they were approximately 20 days old (Galarza *et al.* 2021). Nevertheless, this breeding attempt provided further evidence of how the establishing population will increase metapopulation connectivity between Ospreys in France, the Mediterranean islands and southern Spain.

Figure 8.25. In 2021, the first known albino Osprey chick hatched in a nest at Urdaibai, Basque Country. Sadly it died a few days later (© Urdaibai Bird Center).

VALENCIA

With the Osprey population in Andalucía well established following the first translocation project in Iberia, and the first breeding activity recorded in northern Spain, the third Spanish translocation, a collaboration between Fundación Migres and the Generalitat Valenciana, began in the Valencia Community of eastern Spain in 2019. The key aim of the project, based at Pego-Oliva Marsh Natural Park, is to facilitate connectivity with the Osprey population in the Balearic Islands as well as the expanding population in Andalucía (Generalitat Valenciana 2019). Eight young birds were released during the first three years of the project, four from the Balearic Islands and four from Andalucía. A bird from Menorca, where the population is largely sedentary or makes short-distance movements (see chapter 6), completed a 13-day migration to northern Senegal and settled on the Ferlo River. The satellite tracking also highlighted the risk of electrocution to young Ospreys. Two of the four birds released in the first year were known to have been electrocuted during their first winter. Electrocution is a threat to Ospreys in the region, and following the death of one of the birds at Parc Natural de l'Albufera, 1346 electric poles were protected to prevent repeat occurrences (I. Colodro pers. comm. 2023).

Experiences elsewhere in Europe demonstrate that restoring Osprey populations through translocation requires patience, but plans to expand the project to include hacking at a second coastal site, and to translocate a larger number of birds, including some from Scotland, will enhance its chances of success, and in doing so, help to secure the future of the Osprey in Iberia and the wider region. In 2023, 11 Scottish juveniles were released.

PORTUGAL

It is thought that at the beginning of the twentieth century breeding Ospreys occupied the southern and southwestern rocky coast of Portugal, as well as parts of the central-western sandy coast and some sites further north. However, a combination of persecution and disturbance, principally associated with tourism development, led to a decline in the breeding population from a high of at least 24 breeding pairs to three in 1978 (Palma 2001). These latter pairs eventually disappeared due to low productivity associated with disturbance by anglers, and loss of key breeding adults. The last adult bird disappeared in 2002 (Palma et al. 2019).

An initial plan to restore a breeding population through translocation did not receive the necessary support, but a licence was finally granted in 2011 following the successes of the nearby Spanish project in Andalucía. A total of 56 juvenile Ospreys (33 males and 23 females) were subsequently translocated from Finland (19 males, 10 females) and Sweden (14 males, 13 females) to a hacking site at the vast Alqueva reservoir in eastern Portugal between 2011 and 2015. This huge artificial lake in the Guadiana River basin, close to the Spanish border, covers an area of 250km^2, with 1,160km of shoreline. The reservoir has over 200 islands where the provision of artificial platforms has increased the availability of potential breeding sites.

The first successful breeding at Alqueva occurred in 2015 when a three-year-old male that had been translocated from Finland reared a single chick in a nest on a dead eucalyptus with a German female, translocated to nearby Andalucía in 2012 (Palma et al. 2019). Unusually, a green-ringed bird of unknown origin, and sex, acted as a helper at the nest, assisting with incubation and nest refurbishment, but not food provisioning. This individual did not return the following year but the two translocated birds remained together and reared two chicks. Two other translocated males returned to Alqueva and attempted to breed for the first time in 2018, again on dead eucalyptus trees located in the water. One of these males paired with an unringed female and the other with a female that had fledged from a nest in Andalucía in 2016. Both

Figure 8.26. A programme of artificial nest building has been implemented at Alqueva reservoir and other sites in Portugal to provide robust alternatives to dead eucalyptus trees, which were favoured by breeding Ospreys in the early years of population establishment (© Jorge Safara).

Table 8.7. Breeding attempts in Portugal 2015–2021 (L. Palma unpub. data 2022)

Year	Number of breeding pairs	Number of active nests (i.e. female laid eggs)	Number of chicks hatched	Number of fledged young	Productivity
2015	2	2	3	3	1.5
2016	1	1	2	2	2
2017	3	1	2	1	1
2018	4	3	2	2	0.7
2019	2	2	3	3	1.5
2020	1	1	3	2	2
2021	4	2	6	5	2.5
Total	17	12	21	18	1.5

females laid eggs but neither breeding attempt was successful, with disturbance by boats a factor. The formation of these pairs again demonstrated the ability of translocated males to attract females from other populations. By 2021, a total of 18 chicks had fledged (Table 8.7).

The nests in the dead eucalyptus trees proved to be very unstable and so a programme of artificial nest construction was implemented in order to provide more secure breeding sites (Figure 8.26).

In 2015, a pair of Ospreys also reared two young on the same sea cliffs in Costa Vicentina that had been used by the last breeding pair in 1996, with the nest occupied by White Storks in intervening years. Neither bird was ringed but, as in Spain, it is possible that the presence of translocated birds in the region – at Alqueva and also in nearby Andalucía – helped attract them. The female failed to return the following year.

One significant factor in the early years of population establishment was that all of the returning birds were translocated from Finland and none from Sweden. This may partly be attributable to the fact that more Finnish males departed on migration (18 compared to 11) and that three Swedish males dispersed early, potentially reducing their chances of survival. However, satellite tracking and ringing research has shown that most Swedish Ospreys use a south-westerly route through Europe, as described in chapter 6. Thus, given the likely innate component of migration, it is possible that the Swedish males may have initiated migration in this south-westerly orientation. There is a high chance that they would have perished at sea, unless they made a compensatory change of heading, which juvenile birds, without prior migratory experience, tend not to do (Palma et al. 2019).

Italy

By the early part of the twenty-first century, there had been no breeding Ospreys in mainland Italy for more than 50 years, yet in centuries past they were widely dispersed along the Italian, Corsican and Sardinian coasts, often breeding on spectacular sea cliffs (Monti et al. 2014). In 2006, a project was initiated at Maremma Regional Park, to restore a population on the extensive coastal wetland system in Southern Tuscany and to facilitate geographical expansion to other parts of the species' former range in the region. A total of 33 chicks (21 females and 12 males) were translocated between 2006 and 2010 from Corsica (Figure 8.27), where 25–30 pairs breed each year, mainly on cliffs in the northwest of the island (Monti et al. 2014).

The first pair, comprising a translocated male and an unringed female, bred successfully in 2011. They reared two chicks on an artificial nest in a tree, just 600 metres from the hacking site in Maremma Regional Park. In 2014 a second pair settled on an artificial nest at Diaccia Botrona Nature Reserve, a wetland site approximately 15km to the north, and reared three chicks. The male fledged from the first successful nest at Maremma in 2011, while the female was released in 2010, the last year of the translocation project (Sforzi et al. 2019).

The importance of conspecific attraction in the establishment of Osprey colonies was demonstrated in 2015 when another pair bred at Diaccia Botrona and reared two young. Neither of the breeding birds was ringed, and thus not local translocated or wild-fledged individuals. Instead, it is very likely that they were encouraged to settle by the presence of other breeding Ospreys in the locality. Similarly, in 2018 a fourth pair bred on an artificial nest at WWF Orbetello Reserve, a coastal lagoon 25km south of the hacking site. The male was unringed and the female had been colour-ringed at a nest in Corsica in 2013. They reared three chicks, meaning that eight young fledged from a total of four nests that year. The following summer, in 2019, a fifth breeding pair became established at Orti-Battagone WWF Nature Reserve, a coastal site 48km north of Maremma, and reared three chicks. The male was unringed and the female fledged from Diaccia Botrona in 2016 (Sforzi et al. 2019).

By the end of 2019 a total of 41 chicks had fledged from a total of 23 breeding attempts. Over half of the young during this period were produced by two translocated individuals. The male at

Figure 8.27. A young Osprey makes its first flight at Maremma Regional Park in Italy. A total of 33 chicks (21 females and 12 males) were translocated between 2006 and 2010 from Corsica (© John Wright).

Maremma Regional Park, L7, reared 11 chicks between 2011 and 2019, while the translocated female, S5 fledged a total of 12 chicks at Diaccia Botrona between 2014 and 2019 with one of L7's offspring, IAA (Sforzi *et al.* 2019). As at Rutland Water, this demonstrates the importance of specific individuals in the early years of establishment – particularly as the presence of active nests often encourages non-breeding floaters to join the population. As expected, satellite tracking has shown that the Italian population is partially migratory, with 33.3% of satellite-tagged birds sedentary and 66.7% migratory (n = 15). The migratory birds mainly wintered in the Mediterranean region, but one satellite-tagged juvenile travelled to Mauritania where its satellite-tag stopped transmitting (Monti *et al.* 2018a).

It is hoped that, in time, the central Italy population will recolonise the seven islands in the Tuscan Archipelago National Park, situated midway between Corsica and Tuscany, and thereby further facilitate connections between the two populations. For this reason, nine artificial nests were erected between 2015 and 2019 to encourage young birds dispersing from Corsica and Tuscany to settle and breed (Sforzi *et al.* 2019). It was also notable in the context of metapopulation connectivity in this part of the Mediterranean that in 2020 an unringed female and a colour-ringed male from Corsica became the first Ospreys to breed in Sardinia since 1968 (Fozzi *et al.* 2020). They reared two chicks at a site on the northwestern coast.

SWITZERLAND

Ospreys last bred in Switzerland along the River Rhine in 1914, with the last observation being of a territorial male in 1919. Sightings since have generally been confined to migration periods, even though extensive areas of suitable breeding habitat exist in many parts of the country (Strahm and Landenbergue 2013). With natural recolonisation considered unlikely in the short–medium term, a five-year translocation project was initiated at Bellechasse in the canton of Fribourg, in western Switzerland, in 2015. Six Scottish juveniles (four males and two females) were translocated that summer, and in subsequent years birds were collected from nests in eastern Germany (total 30 birds) and southern Norway (31 birds). Sixty-two of the juveniles subsequently set off on migration (32 males and 30 females) (Strahm and Landenbergue 2021). The sex ratio was therefore much more balanced compared to the first European project at Rutland Water.

The first translocated bird returned in 2018, and as observed in other reintroduction projects, there was a clear difference in the behaviour of returning males and females, with males showing greater fidelity to the hacking area in the Three Lakes region. This was exemplified well in 2021 when two translocated females bred successfully for the first time. PR4, a female translocated from a nest in eastern Germany in 2016, reared two chicks with a German-ringed male in the Moselle region of northeast France, some 200km north of the release site. Meanwhile F02, a female collected from a nest in the German province of Sachsen-Anhalt, paired with a German male and fledged three chicks at a nest in northeast Bavaria in Germany, approximately 450km north-east of Bellechasse. Translocated males, in contrast, were more faithful to the Three Lakes region. One of the returning males, PS7, translocated from Germany in 2017, paired with a seven-year-old female from eastern Germany during summer 2021; this was the first known formation of an Osprey pair in Switzerland for over a century.

France

In 2018, authorities granted a licence to translocate Ospreys from the established population in central France, with strongholds in Orléans Forest and Sologne, to Aquitaine on the Atlantic coast north of Biarritz (Figure 8.28). This, it was hoped, would improve Osprey metapopulation connectivity between the Iberian Peninsula, mainland France and the Mediterranean, and further enhance efforts to restore Ospreys to the region, following the reintroduction project at Urdaibai in the Basque Country (Lesclaux 2019). In all, a total of 38 juvenile Ospreys (24 males and 14 females) were translocated to Marais d'Orx, 105km to the east-north-east of Urdaibai over four summers between 2018 and 2021.

The first indication that the Aquitaine project would enhance wider Osprey conservation efforts was evident even before the first juveniles were translocated in 2018. That year, a male, N4, released at Urdaibai in 2013, reared two young with a four-year-old Corsican female at

Figure 8.28. Translocated Ospreys at Marais d'Orx in 2018. A total of 38 juveniles were released at the site between 2018 and 2021 (© Paul Lesclaux).

Courant d'Huchet Nature Reserve in Les Landes, approximately 31km north of the Marais d'Orx release site, and some 875km from the female's natal site (Lesclaux 2019). Then, during the second year of the translocation project, a two-year-old male from Urdaibai, U18, was identified at the release site interacting with the translocated young. He returned on 10th April the following spring and lingered until late September, attracting at least four different females, including two colour-ringed birds. A female from Andalucía, which in itself was notable, was present for a brief period in the spring, and then a two-year-old German female, BL71, remained with U18 for two-and-a-half months during the latter part of the summer. The two birds were observed provisioning the translocated juveniles on a number of occasions; such behaviour that can greatly strengthen the pair-bond, and attachment to the site. Perhaps encouraged by these interactions with the translocated juveniles, the two birds returned and bred successfully for the first time at Marais d'Orx in 2021, rearing two chicks.

Further pairs became established. In 2021, a third male from Urdaibai, PF – known as Pink Floyd – built a nest and paired with a four-year-old Andalusian female at a site halfway between the other breeding pairs in Aquitaine. Then, in 2022, a male that fledged from the nest at Courant d'Huchet in 2019 began breeding with a translocated female, T13, at a site 58km north of his natal site, still within the Les Landes department (Lesclaux pers. comm. 2022).

Meanwhile on the other side of the Pyrenees a female, T06, translocated to Aquitaine from the Centre-Val de Loire region in 2018, bred for the first time at Urdaibai with an eight-year-old male in 2021, which had been released in the Basque Country in 2013. Although, as described above, the breeding attempt was unsuccessful, the fact that two translocated birds paired in this way was an extremely encouraging sign for the future in the context of regional Osprey conservation efforts, and likely a precursor to further interchange between the two establishing populations, and others in the region, in years to come.

A summary of European Osprey translocation projects is shown in Table 8.8.

THE FUTURE

The evidence from North America and Europe demonstrates that the Osprey is a species that responds well to proactive conservation intervention, and the translocation of young has greatly facilitated the recovery of the species over the past 40 years. Nevertheless, the Osprey remains absent from parts of its former range, particularly in southern and southeastern Europe where populations remain small and disjointed (Dennis 2016). It is clear that translocation is a highly effective tool in these circumstances. The knowledge and capability to run translocation projects successfully, and cost-effectively, is now readily available and, in 2016, the Council of Europe published a report written by Roy Dennis that recommended this ongoing positive approach to Osprey conservation in Europe (Dennis 2016b). As he points out in this report, it is important to recognise that the documented historical range of remnant populations is often a reflection of refugia of low persecution rather than solely an indicator of high-quality habitat or food. As long as people accept them without persecution, Ospreys are able to exploit areas with richer food supplies, often near humans, and thus increase breeding productivity, population size and range more quickly, as exemplified in many parts of North America. It is also important to consider that the translocation of young from colonies close to carrying capacity has little or no discernible effect because the removal of birds in these circumstances reduces competition, including interference by non-breeding floaters at nests, and usually enables translocated individuals to breed at a younger age than they would in an established colony, where age at first breeding is typically higher. In addition, the strong influence of conspecific attraction in the species also means that populations established by translocation are very

Table 8.8. Osprey translocation projects in Europe (1996–2022)

Country	Location	Years	Birds translocated			Donor countries	Year of first breeding by translocated birds	Max no. of pairs
			Total birds	Percentage Male	Percentage Female			
England	Rutland Water, Rutland	1996–2001 and 2005	75	68.0%	32.0%	Scotland	2001	11
England	Poole Harbour, Dorset	2017–		Ongoing		Scotland	2022	1
Spain	Barbate Reservoir and Odiel Marshes, Andalucia	2003–2012	191	54.5%*	45.5%*	Germany, Scotland and Finland	(2005)** 2009	16
Spain	Urdaibai, Basque Country	2013–2017	60	66.7%	33.3%	Scotland	2018	2
Spain	Marina Alta, Valencia	2019–		Ongoing				
Portugal	Alqueva Reservoir	2011–2015	56	58.9%	41.1%	Sweden and Finland	2015	3
Italy	Maremma Regional Park, Tuscany	2006–10	33	36.4%	63.6%	France (Corsica)	2011	5
Switzerland	Bellechasse, Canton of Fribourg	2015–2020	66	53.0%	47.0%	Germany, Norway and Scotland		
France	Marais d'Orx, Aquitaine	2018–2021	38	63.2%	36.8%	France (central)	2018***	3

*number of males and female based on 145 sexed birds (by molecular methods)
**first breeding pair both unringed, but attracted by the presence of translocated juveniles, and so year shown in brackets
***initial pairs not translocated birds, but included individuals attracted by presence of translocated young in area

likely to draw immigrant birds from other populations, as observed at various sites in Europe. It is much less likely that these wandering individuals would settle in the absence of translocated birds.

The methods for the collection, translocation, care and release of young Ospreys are now well understood. Translocated males are key to establishing new populations; and experiences across North America and Europe show that collection should be biased in favour of males – albeit only to a relatively minor degree, given that some translocated females will also return to the release site. High mortality of young birds in their first year, particularly in migratory populations, means that it is essential to release sufficient numbers of individuals to account for these losses. Releasing as large a number of birds as possible in as short a time frame as possible should therefore be the ultimate aim. It is important that all translocations are carefully considered, with proposals evaluated against International Union for Conservation of Nature (IUCN) guidelines for species reintroductions, and appropriate feasibility assessments carried out. Projects should only go ahead if they are in line with these standards.

The conservation of charismatic flagship species such as the Osprey can play a major role in helping to raise awareness of the conservation of wetland ecosystems, and in generating wider public support for the conservation movement. Ospreys create great interest wherever they occur, so aside from the moral responsibility we have to help a species that was lost from many areas entirely through the influence of humans, the return of this charismatic species can have many other far-reaching benefits. Its comeback should be welcomed and celebrated.

Osprey ecotourism

There are few avian species that are able to capture the public imagination in the manner of Ospreys. This has been no more apparent than in the United Kingdom where there is great excitement each spring when Ospreys return to nests around the country, from Poole Harbour (Figure 8.29) on the English south coast to Loch Garten in northern Scotland. The story of the British public's love affair with Ospreys can be traced back to 1959 when George Waterston made the visionary decision to publicise the location of the Loch Garten nest and encourage people to come and view the birds from a special hide. As described earlier, more than 14,000 people took up the invitation that summer, and it created a model for modern-day ecotourism that is replicated at many Osprey viewing sites today (Dennis 2008).

Dickie *et al.* (2006) estimated that 290,000 people visit Osprey watching sites in the UK each year, contributing approximately £3.5 million to local economies in the process. It seems likely that these figures are now even higher given the growing number of Osprey-watching sites around the country (Appendix 2). A study by Natural Resources Wales showed that in 2011, when 40,000 people visited the Cors Dyfi nest in Mid Wales, £180,000 of visitor spend was directly attributable to visits to the site. Furthermore, the presence of Ospreys was a key factor in decisions to visit the area and the overall economic impact of these trips was estimated to be £2.15 million (Munday *et al.* 2015). This shows that while Ospreys are now much more common and widespread than they were in 1959, the bird's popularity is showing no signs of waning and this has significant benefits for the communities in areas where they breed. It is no surprise, therefore, that they are warmly welcomed in these places. Boat of Garten in the Scottish Highlands famously declared itself the Osprey village, but as the species continues to expand, there are other places that might challenge this status. Oakham in Rutland, for example, has a fine Osprey carving on one of its roundabouts in celebration of the successful Osprey translocation at nearby Rutland Water (Figure 8.30), and you can buy a pint of Rutland Osprey from the local Grainstore Brewery. Further afield the

Our relationship with the Osprey

Figure 8.29. Thousands of people travel to see Ospreys at different sites around the UK each year at dedicated visitor centres, viewing hides and even on boat trips, such as this one at Poole Harbour (© Alison Copland).

Osprey was officially chosen as the provincial bird of Nova Scotia, Canada, by an Act of the House of Assembly in 1994, and is also the provincial bird of Södermanland in Sweden (Mackrill 2019).

WATCHING OSPREYS IN THE UK

Breeding Ospreys are now present in many parts of the UK, and dedicated visitor centres and hides provide an excellent way of viewing them. Several sites have also set up hides where you can enjoy close views of fishing Ospreys. Further details of some of the best sites is included in Appendix 2.

LIVE STREAMING AND TRACKING

Today, modern technology means that you do not even need to leave the comfort of your home to enjoy watching Ospreys. Live images from across the UK, Europe, North America and Australia are streamed online in high definition, offering a remarkable insight into Osprey family life, and generating online audiences of thousands. There were at least 30 cameras streaming live in 2014 (Bierregaard *et al.* 2014b), and new ones appear online every year. It is wonderful to watch Ospreys rearing their young in a range of widely contrasting settings, from the remote forests and bogs of Estonia to a roadside nest in urban Long Island, New York. If you ever wanted to appreciate the adaptability of Ospreys then simply spend a few minutes watching these different live streams from all over the world. It is particularly exciting to watch the footage around hatching time, when viewers are able to witness the moment a chick breaks free

237

of its shell for the first time, and the tenderness with which it is fed by its mother a few hours later. These intimate views are not possible in the field because, for the first ten days of their life, the chicks are usually too small to be seen over the rim of the large nest. Such cameras have therefore shed new light on this element of Osprey behaviour which otherwise would go unseen (Figure 8.31).

The publication of Osprey satellite-tracking data has also created considerable interest around the world, enabling people to follow the flights of individual birds during autumn and spring. This work has been pioneered by people like Roy Dennis in Scotland, Pertti Saurola in Finland and Rob Bierregaard in the United States, whose regular updates on the journeys of satellite-tagged Ospreys generated large global online followings. The first Osprey that Roy Dennis tagged with a GPS transmitter, known as Logie, featured on a BBC Radio 4 programme, *World on the Move*, in 2008. Roy also followed three Welsh Osprey chicks from the Cors Dyfi nest on their first migration to West Africa for *BBC Autumnwatch* in 2011.

Figure 8.30. A roundabout on the outskirts of Oakham in Rutland features an impressive Osprey sculpture (© Tim Mackrill).

Osprey education

It is clear that Ospreys appeal to a much wider audience than many of our other, less charismatic, species and so they provide a valuable means with which to inspire the general public about nature, and to raise awareness of specific environmental issues. Visit any Osprey-watching site in the UK, and the awe and excitement that the birds generate is palpable.

Many Osprey projects have education outreach programmes and schools are encouraged to visit. At Rutland Water, an Osprey ambassadors' scheme has been set up where children from local schools are tasked with the job of keeping their fellow students and teachers updated on the progress of the local Ospreys. Encouragingly, the Osprey education network has also spread along the migration flyways, helping to link students from Europe, North America and Africa during World Osprey Week, which myself and colleagues from the three continents established in 2015 (Figure 8.32). This international celebration of Osprey migration, held during March, was timed to coincide with the northward migration of Ospreys from Africa to Europe and from South to North America. Schools could register for World Osprey Week on the Rutland Osprey Project's website and then log-on to an interactive map to follow the journey of nine satellite-tagged Ospreys as they flew north to nests in North America, Corsica, Estonia Finland and the UK. They also had the opportunity to create their own page on the website, and to get in touch with other schools elsewhere on the migratory flyways. In addition, registered schools – both primary and secondary – had free access to a range of teaching resources based around Ospreys and Osprey migration that covered everything from maths and science to English and

Our relationship with the Osprey

Figure 8.31. High-definition nest cameras, such as the one at the Cors Dyfi nest in Wales, offer remarkable views of the first few hours of a newly hatched Osprey's life. They are watched online by thousands of people each day (© Montgomeryshire Wildlife Trust).

art. In The Gambia, an associated environmental education programme, which colleagues from Rutland Water and I set up with Gambian bird guide Junkung Jadama, has also proved a success at several rural schools. Some of students have gone on to help out with beach clean-ups and surveys of wintering Ospreys as a result of their increased knowledge and interest in environmental issues. In fact, the football team at Tanji Lower Basic School, one of the schools involved in the project, situated in a fishing village on the Gambian coast, was renamed Osprey FC. There is considerable scope to extend this work further, particularly in view of the wealth of satellite-tracking data and high-quality nest cameras that are now available, and the fact that since the Covid-19 pandemic people around the world have embraced modern technology and video-calling as a way to communicate.

Figure 8.32. Work produced by children at Provo Primary School in the Turks and Caicos Islands during World Osprey Week (© Sian Jones, Provo Primary School, Turks and Caicos).

Conclusions

- The Osprey's fishing prowess has been well known for centuries and was often referenced in historical texts, including by Shakespeare in *The Tragedy of Coriolanus*.
- Persecution of Ospreys from the Middle Ages onwards led to dramatic population declines in the UK and across Europe. It is thought that early persecution was associated with the protection of fishponds due to the strict religious requirement to eat fish on Fridays, and was actively encouraged later by the paying of bounties.
- The Osprey was given legal protection in European countries during the early–mid-twentieth century, and this allowed local populations to begin to recover. However, egg-collecting remained an issue in Scotland in particular.
- The introduction of organochlorine insecticides, most notably DDT, to agriculture following the Second World War led to thinning of Ospreys eggshells and significantly reduced breeding productivity in North America and Europe. This resulted in severe population declines and retractions in the species' range on both sides of the Atlantic.
- The recovery of Ospreys in the post-DDT era has been facilitated by the provision of artificial nests, which Ospreys readily occupy because prospecting birds prefer to take over established nests.
- Artificial nests facilitate range expansion by encouraging first-time breeders to settle in areas away from core populations. They may also increase productivity in areas that already hold breeding Ospreys, by reducing instances of nests collapsing during the breeding season, which is more likely to occur at natural sites.
- The translocation of Osprey chicks was first undertaken in the United States in 1979 and has since been carried out in 20 states. This has greatly facilitated the recovery of Ospreys in the post-DDT era by restoring breeding populations in areas where they had been lost.
- The first European translocation was undertaken at Rutland Water, central England, in 1996, with other projects following at Poole Harbour in Dorset, Spain (three locations), Italy, Portugal, Switzerland and France. The methods for the collection, translocation, care and release are now well understood as a result. As in the United States, the technique has successfully restored Ospreys to parts of their historical range, although Ospreys remain absent from many areas, particularly in southern and southeastern Europe.
- Osprey ecotourism was pioneered by George Waterston at RSPB Loch Garten in 1959. The model created there is now replicated at many Osprey viewing sites around the UK, which attract thousands of visitors each year and make a significant contribution to local economies.
- Live nest cameras and satellite tracking of individual Ospreys generate large online followings, and have also facilitated education programmes linking schools along Osprey migration flyways.

CHAPTER 9
Longevity and survival

Annual survival

The first year is the most challenging period of an Osprey's life, and significant mortality occurs during the first migration and when searching for a wintering site, as described in chapters 6 and 7. This means that an individual's chances of surviving to two years, when most young Ospreys return to potential breeding areas for the first time, is relatively low. However, as they become older and more experienced, survival rates increase significantly.

Early studies based estimates of survival on recoveries of dead ringed birds. This provided the first evidence of high mortality rates during the first year. Henny and Wight (1969) calculated that first-year survival of individuals from New York and New Jersey in the United States was 42.7–48.5 per cent, and 81.5–83.8 per cent thereafter. Similarly, an extensive analysis of ring recovery data in Sweden over the period 1942–1983 showed that mean survival of first year birds was 53 per cent, and then increased to 72–86 per cent for older age classes (Eriksson and Wallin 1994). Saurola (2005), meanwhile, reported that recoveries of 2,977 Ospreys ringed in Finland indicated that survival of first year birds was approximately 40 per cent.

Studies based on sightings of colour-ringed individuals have provided a more satisfactory, and potentially more accurate, method. This is aided by the fact that adult Ospreys are usually faithful to the same nest – or at least breeding area – each year, and perch prominently at the nest-site on their return. In addition, advances in digital photography and nest cameras mean that returning two- and three-year-old floaters can be identified when they intrude at active nests. The Osprey population in central France has been closely monitored by Rolf Wahl and others since the first breeding attempt in 1985 (Figure 9.1). Rolf initiated a colour-ringing programme in 1995, and this has offered detailed information on many elements of population demographics, which were reviewed by Wahl and Barbraud (2014). Their analysis of a dataset comprising 455 colour-ringed birds, covering a 16-year period between 1995 and 2011, revealed that survival was best modelled as a function of two age classes. Average survival of first-year birds was estimated at 0.495 (±0.043), while the corresponding figure for birds of two years and older was 0.875 (±0.017), which, like the studies in North America and Sweden, illustrated the high mortality of juvenile birds. A similar trend is evident among the partially migratory Mediterranean population. Monti et al. (2014) analysed the survival of juvenile Ospreys translocated from Corsica to Maremma Regional Park on the Tuscany coast in Italy. Survival was initially high during the post-fledging dependence period, when translocated young are supplied with food at the hacking site, to replicate the natural provisioning of adult birds. However, survival dropped to 0.26 (0.13–0.46) after dispersal and during the first winter, even though birds from this region tend to only make short-distance movements within the Mediterranean. As in other studies, annual survival increased thereafter to 0.69 (0.29–0.92) for second-year birds and 0.93 (0.13–0.99) for adults. Väli et al. (2021), meanwhile, found that survival was lower for one-year-old females (0.14) than males (0.27) in the Baltic region, and these sex-based differences were also evident for females (0.64) and males (0.74) in the older age cohort (>1 year).

Interestingly, we have found that the survival of young birds in the Rutland Water population varies considerably between nests, as shown in Table 9.1. Survival rates of birds that fledged

The Osprey

Figure 9.1. Rolf Wahl has closely monitored the Osprey population in Orléans Forest in France since the first pair bred in 1985 (© Rolf Wahl).

from the Site B and Manton Bay nests up to 2019 were very high and approximately twice that of most other breeding sites in the area.

The survival of young during their first year is likely influenced by body condition when they depart on their first migration. This, in turn, will vary according to the provisioning rates of adult males, which itself may be related to access to high-quality foraging sites, and the foraging ability of individual males. The Site B and Manton Bay nests both have easy access to prime

Table 9.1. Variations in known survival to two years of Ospreys from different nests in the Rutland Water area (2001–2019)

Site	Number of young	Known survival to 2 years
B	38	0.39
C	16	0.19
K	18	0.22
MB	27	0.41
N	25	0.24
O	25	0.24
R	11	0.18

Table 9.2. Annual variations in known survival to two years of Ospreys from nests in the Rutland Water area (2001–2019)

Year	Breeding pairs	Fledged young	Known survival to 2 years
2009	4	9	0.44
2010	5	12	0.50
2011	5	10	0.30
2012	4	9	0.33
2013	5	14	0.29
2014	6	11	0.09
2015	8	15	0.27
2016	8	15	0.40
2017	8	16	0.25
2018	9	14	0.07
2019	10	24	0.17

fishing areas, and the birds that have bred in the two locations have proved to be highly productive (Table 8.5).

There is also evidence that annual recruitment of birds aged two years varies significantly between years. Table 9.2 shows known survival to two years of birds that fledged from nests in the Rutland Water area between 2009 and 2019, when the population increased from four to 10 breeding pairs. Eriksson and Wallin (1994) detected a significant relationship between rainfall in West Africa and survival of third-year Swedish Ospreys, with survival higher in wetter years, as described in chapter 7. It is possible that this may also explain the observed fluctuations in survival recorded at Rutland Water, which ranged from a low of 0.09 to a high of 0.50. Survival may also be impacted by the prevelnce of easterly winds during autumn, which can result in inexperienced juveniles being blown out into the Atlantic.

Ospreys have been colour-ringed in Scotland since the 1960s, thanks to the efforts of Roy Dennis and others, and this provides valuable long-term data on the annual survival of adult birds, which typically exceeds 80 per cent (Dennis 2008). Indeed, the figures shown in Table 9.3 are probably an underestimate given that some individuals will go undetected each year and so it is thought that annual survival is probably close to 90 per cent. At Rutland Water, the annual

Table 9.3. Annual survival of adult Ospreys in Scotland (Dennis 2008)

Years	Males	Males at same nest for 3+ years	Females	Females at same nest for 3+ years
1960–1990	82.3%	87.6%	74.6%	87.4%
1991–2004	83.9%	89.1%	82.8%	86.5%

survival of breeding birds over the period 2001–2022 was 89 per cent for males and 93 per cent for females.

Detailed analyses of adult survival and mortality have also been undertaken in North America. Mean annual mortality of adults in two well-studied populations in Massachusetts and Chesapeake Bay ranged from 9.5–17.1 per cent (Poole 1989). Ospreys aged 10 to 13 years were more likely to die than others (15–17.1 per cent) but, conversely, also produced more young at this age than at any other period in their lives. This led Poole (1989) to suggest that there may be a cost of heightened reproductive effort.

Longevity

The data on adult survival demonstrates that an individual Osprey's chances of living from one year to the next increase significantly once they have reached two years of age. Migrating between regular breeding and wintering sites means that some individual Ospreys can live to an old age. In Europe, Ospreys have been recorded living well into their twenties. The oldest Osprey in Finland was 26 years (Saurola 2005), but even that was surpassed by a colour-ringed bird in Scotland that continued breeding until it was 27 years old (R. Dennis pers. comm. 2022) while a breeding male at another nest was thought to be at least 28 years old (Dennis 2008). Another Scottish Osprey, Green J, lived until she was at least 25 years. Roy Dennis originally colour-ringed this bird at a nest in Easter Ross in July 1991, and then in 1999, when she was breeding at a nest near Carrbridge in the Highlands, she became the first Osprey to be satellite-tagged in the UK. The satellite-tracking data showed that she wintered at Gabriel y Galán, a large reservoir in Extremadura in central Spain. It seems likely that the shorter migration to central Spain may well have been a factor in her longevity. She raised two chicks in her last year of successful breeding in 2016.

Similar examples of longevity have been reported in North America. Of breeders in Massachusetts and Chesapeake Bay, 3–3.5 per cent were aged 18+ years (Poole 1989) and Ospreys aged 24–25 years were found breeding on three occasions (Spitzer 1980). Henny and Martell (2017) report that an adult Osprey they satellite-tagged on the Washington side of the Columbia River in June 1996, which subsequently wintered east of Los Mochis in north-central Sinaloa, Mexico, was found electrocuted in April 2016, 3.98km from its 1996 nest-site. Given that it was an adult when it was tagged, they calculated that it was at least 22.83 years old when it was killed. At the time there were only six other Ospreys from the United States that were older, with the oldest being 25.17 years.

Threats to Ospreys

The latter half of the twentieth century saw an upturn in the fortunes of Ospreys in many parts of the species' range, and this trend has continued in most areas in recent decades. Yet, significant threats remain in some localities, many of which are anthropogenic in nature and indicative of wider issues. The high public interest in the Osprey means it can be used as both a sentinel and flagship species to raise awareness of the problems it encounters (Grove et al. 2009).

COLLISION AND ELECTROCUTION

The Osprey's ability to adapt to a human-dominated landscape means that it frequently inhabits areas with power infrastructure. The use of electricity poles and pylons for nesting is covered extensively in chapters 4 and 8, and power poles are often used by birds as perches in other areas, including on the wintering grounds. This inevitably leaves the species prone to electrocution and collisions with powerlines (Washburn 2014). There is also a risk of collision with wind turbines.

Olendorff and Lehman (1986) highlighted how the Osprey is a species at high risk of collisions with powerlines, which can result in neck, head and beak trauma and broken and/or damaged wings. They documented 88 raptor collisions with powerlines, with Ospreys the fourth most numerous species reported. Naïve, recently fledged juveniles are particularly at risk, and at least two young birds are known to have been killed in this way during the post-fledging phase in the Rutland Water area. The vulnerability of the species to powerline collisions was further highlighted by D'Amico et al. (2019) who proposed a species prioritisation method to identify species and areas with potentially highest extirpation risk due to collisions with powerlines, using Spain and Portugal as case studies. They concluded that large, long-lived species and slow-reproducing birds – often with habitat specialisations and specific behavioural traits – were the most in danger, and identified the Osprey as the species with the highest extinction risk in Spain, based on the parameters analysed. The best means of reducing the risk of powerline fatalities is through the installation of line-marking devices which make wires more visible (Washburn et al. 2014).

Collision with wind turbines is another potential source of mortality for Ospreys. Diffendorfer et al. (2021) analysed the potential effects of such collisions on 14 raptor species in the United States. They concluded that although Ospreys are at risk, with an estimated turbine-caused mortality rate of 0.0022, there was a lower potential for population-level impacts than those considered most as risk, which included Golden Eagle, Barn Owl, Ferruginous Hawk *Buteo regalis*, American Kestrel *Falco sparverius* and Red-tailed Hawk *Buteo jamaicensis*.

Electrocution also poses a significant threat to Ospreys, particularly in areas where power poles are used for breeding, or locations that support high densities of wintering birds. Distribution power poles, which support one, two or three energised conductor wires and, sometimes, a grounded neutral wire to provide a return path for electricity, cause electrocution or shock injury when a bird simultaneously contacts two differently energised wires or one energised wire and one grounded area (APLIC 2006). Electrocution risk varies from pole to pole according to various factors including the number, spacing and arrangement of conductor wires, neutral wires, equipment and jumper wires, and the type and extent of grounding (Eccleston and Harness 2018). Hence, some pole configurations pose a greater risk to Ospreys and other raptors. Harness and Wilson (2001) found that three-phase transformer banks in rural habitats were responsible for 53 per cent of detected raptor electrocutions in the western United States but comprised less than 3 per cent of all poles.

Restani (2021) reported that 23 Ospreys (15 adults and eight fledglings) were electrocuted along the Yellowstone River in Montana between 2012 and 2021, where over 95 per cent of the population of 89 pairs nest on power poles or artificial nests. In view of the significant electrocution risk, a team of volunteers coordinated by the Yellowstone Valley Audubon Society (YVAS) monitor breeding activity along a 700km section of the river from Gardiner in south-central Montana to the North Dakota border. Any Ospreys attempting to construct nests on energised power poles are immediately reported to the relevant power company who remove sticks and erect nest deterrents and/or artificial nest platforms which are readily occupied by the birds (Figure 9.2). This proactive approach has been highly successful, and led to a decrease in the proportion of the population electrocuted each year. At the same time the population increased from 48 breeding pairs in 2013 to 89 in 2021. It is clear that without the collaboration between YVAS and the power companies many more Ospreys would be electrocuted, potentially resulting in population-level effects.

Figure 9.2 (Left) An Osprey constructing a nest on an electricity pylon in Montana, USA, behaviour that can lead to fires and power outages and also presents a risk of electrocution of the birds. (Above) Power companies remove nests before egg-laying has taken place and erect artificial nest platforms nearby to obviate such issues. Over 95 per cent of nests along the Yellowstone River are on these platforms (© Marco Restani).

Electrocutions can cause significant mortality in wintering areas as well. Four Ospreys, including a colour-ringed bird from Rutland Water, were electrocuted near the Tagus estuary in Portugal during autumn 2021. This area has become increasingly important for passage and wintering Ospreys from Northern Europe in recent years, and Ospreys regularly perch on power poles in rice fields at Vila Franca de Xira on the eastern side of the estuary. Some poles in the area were already protected; but following the death of the birds, the power company committed to insulate all remaining poles and to instal others as perches for Ospreys away from the powerlines. Similarly, the electrocution of a translocated Osprey near Parc Natural de l'Albufera in eastern Spain led to 1346 poles being protected by the power company to prevent repeat occurrences in the future, as described in the previous chapter (I. Colodro pers. comm. 2023). A range of measures can be used to reduce the risk of electrocution. These include the installation of avian perch deterrents, pole caps, thermoplastic coating, bushing covers – which insulate key electrical infrastructure such as breakers and transformers to prevent incidental contact by animals and birds – and appropriate separation of phase-phase and phase-ground wires (APLIC 2006, Washburn 2014).

Illegal killing and persecution

Historical persecution was one of the main drivers of Osprey population declines in Europe, and although the species now enjoys the highest levels of legal protection, ongoing persecution remains an issue in some areas. Hunting Ospreys can be very conspicuous, and surprisingly approachable, and this makes them relatively easy targets for shooting, which is the main persecution threat.

As described later in this chapter, there is a general trend for Osprey populations to be stable or increasing in most parts of Europe, but this is not the case in Poland where the number of breeding pairs is declining – despite the country's proximity to the rapidly expanding German population. Woźniak et al. (2022) reported that the population has fallen to its lowest point in at least 40 years and numbers less than 25 breeding pairs. The main cause of the decline is thought to be illegal killing of adult birds, particularly those that use fish farms as hunting grounds. The loss of breeding males in this way is especially damaging to small and declining populations. In recent years these losses have been compounded by predation of nestlings and some adults by Northern Goshawk. It is clear that without immigration of birds from neighbouring Germany (700 pairs) and Belarus (150 pairs) the population is at risk of extirpation unless persecution ceases.

In some European countries, including Finland, compensation has been paid to the owners of fish farms that have suffered losses due to Osprey predation. This, however, has been controversial because it is difficult to quantify the extent of any losses, and it has

Figure 9.3. A photographic hide at Horn Mill Trout Farm in Rutland provides an opportunity for incredible close-up views of hunting Ospreys (© Geoff Harries).

been argued that public money would be better spent on the construction of nets, which exclude Ospreys and thus prevent any further damage (Saurola 2005). A different approach has recently been taken in Estonia where fish farm owners can apply to the state Environment Board for annual funds if they are located within 25km of an active Osprey nest (U. Sellis pers. comm. 2022). Under this scheme, the full sum (almost €1,000) is paid if a breeding attempt is successful, with a lesser amount available if it fails. No payments are made if a nest is abandoned. If several nests are located around a specific fish farm then the owners can apply for multiple payments (i.e. one per nest), meaning the more successful Ospreys are in their area, the more the fish farm stands to benefit. These payments are made once the results of annual monitoring of Estonia's Ospreys are published.

A fish farm in Rutland has also benefitted from the presence of Ospreys because the owner took a proactive approach to an increasing number of visits by Ospreys and a range of other piscivorous species. We helped build a sunken hide beside the pond favoured by Ospreys, and photographers now pay for the opportunity to capture shots of hunting Ospreys at very close quarters (Figure 9.3). The hide has been so successful that it now accounts for a significant proportion of the fish farm's annual turnover. A similar approach was also successful at Rothiemurchus Fishery in Aviemore in northern Scotland (Dennis 2008).

In the UK the expansion of the Scottish population during the mid–late twentieth century was limited by egg-collecting, but that has become much less of an issue following the introduction of custodial sentences, as described in the previous chapter (Dennis 2008). Some isolated cases of persecution do continue to occur, though. An Osprey was shot in Lincolnshire in 2010 and earlier that spring two male Ospreys disappeared in suspicious circumstances within a week of each other in the Rutland Water area, with shooting thought to be the likely cause (Mackrill *et al.* 2013). A third breeding male subsequently disappeared in similar circumstances in the same area during May 2011. Losing key breeding birds in this manner has serious implications for a small population and in view of this we satellite-tagged two other adult males later that summer. This, we hoped would act as a deterrent to further persecution, and no further losses occurred in this way.

Like in Europe, shooting of Ospreys occurs in the United States, but usually at a low level. A bigger problem is thought to occur on migration and on the wintering grounds, where Ospreys are considered to be a problem for aquaculture facilities, which have significantly increased in number in many Latin American countries since the 1970s, most notably in Colombia, El Salvador, Panama and Guatemala (Bierregaard *et al.* 2020). In Colombia, where fish farms attract wintering Ospreys in increasing numbers, as many as 270 have been shot in Huila, a department in the south of the country, in a single year alone (Bechard and Márquez-Reyes 2003). Warmer waters in this region are more conducive to the production of tilapia, which are reared in large shallow pools approximately 5ha in size. Aquaculture facilities may hold as many as one million tilapia of various age classes, making them highly vulnerable to predation by Ospreys. In such circumstances netting is the best non-lethal deterrent, but the lack of legal protection in many countries means that shooting is routinely used (Bechard and Márquez-Reyes 2003). Migrant Ospreys are also thought to be at risk of shooting at fish farms in Cuba and Hispaniola, a bottleneck for birds moving south from eastern parts of North America. At least four satellite-tracked birds stopped moving at fish farms in the region, almost certainly due to shooting (Bierregaard *et al.* 2020).

European Ospreys are also at risk during migration and wintering periods. Brochet *et al.* (2016) estimate that the number of Ospreys killed in the Mediterranean region each year ranges from 47 to 349. Many of these birds are shot during migration, with the largest number of deaths occurring in Lebanon. The extent of persecution of Ospreys on African wintering grounds can be more difficult to quantify, but evidence suggest it occurs at a low level relatively widely, particularly in inland areas away from the best wintering habitat on the coast, as described in chapter 7. Saurola (2005), meanwhile, observed that ring recovery data for Finnish Ospreys indicate that hunting pressure on Ospreys has declined significantly across Europe between 1970 and 2005, but remained at the same level in Africa during the same period.

HUMAN DISTURBANCE

One of the most striking differences between *Pandion h. carolinensis* and *P. h. haliaetus* is their differing levels of tolerance to human activity. Ospreys in North America have adapted extremely well to anthropogenic landscapes, and readily build their nests on a range of structures, often in close proximity to humans, as described in chapter 4. In Europe, where historical persecution of Ospreys was much more widespread and had clear population-level effects, Ospreys are considerably more wary of people, even though some individuals may become habituated to certain activities over time. As a result, human disturbance, through a range of different sources, can have an impact on annual breeding success.

As a long-lived species, the Osprey is more sensitive to mortality of adults than reduced breeding performance (Martínez-Abraín 2018), but disturbance can have local impacts – particularly in establishing populations where the production of young has more significant effects on future population demographics. Predictable disturbance that is concurrent with nesting initiation is usually better tolerated than sporadic disturbance or new sources of disturbance that occur during incubation and the early nestling stage, but it varies on a case-by-case basis (Levenson 1979, Poole 1989). Disturbance that causes the adult birds to leave the nest frequently or for extended periods during incubation or when there are small dependent young is particularly serious and can lead to reduced brood size, or complete failure, either through chilling or predation (Levenson and Koplin 1984, Monti *et al.* 2018).

WATER-BASED DISTURBANCE

Activity by boats has been demonstrated to reduce Osprey breeding success in a number of areas. In Corsica, a population of approximately 30 pairs breed on spectacular sea cliffs. Some of these nests lie within the Scandola Marine Protected Area (MPA), a terrestrial and marine reserve of around 2,000 hectares, which attracts around 300,000 visitors each year, predominantly in June–August. Many of these visitors view the coast from licensed boat trips, as described by Monti *et al.* (2018c). Scandola MPA is one of the most pristine environments for marine biodiversity in the Western Mediterranean, and food availability for Ospreys here is high. Ospreys in Corsica are capable of tolerating low–medium human disturbance, but recent major increases in boat traffic, compounded by a lack of restrictions on either the number of visiting vessels or proximity to the coastline, has led to a reduction in breeding success of Ospreys within the MPA. Monti *et al.* (2018c) found that 74.6 per cent of boats passed within 250m of the coast, meaning that nests on frequently used routes experienced regular disturbance. In some cases, boats linger close to nests for several minutes in order to watch the Ospreys with the guide speaking on a megaphone, and this causes particular problems, with females alarm calling and flying off the nests. This leaves eggs or small dependent young vulnerable to predation by Yellow-legged Gulls *Larus michahellis* and Ravens, and also leads to a reduction in food provision by males. It also caused stress, with corticosterone levels in chick feathers three times higher than other locations in the Mediterranean region. Although, as the authors acknowledge, other factors such as density dependence may also be acting on the population, it is clear that tourist boats do have a detrimental impact, indicating that tighter controls are required.

Ironically, the use of high sea cliffs for breeding is thought to be a relict of historical persecution which resulted in selection in favour of individuals that raised young in remote, inaccessible localities. Now, thanks to the legal protection of Ospreys around the Mediterranean, breeding birds may potentially suffer less disturbance if they revert to tree nests (Martínez-Abraín 2018).

Water-based activities have also been identified as a potential threat to the establishing population of Ospreys at Alqueva reservoir in Portugal. As Palma *et al.* (2019) reported, incubating birds are disturbed by recreational boats at some sites, and this resulted in nesting failures of some of the initial breeding pairs. Reduced breeding success can have more serious implications for populations such as this, which are in the early stages of establishment. The implementation of exclusion zones around nests using floating buoys is a simple and important method that can limit disturbance, increase breeding productivity and facilitate the establishment of a self-sustaining population more quickly.

Saurola (2005) estimated that only 15 per cent of occupied Osprey nests in Finland are located close to the shorelines of lakes due to pressure from human recreation. In many cases historical nest-sites have been abandoned, and the Ospreys have been forced to move further into forests to avoid disturbance from activities such as sailing, canoeing, swimming and angling. In Scotland stand-up paddleboards and canoes have also become an increasing issue for nesting Ospreys in some areas (R. Dennis pers. comm. 2022).

Even in North America, where Ospreys are generally more tolerant of human activity, water-based activity can result in disturbance. Rodgers and Schwikert (2002) found that average distances at which Ospreys were disturbed (i.e. took to the air) from active nests in Florida were

49.5m (±21.8m, range 20–159m) and 57.9m (±22.2m, range 30–140m) for jet skis and outboard-powered boats respectively.

TOURISM DEVELOPMENT

Tourism and recreation pose a threat to Ospreys in other areas too. Habib (2019) recommended the installation of artificial nests to replace nest-sites lost to tourism development on the Egyptian Red Sea coast. Similarly, Khan et al. (2008) estimated that close to 80 per cent of former breeding sites in the United Arab Emirates have been lost to tourism development. Ospreys are increasingly restricted to the quieter, more remote islands as a result, and have been helped by the provision of artificial nests. Dennis and Clancy (2014) also reported that expansion of tourism infrastructure is one of the main threats to Ospreys in Australia. Human disturbance during the breeding season is a particular issue in the open and exposed landscapes of the South Australia coastline, where many nests are constructed at or near ground level, causing the birds to leave their nest more readily at critical phases of the breeding cycle (Dennis 2007a, Detmar and Dennis 2018).

FORESTRY AND AGRICULTURE

Forestry operations also have the potential to have detrimental impacts on breeding Ospreys. Saurola (2005) reports that the inappropriate timing of forestry work close to active nests has caused breeding failures in Finland. The construction of logging roads, digging of forest ditches, harvesting, improving of young stands and planting seedlings are all activities that may cause issues during courtship, incubation and brooding periods, and the establishment of buffer zones around nests is therefore of key importance. However, it can be difficult to determine the most appropriate buffer zone size because Ospreys are highly individual in their response to disturbance, while other factors, such as topography, are also important. Ruddock and Whitfield (2007) asked selected experts to record the distance at which Ospreys typically showed a 'static' (e.g. alarm calling) and an 'active' (e.g. taking flight) behavioural response to a single pedestrian observer walking in full view towards an active nest. They received a wide range of observations, but mean values were 329 metres for 'static' (i.e. Osprey alarm calling) and 225 metres for 'active' (Osprey leaving the nest) disturbance during incubation, and 325 metres for 'static' and 221 metres for 'active' during chick-rearing. At Rutland Water we have observed that adult birds become more sensitive to disturbance when young are close to fledging.

There are certain human activities that European Ospreys become habituated to. In Rutland, artificial nests sited on the edge of arable fields have proved highly successful. In these circumstances the breeding birds largely ignore normal farming activities involving heavy farm machinery (see Figure 8.12), although it is important that the operator remains inside the vehicle when they are in close proximity to the nest. In my experience, farmers are always delighted to have a pair of Ospreys on their land, and are careful to adapt their methods to reduce the risk of disturbance. For example, a vehicle driving directly towards a nest can be more threatening to the breeding Ospreys than one moving parallel to them.

As time progresses it seems likely that European Ospreys will become more habituated to human activity, but for now clear differences remain. Forys et al. (2021) found that reproductive success of Ospreys nesting in a highly urban environment in Pinellas County, Florida, where nests are spaced approximately every 400 metres and the mean urban area within 200 metres of nests is 39.9%, was similar to less-developed parts of Florida. A true sign of success of the recovery of Ospreys in Europe will be when they too are nesting successfully in urban environments such as this.

NETTING AND ENTANGLEMENT

Nets are routinely used to protect smaller waterbodies from avian piscivores, but they can present a hazard to fishing Ospreys, who may become entangled and drown. Early spring is one of the most dangerous times of the year for Ospreys in Finland because most fishing grounds are still covered by ice. This forces newly returned birds to widen their search to waterbodies not normally used. Ponds protected by netting have resulted in many cases of drowning during this period (Saurola 2005). In spring 2006 an Osprey from Rutland Water had a lucky escape from a pond in Essex. It was found severely waterlogged on the side of a netted pond, and unable to take off. Fortunately, after a night in a rehabilitation centre it was released again and subsequently returned to Rutland Water. It went on to breed for the first time in 2009 and reared eight chicks over the course of four years (Mackrill *et al.* 2013). Monofilament nets, often used to protect fishponds in Southern Europe, present a particular risk to Ospreys because they are almost invisible. At least one juvenile from Rutland Water has been killed in this way in southern Spain. It is essential, therefore, that nets used to protect fishponds are as visible as possible and fitted well above the water (Dennis 2008).

Nylon line, fishing hooks and even bubble floats are increasingly found on Osprey nests in Scotland. This can lead to chicks becoming tangled in nylon line or even have a hook embedded in their throat or stomach. Breeding adults have also been lost in this way (Dennis 2008). Likewise in the United States, Watts and Paxton (2007) observed that the increased use of Chesapeake Bay by recreational anglers led to discarded monofilament line being increasingly found in nests. Adults often encounter line within floating debris and pick up the material to use as nest lining. This leads to entanglement of both adults and chicks. Similarly discarded fishing net on West African beaches poses a threat to wintering Ospreys (Figure 9.4).

Baling twine, which is used by farmers for storing hay and straw, as well as for a range of other ad-hoc purposes, is another potential entanglement hazard. It becomes frayed easily and

Figure 9.4. Discarded fishing nets pose a threat to wintering Ospreys in West Africa (© John Wright).

Figure 9.5. (Top left) Ospreys may pick up baling twine when they are searching for material to line their nests (© John Wright). (Top right) It frays easily and can lead to adults and chicks becoming entangled, causing constriction, infection and potentially starvation. This juvenile on a nest on the Yellowstone River, Montana, USA was found to be entangled when it was ringed. It had a swollen right leg with superficial skin abrasions. The left leg was deformed, presumably from weeks of restricted movement. Approximately 3–4 per cent of nestlings in the area become entangled in baling twine (© Marco Restani). (Centre) A nestling Osprey that died from entanglement in baling twine. Depending on the year, approximately 40–60 per cent of Osprey nests along the Yellowstone River contain baling twine (© Marco Restani).

is often found discarded in fields where Ospreys collect clumps of grass to line the nest (Figure 9.5a). If it is picked up and taken to the nest, there is potential for the adults and chicks to become entangled, leading to constriction, infection and potentially starvation. On one occasion at Rutland Water, an adult male Osprey was observed hanging from its nest on a strand of baling twine. Fortunately, the bird was able to free itself after a few minutes, before any intervention was required. Similarly, in 2018 Roy Dennis, Ian Perks and I found a chick with a badly swollen leg caused by a thick strand of baling twine at a nest in Moray in northern Scotland. We were able to remove the twine and the bird went on to fledge successfully, but it would undoubtedly have died otherwise.

Baling twine is also a hazard for breeding Ospreys in North America (Figure 9.5b and 9.5c). All nests near agricultural areas in Montana in northwest United States, were found to contain twine, which entangled young in nearly 5 per cent of nests (Blem *et al.* 2002). These findings prompted Seacor *et al.* (2014) to investigate the issue in the Yellowstone River floodplain. Three nestlings at two nests (9.5 per cent of total) were found entangled in twine in 2012: one was already dead, one had to be euthanized and one was cut free and appeared to fledge normally. The work continued in subsequent years and a total of 27 Ospreys were found entangled between 2012 and 2021. Sixteen nestlings and one adult were rescued, each of which would have likely died without such intervention (Restani 2021). The extent to which some pairs collect baling twine in this region was exemplified by the fact that one nest blown down in a storm contained 206 metres of twine, and another 131 metres. Seacor *et al.* (2014) conclude that although mortality resulting from entanglement in baling twine was not sufficient to threaten the stability of the Osprey population in the region, it did present clear welfare concerns because of the potential for adults and nestlings to suffer slow deaths, and they initiated an education programme as a result.

POISONING/CONTAMINATION

DDT had a drastic impact on the population size and distribution of Ospreys in North America and Europe, as described in chapter 8, and although no other pollutants have had such a devastating impact, vigilance is required. Recently, a number of new industrial pollutants have caused some concern. Polybrominated diphenyl ether (PBDE), flame retardants used in thermoplastics, textiles, polyurethane foams and electronic circuitry, are one such example. Lower-brominated PBDEs have been shown to bioaccumulate and biomagnify up the food chain, but unlike organochlorine pesticides and PCBs which have declined in aquatic ecosystems in recent decades, PBDEs have increased since the 1970s (Henny *et al.* 2009). They are considered a potential emerging threat to Ospreys as a result.

Henny *et al.* (2009) studied levels of PBDEs and potential impacts on breeding Ospreys in Oregon and Washington in the United States. They found that PBDEs were ubiquitous in the region, with all 120 eggs analysed between 2002 and 2007 from nests in three major river systems, the Columbia, Willamette and Yakima, containing quantifiable concentrations. There was no evidence of a reduction in reproductive performance when levels were below 1,000 ng/g ww. However, a small number of Osprey eggs from the Willamette River and Columbia River contained PBDE concentrations above this threshold, with breeding productivity reduced by almost half. A subsequent study in the Columbia River Basin in 2008–2009 found no such relationship between productivity and PBDE (Henny *et al.* 2011), but the authors conclude that long-term monitoring is required. Chen *et al.* (2010), meanwhile, examined PBDE contamination in Osprey, Double-crested Cormorant *Nannopterum auritum*, Brown Pelican *Pelecanus occidentalis* and Peregrine at Chesapeake Bay. They estimated that the biomagnification factor of PBDEs in the fish–Osprey food chain was 25.1, and concluded that PBDEs pose a moderate threat to Ospreys through impairment of reproductive performance. It is clear from these studies that further monitoring of this emerging threat is essential, particularly as PBDEs are also known to be prevalent in the environment in areas inhabited by Ospreys in Europe (Nordlöf *et al.* 2010).

As a well-studied species at the top of the aquatic food chain, it is clear that Ospreys can play a key role in highlighting new threats. Grove *et al.* (2009) proposed the Osprey as a sentinel species for the monitoring of environmental contaminant exposure and effects in aquatic ecosystems. They argue that various key characteristics support this, including the following:

1. The exclusively piscivorous diet, and associated position at the top of freshwater and saltwater food chains.
2. Localised feeding habits close to nests, with prey readily identifiable, especially using nest cameras.
3. Ease of long-term monitoring due to longevity and site fidelity of individuals and the highly visible nature of nests.
4. Adaptation to the human landscape where contamination is most prevalent.
5. Tolerance of short-term nest disturbance for monitoring purposes, with no effect on breeding success.
6. Regular spatial distribution of nests, enabling evaluation of spatial changes in contamination levels along rivers, including upstream and downstream of known or suspected point sources.
7. The ability to accumulate most, if not all, lipophilic contaminants.
8. Known sensitivity to many contaminants; and, crucially, a cosmopolitan distribution that facilitates worldwide study.

It says much about the recovery of Osprey populations in the post-DDT era, and its ability to adapt to highly anthropogenic landscapes, that a species that was so badly affected by environmental contaminants can now play a central role in identifying and monitoring emerging threats in this way.

Overfishing

The Osprey's ability to catch a broad range of freshwater and saltwater fish means that it is highly adaptable and can adjust its diet according to local food availability through the year. There are, nonetheless, concerns that overfishing of favoured prey species may become an issue in some areas. In North America the Atlantic Menhaden *Brevoortia tyrannus* is one of the Osprey's preferred prey species along the northeastern coast. However, its population has been dramatically reduced by overfishing in recent decades. The Atlantic States Marine Fisheries Commission (ASMFC) estimates show a severe population decline during the late 1980s with no recovery over the next 25 years (Bierregaard *et al.* 2014a). Similarly, Winter Flounder *Pseudopleuronectes americanus* harvest declined by 90 per cent from 1981 to 1991 (ASMFC 1993). It is though that these declines may have contributed to reduced Osprey productivity, especially when combined with increased nesting density following the DDT era (Bierregaard *et al.* 2014a). It is hoped that new management measures implemented in the fishery will prevent further declines and prevent any associated issues for local Osprey populations.

In West Africa increasing demand for fishmeal and fish oil from major markets is causing concern regarding overexploitation of small pelagic fish, and the associated socioeconomic and environmental impacts. In 2016, West African countries were responsible for producing 7 per cent of the world's fishmeal, and some countries have experienced a particularly steep rise in production (Gorez and Foday Saine 2021). In The Gambia three industrial fishmeal factories, each of Chinese ownership, located in Sanyang, Gunjur and Kartong, have significantly increased production of fishmeal in recent years, and this led to the highest level of small pelagic fish catch (46,662 tonnes) ever recorded in 2017 when the three factories were fully operational for the first time. Data from the Working Group of the Fishery Committee for the Eastern Central Atlantic (CECAF) confirmed that the feed industry stimulated an increase in fishing effort. This intense fishing effort has likely been contributing to the exploitation of small pelagics, whose total catch had dropped to a very low level in 2019, at 26,213 tonnes (Thiao and Bunting 2022). The species caught include those such as the Bonga Shad that are also targeted by Ospreys, causing concerns that unsustainable harvesting of these species for fishmeal production could have an impact on overwintering Ospreys and other piscivorous species, in a manner that Gambian artisanal fisheries never did (Figure 9.6). Furthermore, concerns have also been raised about pollution from the factories and its associated impact on the local environment, as well as various socioeconomic and food-security issues (Gorez and Foday Saine 2021).

In Morocco the breeding population of Ospreys at Al Hoceima National Park, where pairs nest on high calcareous cliffs, declined by 35.7 per cent between 1983–90 and 2012–13, as described by Monti *et al.* (2013). Several threats have been identified, including illegal fishing practices such as use of dynamite, which is tossed from a cliff in order to kill fish that are then collected by a swimmer, and poison where copper sulphate is used to force Octopus *Octopus vulgaris* out of their holes, leading to poisoning of other marine organisms. These practices not only deplete fish stocks, but may also lead to disturbance of breeding Ospreys, especially when combined with the effects of other fishing activities and tourism.

Figure 9.6. Traditional artisanal fisheries pose no threat to Ospreys in West Africa, but industrial-scale fishing, particularly in relation to the production of fishmeal, is a cause for concern (© John Wright).

DISEASE AND PARASITES

Evidence suggests that disease is relatively unimportant in Osprey population dynamics, but some instances have been recorded. At Rutland Water an underweight nestling died of aspergillosis, and the same fate befell a translocated Osprey at Poole Harbour. Wheezing, a common symptom of aspergillosis, is occasionally found in nestlings in poor condition in Scottish nests (Dennis 2008). Cases of this respiratory infection caused by exposure to *Aspergillus* fungi have also been recorded in Florida (Deem *et al.* 1998). *Aspergillus* fungi are ubiquitous in the environment, and can proliferate in a large number of environments on organic substances. Immunosuppressed individuals are particularly at risk (Arné *et al.* 2021).

In addition, avian cholera was found in four adult Ospreys at Chesapeake Bay in the United States. This occurred during an outbreak of the disease among waterfowl, and it is thought that transmission to Ospreys occurred when bones of diseased waterfowl were used for nest-building (Hindman *et al.* 1997). Meanwhile, four juvenile Ospreys that were translocated to Rutland Water in 1996 died of a *Salmonella* infection. That year the project's licence stipulated that only the smallest chick in a brood could be collected, and it is thought that these runt chicks were more vulnerable to infection. No cases were recorded in subsequent years, when stronger, healthier young were translocated (Mackrill *et al.* 2013).

Feather mites, which feed on feather fragments, lipids, and feather fungi, bacteria and algae, are sometimes present on Ospreys when they are ringed as nestlings. They usually do not cause problems for the host but can irritate if present in very large numbers, which may result in feather pulling and associated damage (Philips 2000). *Bonnetella* feather mites only occur on Ospreys, while *Analloptes* mites have also been recorded (Miller *et al.* 1997, Philips 2000). *Kurodaia haliaeeti* is specific to the Osprey and has never been recorded on any other hosts (Nasser *et al.* 2019).

Ospreys, like most birds of prey, host a wide variety of endoparasites that often have complicated developmental cycles in which the Osprey is the final host, and prey species are intermediate hosts (Krone 2000). Three Ospreys recovered dead in Germany were all found to be infected with *Nematoda* and/or *Trematoda* helminths (Krone 2000). Meanwhile, a total of 28 species of helminths (17 trematodes, 3 cestodes, 7 nematodes and 1 acanthocephalan) were found in 17 dead Ospreys from the United States. These included seven species thought to be Osprey specialists, including three species (*Neogogatea pandionis, Paradilepis rugovaginosus, Paradilepis simoni*) reported only from North America and four recorded in both North America and Europe (*Pandiontrema ryjikovi, Scaphanocephalus expansus, Contracaecum pandioni* and *Sexanocara skrjabini*) (Kinsella *et al.* 1996). The authors of the research suggest that this number of specialists is large in comparison to other avian hosts and may reflect the Osprey's reproductive and ecological isolation from other raptors since the Pleistocene (Poole 1989, Kinsella *et al.* 1996). There were no indications that the presence of the helminths in the digestive tracts of the Ospreys contributed to their death. *Scaphanocephalus expansus* was also found in the small intestine of an Osprey found dead in the Canary Islands (Foronda *et al.* 2009).

INTERSPECIFIC PREDATION AND COMPETITION

Although the Osprey is at the top of the aquatic food chain, nestlings and occasionally adults can be vulnerable to predation in some circumstances. The persecution of Ospreys in Poland has led to the recent population declines referred to earlier in this chapter, but in recent years this has been exacerbated by predation of nestlings, and sometimes adults, by Northern Goshawk. During 2018–20, 27 Osprey nests were equipped with cameras to monitor breeding success and causes of mortality (Woźniak *et al.* 2022). There were five cases of total nest failure during this period, two of which were caused by Goshawk predation. Three juveniles were also predated by a Goshawk during the post-fledging period. When other nests, without cameras, were also factored in, Osprey nestlings were known to have been predated at a total of 10 nests during the period, with 50 per cent of losses caused by Goshawks. In addition, a breeding male, and probably the adult female from the same nest, were also killed by a Goshawk at one of the sites. Predation by Goshawks has been recorded in other European countries, including Estonia, Finland and Latvia (Väli *et al.* 2021), but in Poland it has much more serious implications because Osprey numbers there are already in decline, potentially leaving the population vulnerable to local extinction (Woźniak *et al.* 2022). Even juvenile Goshawks are capable of predating Ospreys. A recently-fledged juvenile Osprey was killed by a young Goshawk at Poole Harbour on the South Coast of England in 2022 (Figure 9.7).

Osprey nestlings are also known to be vulnerable to predation by Eurasian Eagle Owl *Bubo bubo* (Odsjö and Sondell 2001, Väli *et al.* 2021) at night, and a case of White-tailed Eagle predation was documented in Poland in 2020 (Woźniak *et al.* 2022), having previously been recorded in Russia (M. Babushkin pers. comm. 2019). In Scotland, a Common Buzzard was filmed snatching a three-week-old nestling out of an Osprey nest at Lochter in Aberdeenshire in 2012, although such behaviour is very rare.

In North America Bald Eagles are known to predate nestlings (Flemming and Bancroft 1990, Liston 1997, Poole 2019), and very occasionally adult birds (MacDonald and Seymour 1994). However, there is no evidence that this has population-level impacts (Poole 2019).

Like the closely related White-tailed Eagle, instances of kleptoparasitism are more common than predation (Ogden 1975, Poole 2019). As described in chapter 4, Bald Eagles have sometimes been shown to displace Ospreys from the best breeding sites close to water (Ogden 1975, Gerrard *et al.* 1975), although this tends to be when there are isolated pairs of Ospreys which

Figure 9.7. Goshawks are capable of predating both juvenile and adult Ospreys. This juvenile female attacked and killed a recently fledged juvenile female Osprey at Poole Harbour in Dorset, southern England (© Birds of Poole Harbour).

may be more vulnerable to interspecific competition. In areas such as Chesapeake Bay where Ospreys nest in very high densities, the species coexist well (Poole 2019).

Interspecific competition with the White-tailed Eagle is thought to influence the distribution of Ospreys in Japan (Shoji et al. 2011) and Russia (Babushkin pers. comm. 2019), in the same way as the closely related Bald Eagle, with Ospreys choosing sites further away from water potentially to reduce the risk of kleptoparasitism.

White-tailed Eagles have also been known to take over Osprey nests on some rare occasions (Treinys et al. 2011), including in Scotland (R. Dennis pers. comm. 2022). Treinys et al. (2011) concluded that the expansion of White-tailed Eagles in Lithuania may limit the future distribution of Ospreys, but they did acknowledge that changes in habitat use in space and time, as observed in other areas, may change the dynamic – particularly given the plasticity in Osprey nest-site choice. Ospreys and White-tailed Eagles are known to coexist in many parts of Europe.

The Great Horned Owl *Bubo virginianus* is considered to be a significant predator of Ospreys in North America, particularly in the northeastern United States (Bierregaard et al. 2020). Like Eagle Owls in Europe, they have the advantage of attacking Osprey nests at night. For this reason, it can be difficult to document the extent of predation, but at least 20 per cent of 101 nestlings lost from a southeastern Massachusetts colony over the course of five years were thought to have been killed by Great Horned Owls (Bierregaard et al. 2020). Adult Ospreys may also be taken on some occasions (Cold 1993), and satellite data indicate that Great Horned Owl predation may also be a hazard for migrant Ospreys (Bierregaard et al. 2020).

Some mammalian predators also pose a threat. Pine Martens are known to predate Osprey eggs and also small nestlings in various European forests, including in Sweden (Odsjö and Sondell 2001) and Poland (Woźniak et al. 2022). Recent population increases of Pine Martens in Scotland mean that this phenomenon is also increasingly recorded there, with claw marks visible on the trunks of trees with failed nests (Dennis 2008). On one occasion Roy Dennis found a Pine Marten with young in a regularly used Osprey nest in Moray. Meanwhile, Raccoons predate Osprey eggs and small nestlings in North America, in the same way as Pine Martens in Europe. It is thought that they have contributed to a shift in Osprey nest-site preference at Chesapeake Bay, where the majority of nests are now built over the water, whereas historically they tended to breed in woodlands around the water's edge (Henny et al. 1974, Bierregaard et al. 2020).

Ospreys are also vulnerable to predation on their wintering grounds. Nile Crocodiles *Crocodylus niloticus* are a particular threat in West Africa when birds are perched on the ground near water, or bathing in the shallows, with naïve juveniles most at risk (Prevost 1982). In 2018 an Osprey that we translocated to Poole Harbour from northern Scotland was found dead at Kartong in the south of The Gambia. It appeared to have died of injuries sustained when it was attacked while perched on the ground because it was missing most of its tail. Crocodiles are widespread in this area, but an attack by a feral dog is also possibility. Similarly, it is thought that Black Caiman *Melanosuchus niger* and other caiman species may predate wintering Ospreys in South America (Bierregaard et al. 2020).

Climate change

The broad global distribution of the Osprey and, in particular, its ability to exploit a range of environments for breeding and wintering mean that it should be fairly resilient to environmental change caused by global warming (Gilroy et al. 2016). Indeed, the increasing tendency for Ospreys to winter on the Iberian Peninsula rather than continue south across the Sahara (Martín et al. 2019), described in chapter 7, may be related to climate change. Nevertheless, there is broad consensus that the frequency and intensity of storms, cyclones and high-impact wind speed will increase over the next two centuries (Molter *et al.* 2016), and this may have implications for Ospreys. An uptick in the frequency and severity of storms related to climate change could result in increased mortality during migratory periods, particularly among first-year birds. It may also lead to an increase in cases of brood reduction or nest failure if storms occur during key points in the breeding season, such as hatching, when small young are known to be particularly vulnerable to such conditions (Dennis 2008).

Current status of the Osprey

BirdLife International estimates the current world population of Ospreys to number 100,000–120,000 mature individuals (BirdLife International 2022). The accuracy of population estimates varies from country to country, but it is encouraging that the species is either stable or increasing in most areas it occurs – although, as detailed below, there are some exceptions.

Europe, North Africa and the Middle East

In Europe, where persecution and DDT severely impacted population size and distribution, the Osprey is recovering strongly in many areas. Schmidt-Rothmund *et al.* (2014) estimated that the number of breeding pairs in Europe, northern Africa and the Middle East almost doubled between the 1980s (approx. 5,500 pairs) and the early twenty-first century when it reached between 9,500 and 11,500 pairs.

Scandinavia and Russia

The northern forests and bogs of Scandinavia and Russia are the stronghold of Ospreys in Europe. In Sweden, where Ospreys breed in large numbers near inland lakes and along the Baltic Sea coast, there are an estimated 4,100 breeding pairs. The population is thought to be stable in most areas (Ottosson *et al.* 2012) and increasing in Scania in the south (Schmidt-Rothmund *et al.* 2014). The closely monitored Finnish population numbers approximately 1,400 pairs. The population increased by 80 per cent between 1972 to 1994, and then by 18.9 per cent between 1994 and 2020 (Saurola 2021). Meanwhile, the neighbouring Norwegian population has increased from an estimated 150–200 pairs in 2000 (Steen and Hansen 2001) to 500 in 2012 (Schmidt-Rothmund *et al.* 2014). Somewhat surprisingly, there are only a small number of breeding pairs in Denmark, with a maximum of three pairs in recent years (Schmidt-Rothmund 2014), but this is likely to increase in future years. Interestingly, the breeding male at one of the sites in northwest Jutland in 2018 was a Welsh male 900km from its natal nest at Cors Dyfi.

There are thought to be at least 2,000–4,000 pairs in western Russia, but there are huge areas that are not surveyed due to lack of funding and staff (Mischenko 2004). However, Babushkin and Kuznetsov (2013) estimated that the population in their study area in northwest Russia is 600 pairs, of which 55–60 nest in the Darwin Nature Reserve and its buffer zone. The Darwin Nature Reserve covers a total area of 1,126km^2 in the upper reaches of the Volga River, in the northwestern part of the Rybinsk reservoir, and consists predominantly of raised bog with pine and mixed forests. The Osprey population increased in density in this area during the 1950s–1970s, which facilitated expansion to other lakes and reservoirs of the Vologda Lake District and southeastern Onega area, where habitats are similar (Kuznetsov and Babushkin 2005). Meanwhile, Shukov (2019) reported 50–60 pairs in the Nizhny Novgorod Region.

BALTIC STATES AND EASTERN EUROPE

Latest data indicate that the Osprey populations in the Baltic are faring well in the post-DDT era. The breeding population in Latvia has increased from 45–50 pairs in the 1980s to 220–240 pairs in 2019, while in Estonia the number of breeding pairs increased from 10 pairs to 90–100 pairs during the same time period (Väli *et al.* 2021). In neighbouring Lithuania, there is a smaller population of 20–30 pairs, with most located in eastern parts of the country near the Baltic Sea coast (Schmidt-Rothmund *et al.* 2014). Across the border in Belarus, there is a population of 150–180 breeding pairs, mainly in the north (Dombrovski and Ivanovski 2005) (Figure 9.8).

As described earlier in this chapter, Poland is the only European country with a breeding population of Ospreys that is in decline. Illegal persecution has left the small population vulnerable to stochastic events and predation by Northern Goshawk, with the latest estimate being fewer than 25 breeding pairs (Woźniak *et al.* 2022). In Ukraine there are just one or two pairs (Dennis 2016).

CENTRAL AND WESTERN EUROPE

The situation in Poland contrasts greatly with neighbouring Germany, where the recovery of Ospreys has been spectacular. There are now an estimated 750 breeding pairs in Germany, with the population more than tripling in recent decades (Schmidt-Rothmund *et al.* 2014, Schmidt-Rothmund pers. comm. 2022). The largest German population is in the federal state of Brandenburg, in the east, where there were 314 territorial pairs in 2008, with a nesting density of 1.06 pairs per 100km². The success of Ospreys in this core area facilitated dispersal to other parts of the country from the early 1990s onwards, aided in particular by the provision of artificial nests on electricity pylons. This led to recolonisation of northwestern Germany (Lower Saxony) and southern Germany (northeastern Bavaria). Meanwhile, it is hoped that the translocation project based in the Canton of Fribourg in western Switzerland will result in the

Figure 9.8. There are approximately 150–180 pairs of breeding Ospreys in Belarus, mainly in forests and bogs (© Denis Kittel).

recolonisation of Switzerland for the first time in more than a century. Early results have been encouraging, as described in chapter 8 (Strahm and Landenbergue 2021).

Ospreys recolonised mainland France in 1985, when a pair bred near Orléans. Orléans Forest has since become a stronghold for the species, aided by the provision of artificial nests and protection of breeding pairs (Wahl and Barbraud 2014). It is thought that the French population now exceeds 100 pairs, with 40 pairs in Orléans Forest and surrounding private land, and a further 40 pairs in the Sologne area, including the Parc de Chambord. In addition, 8–10 pairs breed on electricity pylons further down the Loire River and there are 10–12 pairs in northeast France (R. Wahl pers. comm. 2022). Ospreys bred in southwest France for the first time in 2018, aided by translocation projects in Aquitaine (Lesclaux 2019) and at Urdaibai on the Basque Country coast in Spain, as described in chapter 8.

Ospreys have recently colonised the Netherlands, with the first pair breeding at the Biesbosch in 2016. By 2021 there were five pairs in the area, with a total of eight young fledging that year (P. Voskamp pers. comm. 2022) (Figure 9.9). Several of the breeding birds are colour-ringed individuals from the large German population, while another is from Rutland Water as described in chapter 3. It is likely that the Low Countries will support a growing population of Ospreys in future years and there have been recent breeding attempts in Limburg, Flanders, Belgium (P. Voskamp pers. comm. 2022).

The continuing population expansion of Ospreys in the United Kingdom is a conservation success story, as described in chapter 8. The latest population estimates indicate that there are approximately 300 breeding pairs in Scotland (R. Dennis pers. comm. 2022) and in excess of 35 pairs in England and Wales (see Figure 8.14). Although some populations in northern Scotland appear to have reached carrying capacity, those further south are continuing to increase and

Figure 9.9. Ospreys began breeding in the Netherlands in 2016 and there were five pairs at De Biesbosch in 2021 (© Rob Braat).

there is considerable scope for further expansion through England and Wales. In 2023, Ospreys returned to breed in Northern Ireland, with a pair rearing young in County Fermanagh: the first successful nest on the island of Ireland for more than 200 years.

SOUTHERN EUROPE AND THE MEDITERRANEAN

The Osprey population in southern Europe and the Mediterranean region is small and fragmented compared to historical times, but the situation has been improved in recent decades by reintroduction projects (Monti *et al.* 2018). Mainland Spain (Morandini *et al.* 2019, Galarza *et al.* 2021), Portugal (Palma *et al.* 2019) and Italy (Sforzi *et al.* 2019) now have small breeding populations established through translocation projects, as described in chapter 8. The well-studied population in Corsica remains stable at approximately 30 breeding pairs and there are in excess of 20 pairs on the Balearic Islands (Monti *et al.* 2018). It was also encouraging that, in 2020, Ospreys nested on Sardinia for the first time since 1968 (Fozzi *et al.* 2020). A small number of Ospreys also breed along the north coast of Africa, but face local extinction risks. The breeding population at Al-Hoceima National Park in Morocco declined from a mean 14 breeding pairs in 1983–1990 to nine pairs in 2012–13 (Monti *et al.* 2013). Meanwhile, there is another small population of fewer than 10 pairs on the Mediterranean coast of western Algeria (Monti *et al.* 2018).

In the Atlantic Ocean, the isolated breeding population in Cape Verde has increased by approximately 30 per cent since around the turn of the millennium, and now stands at approximately 115 pairs (Palma *et al.* 2020) (Figure 9.10). The situation in the Canary Islands is less positive, where the local breeding population has declined to 4–7 pairs (Figure 9.11) (Schmidt-Rothmund *et al.* 2014, M. Siverio, B. Rodríguez and F. Siverio pers. comm. 2022).

Figure 9.10. There are approximately 115 pairs of breeding Ospreys in Cape Verde, with nests in a variety of locations, including on the ground (© Pedro López).

Longevity and survival

Figure 9.11. Ospreys at a nest in Tenerife in the Canary Islands, where the species has become a rare breeder, with 4–7 pairs (© Beneharo Rodríguez).

SOUTHEAST EUROPE

The Osprey is very rare in southeastern Europe, where it remains absent from many parts of its historical range. There were 10–20 pairs of breeding Ospreys in Turkey in the mid-1980s but there have been no confirmed breeding attempts in recent years (Schmidt-Rothmund *et al.* 2014). Likewise, recent breeding activity in Bulgaria has been limited to a few isolated pairs, mainly on the Trakia Plain and in the east of the country (Iankov 2007), while only two breeding pairs were reported from Moldova in the 1990s (Schmidt-Rothmund *et al.* 2014) and no confirmed breeding attempts more recently (Zubcov *et al.* 2014). In addition, recent data has shown there are 2–5 pairs in Armenia and fewer than five pairs in Azerbaijan (Dennis 2016).

MIDDLE EAST AND NORTHEAST AFRICA

Ospreys breed along the Red Sea coast of northeast Africa, from Egypt south through Sudan and Eritrea to Djibouti (Shobrak *et al.* 2003) (Ali *et al.* 2015) (Figure 9.12). Habib (2019) located 40 breeding pairs on islands in the Red Sea and South Sinai provinces of Egypt during surveys carried out between 2012 and 2018. Nests were mainly on sand dunes, but some were found on rooftops. This represents a significant decline from approximately 150–180 pairs in the in the early 1990s (Fisher *et al.* 2001), thought to be due to increasing tourism pressure. Further south, Semere *et al.* (2008) estimated that 220 pairs of Ospreys breed on the Eritrean Islands in the Red Sea, with nests located on 153 of the 325 islands surveyed.

Breeding Ospreys are numerous along the Red Sea coast of Saudi Arabia, with the highest breeding concentrations on islands, including Farasan Island and Al Wajh Archipelago, where there are an estimated 500 pairs (Jennings 2010). Meanwhile, Shobrak and Aloufi (2014) located at least one pair of Ospreys on all 16 northern Red Sea islands they surveyed between 2010 and

Figure 9.12. An adult (right) and juvenile (left) in Sudan. Ospreys breed along the Red Sea coast in northeast Africa, from Egypt south to Djibouti (© Tim Sarney).

2011, with the larger islands often holding two active nests. Al-Saghier (2002) recorded 66 pairs in the Yemeni Red Sea.

Elsewhere in the Middle East, the United Arab Emirates, Bahrain and Oman also hold breeding populations. Approximately 75 per cent of breeding Ospreys in the region occur in the United Arab Emirates, where the population was estimated to be 70–75 pairs by Khan *et al.* (2008) after an extensive survey of most of the coastline of the Abu Dhabi Emirate and islands in the Persian Gulf. Forty-one nests were found on islands, and 15 at coastal sites, with 27 per cent of nests on artificial platforms. Quieter, more remote islands are increasingly favoured due to the lack of human disturbance (Khan *et al.* 2008). As in other parts of the region, it is clear that development for tourism has had an impact, with the loss of nearly 80 per cent of former key island breeding sites such as Yasat, Judairah and Ghagah.

Elsewhere in the region, a population of approximately 20 pairs nest on Hawar Island in Bahrain (Aspinall *et al.* 2003). Small breeding populations are also found in Oman, on the Gulf of Oman coast and also on the shores of the Arabian Sea (Khan *et al.* 2008).

Japan and Eastern Eurasia

The Osprey breeds across the whole of Russia, but little is known about its status in the more remote regions, particularly in central and eastern areas (Kuchin 2004). Shikalova and Vinogradov (2011) located nine active nests during summer 2009, and seven the following year, in the Sayano-Shushensky Nature Reserve, on the southern bank of the Yenisei River along the Sayano-Shushenskoye reservoir in the West Sayan mountains of south Siberia. Mitrofanov (2016) reported that the breeding population in the wider southern Siberia region is declining.

In July 2011, ornithologists visiting the remote Mys Shmidta on the Arctic coast of Chukotka to study Spoon-billed Sandpipers *Calidris pygmaea* recorded a fishing Osprey at Erokymanky lagoon (Arkhipov *et al.* 2013). This is the most north-easterly record of an Osprey in Eurasia and approximately 600km from its known breeding range. While this may have a been wandering non-breeding third- or fourth-calendar-year bird, the presence of breeding Ospreys at a similar latitude in North America indicates that breeding in such a location cannot be excluded.

The status and distribution of the Osprey in Japan is poorly known. Shoji *et al.* (2011) located 62 Osprey nests in eight of 14 districts of the northern island of Hokkaido, with most nests in the west. More than half of the nests they located were on rocks or cliffs. Overall, the population of the whole of Japan is thought to number fewer than 1,000 individuals and is classified as Near Threatened by the Japanese Ministry of the Environment (Nagai *et al.* 2021). There seems to have been a recent shift to inland areas, with an increasing number of pairs breeding around reservoirs (Nagai *et al.* 2021).

Longevity and survival

NORTH AMERICA AND MEXICO

North America accounts for almost two-thirds of the world's Ospreys, and Poole (2019) and Bierregaard *et al.* (2020) provide a detailed overview of the most recent population changes of *P. h. carolinensis*. Significant increases of the order of 50–100 per cent have been recorded in most areas since 1990, although there are some more remote parts of Canada where population estimates are very difficult.

UNITED STATES

The United States holds approximately 25,000–30,000 breeding pairs, with the population widely distributed across the country. Indeed, by 2011 nesting Ospreys were absent from just two or three states (Poole 2019). The core of the United States population nest along the East Coast, where the best estimates of current population size have been made. Florida, in the south, supports at least 5,000 pairs, with particularly high concentrations around shallow inland lakes where flat-topped cypress trees (*Taxodium* spp.) sometimes hold multiple nests (Poole 2019). Moving north, Ospreys are widely distributed around inland reservoirs in South Carolina and North Carolina, and also breed along the coast, with 200 pairs within 25km of Parris Island (Bierregaard *et al.* 2020).

The stronghold for North American Ospreys is found around Chesapeake Bay, where there

Figure 9.13. The Osprey's ability to adapt to human-dominated landscapes has enabled it to make a spectacular recovery in the United States. This photograph shows nestling Ospreys on an artificial nest platform along the Yellowstone River, Montana. The nest was located within an industrial area adjacent to an active railroad yard (© Marco Restani).

are at least 9,000 breeding pairs (Poole 2019). Further north along the Atlantic coast, New Jersey holds at least 420 breeding pairs (Clark and Wurst 2014), and the Osprey population in southern New England and Long Island, New York, now exceeds pre-DDT levels, with more than 1,200 breeding pairs (Bierregaard *et al.* 2014a). Continuing north, the breeding population in Vermont and New Hampshire stands at approximately 275 pairs, with a further 1,500 pairs in Maine (Bierregaard *et al.* 2020).

Elsewhere, breeding populations are widely scattered, with translocations, artificial nests and the species' ability to exploit a range of anthropogenic structures for breeding facilitating population expansion in many areas (Figure 9.13), including around inland reservoirs. For example, reservoirs at Flathead Lake and along the upper reaches of the Missouri River support up to 200 pairs, with many breeding on artificial nests (Poole 2019). Further west, the breeding population of Ospreys along the Willamette River in Oregon, where electricity pylons are frequently exploited for breeding, now numbers over 250 pairs. Meanwhile, the provision of artificial nests has also facilitated the establishment of breeding pairs around San Francisco Bay in California (Poole 2019).

CANADA

Poole (2019) estimated that the Canadian population is approximately half that of the United States, thus 12,500–15,000 pairs, with breeding Ospreys widely distributed across the country. Like further south, the east coast of Canada is a stronghold, with pairs breeding as far north as 59° latitude in Labrador. Ospreys are also numerous in the Great Lakes basin, where the population was estimated to number 900 pairs in 1988–1993 by Ewins (1996), including well-monitored populations of approximately 60 pairs each in the Kawartha Lakes and Georgian Bay regions (Poole 2019). Further west, breeding Ospreys are widely distributed in the Rocky Mountains and along the Pacific Coast, although not in the densities that occur on the East Coast (Bierregaard *et al.* 2020). The most northerly breeding Ospreys occur at 68° latitude in the province of Nunavut, where the lack of trees means that breeding pairs depend on human structures for nesting. Ospreys also occur in the Boreal Forests of northern Canada, although generally in low densities (Poole 2019).

MEXICO

The ability of the Osprey to exploit a wide range of environments in exemplified by the fact that Mexico supports significant breeding populations in the northwest of the country, centred on the Gulf of California. Breeding pairs are widely distributed on the coastlines of the Baja Peninsula, along the mainland coast of Sonora and Sinaloa, and on islands in the Gulf of California itself. Aerial surveys undertaken by Henny *et al.* (2007) identified a minimum of 1,342 nesting pairs in this region, which was an 81% increase on surveys undertaken in 1997, but only a 3 per cent increase compared to 1992/93, indicating that the population may have plateaued. As in other areas, there has been a recent shift to the use of artificial structures by Ospreys in Mexico.

CARIBBEAN

The Caribbean race of the Osprey, *P. h. ridgwayi*, exists in the Bahamas, the Turks and Caicos Islands, Cuba, the Yucatan Peninsula and Belize. It generally occurs in fairly low numbers and is the least studied of the four subspecies (Wiley *et al.* 2014). It is thought likely that its population is severely reduced compared to historical times (Poole 2019).

In the Bahamas, breeding Ospreys are widespread south of the Northeast Providence

Channel, although generally in low densities. Wiley *et al.* (2014) report breeding from a total of 25 islands and cays. Meanwhile, Ospreys are fairly common and widespread on the Turks and Caicos Islands, with breeding records for 12 islands and cays (Wiley *et al.* 2014). There have been no thorough surveys of breeding Ospreys in the region, but Poole (2019) estimated there are probably 100–200 pairs overall, with nests on low rocky outcrops, mangroves and navigational structures.

In Cuba, Ospreys are associated with coastal bays, cays and saltwater areas, as well as coastal lakes and river mouths (Wiley *et al.* 2014). Most observers consider the Osprey to be a rare permanent resident, except in Oriente province where it is more numerous (Garrido and García Montaña 1975, Wotzkow 1985, Wiley *et al.* 2014). However, as Poole (2019) points out, the coastline is extensive and remains largely unsurveyed, meaning there are no population data. An estimate based on the population density of Ospreys elsewhere in the Caribbean, and the extent of the Cuban coastline, suggest there could be 500–1,000 pairs, making Cuba the likely stronghold for *P. h. ridgwayi* (Poole 2019). In recent years *P. h. carolinensis* has also been recorded breeding in Cuba, and there have been some reports of mixed pairs reproducing (Wiley *et al.* 2014).

Sprunt (1977) undertook aerial surveys of the Yucatan coast and found nine nests at Bahía de la Ascensión and five at Bahía del Espírito Santo, but only one or two other nests outside this region in Caribbean Mexico. More recent population data are available for Belize, where Poole (2019) reports 60–70 breeding pairs on mangrove cays that have formed behind the fringing barrier reef. Smaller numbers of nests are also found on offshore islands. Surprisingly, there are no breeding Ospreys in Honduras, immediately to the south of Belize, despite the fact that potential breeding habitat is unchanged (Poole 2019).

Australia and Pacific Islands

The Australasian Osprey is the only *Pandion* subspecies with its primary distribution in the Southern Hemisphere. It occurs over a broad region between New Caledonia in the South Pacific, across Australia and New Guinea, north to central Indonesia.

Dennis and Clancy (2014) provided a detailed overview of the status and distribution of the Osprey in Australia, where the population is largely confined to coastal areas, including estuaries to their tidal limit (Marchant and Higgins 1993). The largest population densities are found across the northern tropical and subtropical regions, particularly on the northwestern coastline and on the many islands in the Gulf of Carpentaria (Chatto 2006, Dennis and Clancy 2014). Ospreys are also numerous in tropical regions of Western Australia, where they breed on islands and on the lower reaches of major rivers. It is thought there could be as many as 500–1,000 pairs along this vast and remote section of coastline (Poole 2019). Dennis and Clancy (2014) report that a survey by the Western Australia Department of Environment and Conservation located 10 occupied territories on the North West Cape peninsula, with nests approximately 17.7km apart. The populations in this northern part of the species' range in Australia appear relatively stable and secure (Debus 2012, Dennis and Clancy 2014).

Further south, Ospreys are more sparsely distributed. There are no breeding Ospreys on a 500km stretch of coastline along the Great Australian Blight in South and Western Australia, but then a well-monitored population further east in South Australia (Dennis and Clancy 2014) (Figure 9.14). Comprehensive surveys in South Australia, first undertaken between 2008 and 2010 and then repeated again in 2015–17, revealed a decline in the breeding population from 58–43 pairs (Detmar and Dennis 2018). The declines were most evident in the west of the state, where the number of occupied territories decreased from 33 in 2010 to 22 in 2017, and on

Figure 9.14. Ospreys breed in many coastal areas of Australia (© Andrew Brooks).

Kangaroo Island, situated off the coast near Adelaide, where the population fell from 14 to eight pairs. In South Australia's open coastal landscape, Osprey nest-sites are particularly vulnerable to disturbance from human activity and it is thought that this may be a key contributing factor in the recent declines (Detmar and Dennis 2018).

Ospreys are absent from the cold coastal waters of Tasmania and Victoria, but the population in New South Wales has increased markedly in recent decades. There has been a significant southern expansion of more than 800km beyond the historical breeding range in the region. The southernmost breeding record prior to 1996 was Lake Macquarie, approximately 400km north of Sydney; but in 2008 nesting was confirmed approximately 620km further south. The number of breeding pairs in the state is now estimated to be more than 125 (Ekert and Brady 2007, Dennis and Clancy 2014). It is thought that this southern expansion of range may be linked to changing climatic conditions associated with the strengthening of the tropical Eastern Australia Current (EAC) in recent decades, resulting in warmer sea temperatures (Ridgeway and Hill 2009, Dennis and Clancy 2014). This expansion has also been assisted by pairs choosing to nest on communication towers and other structures, including a series of strategically placed artificial nest platforms, which approximately 30% of breeding pairs now utilise (Moffatt 2009).

Further north, the Osprey breeds along the entire coastline of Queensland, particularly in river estuaries and on the islands of the Great Barrier Reef Marine Park, with the population there estimated to number 250–500 pairs (Dennis and Clancy 2014, Poole 2019). In south-eastern Queensland breeding pairs can be found in heavily urbanised areas, where they readily use artificial structures, as well as natural landscapes with little human presence (Thomson *et al.* 2019).

Elsewhere, the population status of the Eastern Osprey is less well known. The population on New Caledonia was estimated to number 250 pairs by Bretagnolle *et al.* (2001).

Longevity and survival

Figure 9.15. The sight of a hunting Osprey is one that more people can now enjoy thanks to the continued recovery of the species in recent decades (© John Wright).

Baudat-Franceschi *et al.* (2008) recorded 21 pairs of Ospreys in the Northwest Lagoon, between Koumac and Yandé, while Baling *et al.* (2008) located breeding Ospreys on five of 23 islands they surveyed in the Southern Lagoon, situated off the southern part of Grande Terre. As in other parts of their range, Ospreys in New Caledonia are well integrated with the human population, with some nests on artificial structures.

Other breeding populations occur, typically in relatively low densities, along the coasts of New Guinea (River *et al.* 2007), and west to the Moluccas and Lesser Sundas islands of Indonesia (Mees 2006, Mittermeier *et al.* 2013).

The future

It is encouraging that the Osprey is either stable or increasing in most areas, but this wide global distribution should not hide the fact that the species remains absent or very rare in parts of its historical range and is at risk of local extinction in some places. The severe population declines that occurred as a result of the impacts of DDT demonstrate that any new or emerging threats should be treated seriously, and that ongoing efforts to restore the species to its former range must be encouraged. The success of artificial nest building and translocation programmes show that the Osprey is a species that responds well to proactive conservation interventions. In my view, restoring this iconic species to areas where it has been lost due to anthropogenic actions is a moral responsibility. At a time of great concern for our planet due to the climate and biodiversity crises, the Osprey can bring hope and, crucially, inspire more people to take an interest in the natural world and to help safeguard its future (Figure 9.15).

Conclusions

- Significant mortality of Ospreys occurs during their first year, particularly when migrating south for the first time and when searching for a suitable wintering site.
- Survival increases once young birds reach two years. By this age they have usually established a wintering site, completed one full migration cycle (i.e. one flight south and one north) and returned to their natal area for the first time. This experience greatly increases their survival prospects.
- Annual survival of breeding adults may exceed 90 per cent and, in some cases, adult Ospreys can live into their late twenties.
- Although Osprey populations recovered well during the late twentieth and early twenty-first centuries, local threats do remain, which are often anthropogenic in nature. These include electrocution and collisions; illegal persecution; disturbance, particularly in relation to tourism and recreational activities; netting and entanglement; pollutants, including PBDEs, which are considered a potential emerging threat; and overfishing.
- Interspecific competition with some species, including *Haliaeetus* eagles such as the White-tailed Eagle and Bald Eagle may change individual Ospreys' habits in some areas, but they generally coexist well.
- Osprey nestlings, and occasionally adults, can be vulnerable to predation by a range of species, particularly Northern Goshawk, Eagle Owl, Great Horned Owl, Bald Eagle and, in some cases, White-tailed Eagle.
- The world population of Ospreys now likely exceeds 50,000 pairs with close to two-thirds in North America. Although some populations are vulnerable to local extinction, most are stable or increasing.

APPENDIX 1
Osprey breeding attempts in the Rutland Water area 2001–2022

Year	Pairs	Origin of breeding birds			Young	Prod.	No. survived to two years
		Tran.	Scot.	Rut.			
2001	1	1♂	1♀		1	1.0	0
2002	1	1♂	1♀		0	0.0	0
2003	2	2♂ & 2♀			5	2.5	0
2004	1	1♂ & 1♀			2	2.0	2
2005	1	1♂ & 1♀			3	3.0	2
2006	1	1♂ & 1♀			3	3.0	1
2007	2	2♂ & 1♀		1♀	5	2.5	0
2008	3	3♂ & 1♀	1♀	1♀	3	1.0	2
2009	4	4♂	2♀	2♀	9	2.25	4
2010	5	3♂	3♀	2♂ & 2♀	12	2.4	6
2011	5	3♂	3♀	2♂ & 2♀	10	2.0	3
2012	4	3♂	2♀	1♂ & 2♀	9	2.25	3
2013	5	1♂	3♀	4♂ & 2♀	14	2.8	4
2014	6	1♂	3♀	5♂ & 3♀	11	1.83	1
2015	8	1♂	3♀	7♂ & 5♀	15	1.88	4
2016	8		2♀	8♂ & 6♀	15	1.88	6
2017	8		3♀	8♂ & 5♀	16	2.0	4

The Osprey

Year	Pairs	Origin of breeding birds			Young	Prod.	No. survived to two years
		Tran.	Scot.	Rut.			
2018	9		3 ♀	9 ♂ & 6 ♀	14	1.55	2
2019	10		3 ♀	10 ♂ & 7 ♀	24	2.4	4
2020	10		3 ♀	10 ♂ & 7 ♀	19	1.9	2
2021	8		3 ♀	8 ♂ & 5 ♀	19	2.38	7
2022	11		6 ♀	11 ♂ & 5 ♀	22	2.0	?
Total	113				231	2.04	49

Tran. = Translocated; Scot. = Scotland or unringed; Rut. = Rutland; Prod. = Productivity.
No. survived to two years should be considered a minimum given that birds may have gone unrecorded, particularly in recent years

APPENDIX 2
Selected Osprey viewing sites in the UK

Scotland

Loch Garten, Highlands

More than 60 years after Ospreys first bred here, Loch Garten remains one of the best Osprey-viewing locations in the UK. The RSPB visitor centre in the heart of Abernethy Forest provides excellent views of the long-established nest, and staff and volunteers are on hand to give you all the latest news.

www.rspb.org.uk/reserves-and-events/reserves-a-z/loch-garten

Loch of the Lowes, Perthshire

Like Loch Garten, the Scottish Wildlife Trust Reserve at Loch of the Lowes is synonymous with breeding Ospreys. A purpose-built viewing hide, just a short walk from the visitor centre, offers panoramic views across the loch to the Osprey nest on its shore.

www.scottishwildlifetrust.org.uk/reserve/loch-of-the-lowes

England

Foulshaw Moss, Cumbria

Ospreys first bred at Foulshaw Moss – a lowland raised peatbog of international importance south of Kendal – in 2014, and they now return each year. You can observe the birds from a special viewpoint on this Cumbria Wildlife Trust reserve.

www.cumbriawildlifetrust.org.uk/reserves/foulshaw-moss

Kielder Forest, Northumberland

Ospreys returned to breed in Kielder Forest in 2009, and a team of knowledgeable volunteers share their expertise with visitors from an Osprey viewing point and wildlife cabin at Tower Knowe Visitor Centre. The Kielder Osprey Project is a partnership between Northumberland Wildlife Trust, Kielder Water and Forest Park Development Trust and Forestry England.

www.kielderospreys.wordpress.com

Rutland Water, Rutland

The Lyndon Visitor Centre forms the base of the Rutland Osprey Project, a partnership between the Leicestershire and Rutland Wildlife Trust and Anglian Water. Live images are relayed to the visitor centre, and two hides offer spectacular views of the Manton Bay nest. In addition, special Osprey cruises run on the *Rutland Belle* offer a way to see fishing Ospreys in action.

www.ospreys.org.uk

POOLE HARBOUR, DORSET

The site of the second English Osprey translocation is a great place to view these birds, particularly from boat trips organised by Birds of Poole Harbour. Fishing Ospreys can be seen anywhere around the harbour, including from RSPB Arne.

www.birdsofpooleharbour.co.uk

www.rspb.org.uk/reserves-and-events/reserves-a-z/arne/

Wales

CORS DYFI, POWYS

Ospreys first nested at the Montgomeryshire Wildlife Trust's Cors Dyfi Reserve in 2011, and you can enjoy superb views of the nest, along with live CCTV images from the purpose-built 360° Observatory, at the Dyfi Wildlife Centre.

www.dyfiospreyproject.com

GLASLYN, GWYNEDD

Ospreys have bred in the Glaslyn Valley each year since 2004. Bywyd Gwyllt Glaslyn Wildlife operates a viewing point in a stunning location at Pont Croesor, with a backdrop of Snowdonia.

www.glaslynwildlife.co.uk

Osprey photography

There are now several places where wildlife photographers have the opportunity to photograph fishing Ospreys at very close quarters from specially designed hides.

AVIEMORE, HIGHLANDS

Wildlife photographers have the opportunity to enjoy stunning close-up views of fishing Ospreys from specially designed photographic hides at Rothiemurchus Fishery and also at nearby Aviemore Ospreys.

www.rothiemurchus.net/outdoor-activities-at-aviemore/wildlife-watching-photography/

www.aviemoreospreys.co.uk

HORN MILL TROUT FARM, RUTLAND

A photographic hide at this working trout farm provides the best opportunity of photographing fishing Ospreys anywhere in England.

www.rivergwashtroutfarm.co.uk/horn-mill-osprey-hide/

References

Agostini, N., M. Panuccio, and C. Pasquaretta. 2015. Morphology, flight performance, and water crossing tendencies of Afro-Palearctic raptors during migration. *Current Zoology* 61: 951–958.

Al-Saghier, O. 2002. Survey of the Breeding Seabirds in Red Sea of the Republic of Yemen. Report for PERSGA, Jeddah.

Alerstam, T. 1990. *Bird Migration*. Cambridge: Cambridge University Press.

Alerstam, T. 2000. Bird migration performance on the basis of flight mechanics and trigonometry. In Domenici, P. and R.W. Blake (eds) *Bio-mechanics in Animal Beh*aviour (pp. 105–124). Oxford: Bios Scientific.

Alerstam, T. 2006. Strategies for the transition to breeding in time-selected bird migration. *Ardea* 94: 347–357.

Alerstam, T. 2011. Optimal bird migration revisited. *Journal of Ornithology* 152: S5–S23.

Alerstam, T., and A. Lindström 1990. Optimal bird migration: the relative importance of time, energy, and safety. In Gwinner, E. (ed.) *Bird Migration: Physiology and Ecophysiology* (pp. 331–351). Heidelberg, Germany: Springer.

Alerstam, T., and A. Hedenström. 1998. The development of bird migration theory. *Journal of Avian Biology* 29: 343–369.

Alerstam, T., M. Hake, and N. Kjellén. 2006. *Temporal and spatial patterns of repeated migratory journeys by Ospreys*. Animal Behaviour 71: 555–566.

Ali, A.A.M., Z.N.E. Mahmoud, and S.E.M. Elamin. 2015. Nesting, Hatchling Breeding and Feeding of Osprey in Um El Sheikh Island, Dongonab Bay, Sudan. *Advances in Environmental Biology* April: 226–229.

Allen, L.L., K.L. Morrison, W.A.E. Scott, S. Shinn, A.M. Haltiner, and M.J. Doherty. 2018. Differences between stance and foot preference evident in Osprey fish holding during movement. *Brain and Behavior* 8: e01126.

Ames, Peter L. 1966. DDT residues in the eggs of the osprey in the north-eastern United States and their relation to nesting success. *Journal of Applied Ecology* 3: 87–97.

Anderson, D.W., T.H. Suchanek, C.A. Eagles-Smith, and T.M. Cahill Jr. 2008. Mercury residues and productivity in Osprey and grebes from a mine-dominated ecosystem. *Ecological Applications* 18(8): A227–A238.

Arkhipov, V.Y., T. Noah, S. Koshkar, and F.A. Kondrashov. 2013. Birds of Mys Shmidta, north Chukotka, Russia. *Forktail* 29: 25–30.

Arné, P., V. Risco-Castillo, G. Jouvion, C. Le Barzic, and J. Guillot. 2021. Aspergillosis in wild birds. *Journal of Fungi* 7: 241.

Aspinall, S., I. Al Madany, H. King, *et al.* 2003. Hawar Islands Biosphere Reserve Study, Bahrain. National Commission for Wildlife Protection, Kingdom of Bahrain.

Atlantic States Marine Fisheries Commission (ASMFC) 1993. *Stock assessment update and overview of interstate fishery management activities for inshore stocks of winter flounder Pseudopleuronectes americanus*. E.T. Christian (ed.). Atlantic States Marine Fisheries Commission, Baltimore, MD.

Avian Power Line Interaction Committee (APLIC). 2006. *Suggested practices for avian protection on power lines: the state of the art in 2006*. Edison Electric Institute, APLIC, and the California Energy Commission, Washington, DC/Sacramento.

Babuskin M.V., and A.V. Kuznicov. 2012. Diet of the Osprey in the Darwin Nature Reserve (North-Western Russia). *Proceedings of the VI International Conference on Raptors and Owls of Northern Eurasia*: 330–334. [In Russian]

Babushkin, M.V., and A.V. Kuznetsov. 2013. The Current Number and Distribution of the Osprey and the White-tailed Eagle in North-West Russia. *Raptors Conservation* 27: 32–39.

Babushkin, M.V., A. Kuznetsov, and M. del Mar Delgado. 2019. Autumn migratory patterns of north-west Russian Ospreys. *Ardeola* 66: 119–128.

Backman, J., and T. Alerstam. 2003. Orientation scatter of free-flying nocturnal passerine migrants: components and causes. *Animal Behaviour* 65: 987–996.

Bai, M.L., D. Schmidt, E. Gottschalk, and M. Mühlenberg. 2009. Distribution pattern of an expanding Osprey population in a changing environment. *Journal of Ornithology* 150: 255–263.

Balbontín, J., and M. Ferrer. 2005 Factors affecting the length of the post-fledging period in the Bonelli's Eagle *Hieraaetus fasciatus*. *Ardea* 93(2): 189–198

Baling, M., D.H. Brunton, and D. Jeffries. 2008. Marine and coastal bird survey of islands in the Southern Lagoon, New Caledonia. *Notornis* 55: 111–113.

Bartosik, M.B. 2009. Osprey: notes on unknown and poorly studied behaviors. *Bulletin of the Texas Ornithological Society* 42: 18–36.

Baudat-Franceschi, J., J. Spaggiari, and N. Barré. 2008. Breeding seabirds of conservation interest. In *A Rapid Marine Biodiversity Assessment of the Coral Reefs of the Northwest Lagoon, between Koumac and Yandé, Province Nord, New Caledonia* (p. 64). Conservation International.

Beddow, T. 1990. The recovery of the East Tennessee Osprey population. *The Migrant* 61: 92–94.

Beech, M. 2003. The diet of Osprey on Marawah island, Abu Dhabi Emirate, UAE. *Tribulus* 13: 22–25.

Bechard, M.J., and C. Márquez-Reyes. 2003. Mortality of wintering Ospreys and other birds at aquaculture facilities in Colombia. *Journal of Raptor Research* 37: 292–298.

Bent, A.C. 1937. Life histories of North American birds of prey, Part 1. *United States National Museum Bulletin* 167.

Bider, J.R., and D.M. Bird. Distribution and densities of Osprey populations in the Great Whale region of Quebec. In *Proceedings of the 1st international symposium on bald eagles and ospreys*, Montreal, QC.

Bierregaard, R.O., A.B.D.L. Gibson, R.S. Kennedy, A.F. Poole, M.S. Scheibel, and J. Victoria. 2014a. Post-DDT recovery of Osprey populations in southern New England and Long Island, New York, 1970–2013. *Journal of Raptor Research* 48: 361–374.

Bierregaard, R.O., A.F. Poole, and B.E. Washburn. 2014b. Ospreys (*Pandion haliaetus*) in the 21st century: Populations, migration, management, and research priorities. *Journal of Raptor Research* 48: 301–308.

Bierregaard, R.O., A.F. Poole, M.S. Martell, P. Pyle, and M.A. Patten. 2020. *Osprey (Pandion haliaetus)*, version 1.0. In Rodewald (ed.) *Birds of the World*. Ithaca, NY: Cornell Lab of Ornithology

Bijleveld, M. 1974. *Birds of Prey in Europe*. MacMillan Press, London.

BirdLife Australia. 2019. Working List of Australian Birds. https://birdlife.org.au/conservation/science/.

BirdLife International (2022) Species factsheet: *Pandion haliaetus*. BirdLife International (2022) IUCN Red List for birds. Downloaded from http://www.birdlife.org on 22/04/2022.

Birkhead, T.R., and C. M. Lessells. 1988. Copulation behaviour of the osprey *Pandion haliaetus*. *Animal Behaviour* 36: 1672–1682.

Blem, C.R., L.B. Blem, and P.J. Harmata. 2002. Twine causes significant mortality in nestling ospreys. *The Wilson Bulletin* 114: 528–529.

Bohrer, G., D. Brandes, J.T. Mandel, *et al.* 2012. Estimating updraft velocity components over large spatial scales: contrasting migration strategies of golden eagles and turkey vultures. *Ecology Letters* 15: 96–103.

Bretagnolle, V., and J-C. Thibault. 1993. Communicative behaviour in breeding Ospreys (*Pandion haliaetus*): description and relationship of signals and life history. *The Auk* 110: 736–751.

Bretagnolle, V., J-C. Thibault, and J-M. Dominici. 1994. Field identification of individual Ospreys using head marking pattern. *Journal of Wildlife Management* 58: 175–178.

Bretagnolle, V., M. Pandolfi, V. Lecoq, and J. Broudissou. 2001. Le balbuzard pêcheur *Pandion haliaetus* en Nouvelle-Calédonie: effectif, eintroduct et menaces. *Alauda* 69: 491–501.

Bretagnolle, V., F. Mougeot, and J-C. Thibault. 2008. Density dependence in a recovering osprey population: demographic and behavioural processes. *Journal of Animal Ecology* 77: 998–1007.

Brochet, A-L., W. Van den Bossche, S. Jbour, *et al.* 2015. Preliminary assessment of the scope and scale of illegal killing and taking of birds in the Mediterranean. *Bird Conservation International* 26: 1–28.

Brown, P., and G. Waterston. 1962. *The Return of the Osprey*. London: Collins.

Brown S.P., G. Loot, A. Teriokhin, A. Brunel, C. Brunel, and J.F. Guégan. 2002. Host manipulation by *Ligula intestinalis*: a cause or consequence of parasite aggregation? *International Journal for Parasitology*. 32: 817–824.

Bryan, G.W. 1984. Pollution due to heavy metals and their compounds. *Marine Ecology* 5: 1289–1431

Bustamante, J. 1993. Post-fledging dependence period and development of flight and hunting behaviour in the red kite *Milvus milvus*. *Bird Study* 40: 181–188.

Bustamante, J. 1994. Family break-up in Black and Red Kites *Milvus migrans* and *M. milvus*: is time of independence an offspring decision? *Ibis* 136: 176–184.

Bustamante, J. 1995. Duration of the post-fledging period of Ospreys at Loch Garten, Scotland. *Bird Study* 42: 31–36.

Canal, D., V. Morandini, B. Martín, *et al.* 2018. Productivity is related to nest-site protection and nesting substrate in a German Osprey population. *Journal of Ornithology* 159: 265–273.

Carpenteri, S. 1997. *The Fish Hawk: Osprey*. Minnetonka, Minnesota: North Word Press.

Carrete, M., J.A. Donázar, and A. Margalida. 2006. Density-dependent productivity depression in Pyrenean bearded vultures: implications for conservation. *Ecological Applications* 16: 1674–1682.

Carson, R. 1962. *Silent Spring*. Boston, MA: Houghton Mifflin Harcourt.

Carss D.N., and K. Brockie 1994. Prey remains at Osprey nests in Tayside and Grampian 1987–1993. *Scottish Birds* 17: 132–145.

References

Carss, D.N., and J.D. Godfrey 1996. Accuracy of estimating the species and sizes of Osprey prey: a test of methods. *Journal of Raptor Research* 30: 57–61.

Cartron, J-L.E., and M.C. Molles. 2002. Osprey diet along the eastern side of the Gulf of California, Mexico. *Western North American Naturalist* 62: 249–252.

Casado, E., and M. Ferrer. 2005. Analysis of reservoir selection by wintering ospreys in Andalusia, Spain: A potential tool for reintroduction. *Journal of Raptor Research* 39: 168–173.

Cash, C.G. 1914. History of Loch an Eilein Ospreys. *Scottish Naturalist* 25: 149–58.

Chatto, R. 2006. The distribution and status of waterbirds around the coast and coastal wetlands of the Northern Territory. *Technical Report 76/2006*. Parks and Wildlife Commission of the Northern Territory, Palmerston, Australia.

Chevallier, D., Y. Handrich, J-F. Georges, *et al.* 2010. Influence of weather conditions on the flight of migrating black storks. *Proc. R. Soc. B.* 277: 2755–2764.

Chen, Da, R.C. Hale, B.D. Watts, M. J. La Guardia, E. Harvey, and E.K. Mojica. 2010. Species-specific accumulation of polybrominated diphenyl ether flame retardants in birds of prey from the Chesapeake Bay region, USA. *Environmental Pollution* 158: 1883–1889.

Christidis, L., and Boles W.E. 2008. *Systematics and Taxonomy of Australian Birds*. Collingwood, Victoria: CSIRO Publishing.

Clancy, G.P. 2005. The diet of the Osprey (*Pandion haliaetus*) on the north coast of New South Wales. *Emu* 105: 87–91.

Clancy, G.P., 2006. The breeding biology of the Osprey *Pandion haliaetus* on the north coast of New South Wales. *Journal of the Australian Bird Study Association* 30: 1–8.

Clark, K.E., W. Stansley, and L.J. Niles. 2001. Changes in contaminant levels in New Jersey osprey eggs and prey, 1989 to 1998. *Archives of Environmental Contamination and Toxicology* 40: 277–284.

Clark, K.E., and B. Wurst. 2014. The 2014 Osprey Project in New Jersey. Endangered and Nongame Species Program, N.J. Division of Fish and Wildlife.

Clobert, J., E. Danchin, A.A. Dhondt, and J.D. Nichols. 2001 *Dispersal*. New York: Oxford University Press.

Cold, C.W. 1993. Adult male Osprey killed at nest by Great Horned Owl. *Passenger Pigeon* 55: 269–270.

Creswell, R.T., 1902. *Aristotle's History of Animals in Ten Books*. London: George Bell.

D'Amico, M., R.C. Martins, J.M. Álvarez-Martínez, M. Porto, R. Barrientos, and F. Moreira. 2019. Bird collisions with power lines: Prioritizing species and areas by estimating potential population-level impacts. *Diversity and Distributions* 25: 975–982.

Dailey, J. 2021. Kielder Ospreys Blog. Additional Information. https://kielderospreys.wpcomstaging.com/additional-information/

Dailey, J. 2022. Kielder Ospreys Blog. Nest 1A. https://kielderospreys.wpcomstaging.com/category/nest-1a/.

Dale, S. 2001. Female-biased dispersal, low female recruitment, unpaired males, and the extinction of small and isolated bird populations. *Oikos* 92: 344–356.

Danemann, G.D. 1994. Biologia eintroducti del Aguila Pescadora (*Pandion haliaetus*) en Isla Ballena, Laguna San Ignacio, Baja California Sur, Mexico. Instituto Politecnico Nacional, Departamento de Biologia Marina, CICIMAR La Paz, Baja California Sur, Mexico

Danemann, G.D., and J.R. Guzman-Póo. 1992. Notes on the birds of San Ignacio Lagoon, Baja California Sur, Mexico. *Western Birds* 23: 11–19.

D'Arcy, G. 1999. *Ireland's Lost Birds*. Dublin: Four Courts Press.

Debus, S. 2012. *Birds of Prey of Australia* (second edition). Collingwood, Victoria, Australia: CSIRO Publishing.

Deem, S.L., S.P. Terrell, and D.J. Forrester. 1998. A retrospective study of morbidity and mortality of raptors in Florida: 1988–1994. *Journal of Zoo and Wildlife Medicine* 29: 160–164.

Del Hoyo, J., A. Elliott, and J. Sargatal. 1992. *Handbook of the Birds of the World. Vol. 1, no. 8*. Barcelona: Lynx edicions.

Dennis, R. 2008. *A Life of Ospreys*. Dunbeath, Caithness, Scotland: Whittles Publishing.

Dennis, R. 2015. Green J. Roy Dennis Wildlife Foundation website: https://www.roydennis.org/category/.

Dennis, R. 2016a. Beatrice. Roy Dennis Wildlife Foundation website: https://www.roydennis.org/.

Dennis, R. 2016b. *Plan for the recovery and conservation of Ospreys in Europe and the Mediterranean Region in particular*. Strasbourg: Council of Europe.

Dennis, R. 2019. Deshar. Roy Dennis Wildlife Foundation website: www.roydennis.org/category/osprey/.

Dennis, T.E., 2007a. Distribution and status of the Osprey (*Pandion haliaetus*) in South Australia. *Emu-Austral Ornithology* 107: 294–299.

Dennis. T.E. 2007b. Reproductive activity in the Osprey (*Pandion haliaetus*) on Kangaroo Island, South Australia. *Emu-Austral Ornithology* 107: 300–307

Dennis, T.E., S.A. Detmar, A.V. Brooks, and H.M. Dennis. 2011. Distribution and status of White-bellied Sea Eagle, *Haliaeetus leucogaster*, and Eastern Osprey, *Pandion cristatus*, populations in South Australia. *South Australian Ornithologist* 37: 1–16

Dennis, T.E., and G.P. Clancy. 2014. The status of the Osprey (*Pandion haliaetus cristatus*) in Australia. *Journal of Raptor Research* 48: 408–414.

DesGranges, J.L., J. Rodrigue, B. Tardif, and M. Laperle. 1998. Mercury accumulation and biomagnification in Ospreys (*Pandion haliaetus*) in the James Bay and Hudson Bay regions of Québec. *Archives of Environmental Contamination and Toxicology* 35: 330–341.

Detmar, S.A., and T.E. Dennis. 2018. A review of Osprey distribution and population stability in South Australia. *South Australian Ornithologist* 43: 38–54.

Diffendorfer J.E., J.C. Stanton, J.A. Beston, et al. 2021. Demographic and potential biological removal models identify raptor species sensitive to current and future wind energy. *Ecosphere* 12: e03531. 10.1002/ecs2.3531

Dombrovski, V., and V. Ivanovski. 2005. New data on numbers and distribution of birds of prey breeding in Belarus. *Acta Zoologica Lituanica* 15: 218–227.

Duerr, A.E., T.A. Miller, M. Lanzone, et al. 2012. Testing an emerging paradigm in migration ecology shows surprising differences in efficiency between flight modes. *PloS ONE*: e35548.

Dupart, J.-M. 2021. Rapport du comptage et de l'observation des Balbuzards Pêcheurs dans la moitie nord du eintro pour la saison 2020–2021. [In French]

Dupart, J.-M. 2022. Rapport du comptage et de l'observation des Balbuzards Pêcheurs au Senegal pour la saison 2021–2022. [In French]

Durand, J-D., K-N. Shen, W-J. Chen, et al. Systematics of the grey mullets (Teleostei: Mugiliformes: Mugilidae): molecular phylogenetic evidence challenges two centuries of morphology-based taxonomy. *Molecular Phylogenetics and Evolution* 64: 73–92.

Duriez, O., G. Peron, D. Gremillet, A. Sforzi, and F. Monti, F. 2018. Migrating ospreys use thermal uplift over the open sea. *Biology Letters* 14: p.20180687.

Dyfi Osprey Project. 2022a. *Dyfi Osprey family tree.* https://www.dyfiospreyproject.com/dyfi-osprey-family-tree

Dyfi Osprey Project. 2022b. *Dyfi Osprey egg and hatching data.* https://www.dyfiospreyproject.com/dyfi-osprey-egg-and-hatching-data

Dyfi Osprey Project. 2022c. *Dyfi Osprey migration data.* https://www.dyfiospreyproject.com/.

Eccleston, D.T. and R.E. Harness. 2018. *Raptor electrocutions and power line collisions.* In Mendez, M. *Birds of Prey Biology and Conservation in the XXI Century* (pp. 273–302). Cham, Switzerland: Springer.

Edwards, Jr., T.C. 1989. Similarity in the Development of Foraging Mechanics among Sibling Ospreys. *The Condor* 91: 30–36.

Edwards, A.J., C.G. Anthony, and P.O. Abohweyere. 2001. A revision of Irvine's marine fishes of tropical West Africa. *Darwin Initiative Report.*

Ekert, P.A. and A.M. Brady. 2007. A review of the status of breeding Osprey (*Pandion haliaetus cristatus*) in NSW (2006). Unpublished report to the Foundation for National Parks and Wildlife. Ekerlogic Consulting Services, Wallsend, NSW Australia.

Elkins, N. 1995. *Weather and Bird Behaviour* (second edition). London: T & A D Poyser.

Elliott, J.E., M.M. Machmer, L.K. Wilson, and C.J. Henny. 2000. Contaminants in ospreys from the Pacific Northwest: II. Organochlorine pesticides, polychlorinated biphenyls, and mercury, 1991–1997. *Archives of Environmental Contamination and Toxicology* 38: 93–106.

Englund, J.V. and V.L. Greene. 2008. Two-year-old nesting behavior and extra-pair copulation in a reintroduced Osprey population. *Journal of Raptor Research* 42: 119–124.

Eriksson, M.O. 1985. Fish delivery, production of young, and nest density of Osprey (*Pandion haliaetus*) in southwest Sweden. *Canadian Journal of Zoology* 64: 1961–1965.

Eriksson, M.O. and K. Wallin. 1994. Survival and breeding success of the Osprey *Pandion haliaetus* in Sweden. *Bird Conservation International* 4: 263–277.

Evans, E. 2014. *Ospreys in Wales: The First Ten Years.* Aberystwyth: Cambrian Printers.

Evans, A.H., and W. Turner. 1903. *Turner on Birds: A Short and Succinct History of the Principal Birds Noticed by Pliny and Aristotle First Published by Doctor William Turner 1544*: Edited, with Introduction, Translation, Notes and Appendix.

Ewins, P.J. 1996. The use of artificial nest-sites by an increasing population of Osprey in the Canadian Great Lakes basin. In: Bird, D.M., D.E. Varland, and J.J. Negro (eds) *Raptors in Human Landscapes* (pp. 109–123). San Diego, CA: Academic Press Inc.

Ewins, P.J. 1997. Osprey (*Pandion haliaetus*) populations in forested areas of North America: changes, their causes and management recommendations. *Journal of Raptor Research* 31: 138–150.

Ewins, P.J., M.J. Miller, M.E. Barker, S. Postupalsky. 1994. Birds breeding in or beneath osprey nests in the Great Lakes basin. *The Wilson Bulletin* 106: 743–749.

Falkenberg, I.D., T.E. Dennis, and B.D. Williams. 1994. Organochlorine pesticide contamination in three species of raptor and their prey in South Australia. *Wildlife Research* 21: 163–173.

Ferguson-Lees, J., and D.A. Christie. 2001. *Raptors of the World*. London: Christopher Helm.

Ferrer, M., and E. Casado. 2004. Osprey (*Pandion haliaetus*) reintroduction project in Andalusia (Southern Spain). Centro Superior de Investigaciones Científicas. www.fundacionmigres.org/documentos.htm.

Fields, L.E., and J.E. Pagel. 2016. Osprey Occupancy of Mono Lake—Unique Habitat in Eastern California. *Journal of Raptor Research* 50: 97–102

Fisher, P.R. 2001. Ecology and behaviour of the Osprey *Pandion haliaetus* on the Farasan Islands, Red Sea, Saudi Arabia. Doctoral dissertation, Manchester Metropolitan University.

Fisher, P.R., S.F. Newton, H.M.A. Tatwany, and C.R. Goldspink. 2001. The status and breeding biology of the Osprey *Pandion haliaetus* in the Middle East. *Vogelwelt* 122: 191–204.

Flemming, S.P., and R.P. Bancroft. 1990. Bald Eagle attacks Osprey nestling. *Journal of Raptor Research* 24: 26–27.

Flemming, S.P., and Smith, P.C. 1990. Environmental influences on Osprey foraging in northeastern Nova Scotia. *Journal of Raptor Research* 24: 64–67.

Flemming, S.P., R.P. Bancroft, and N. Seymour. 1991. Enhanced fledging success by colonially nesting Ospreys in Nova Scotia coastal habitat. *The Wilson Bulletin* 103: 664–668.

Flemming, S.P., P.C. Smith, N.R. Seymour, and R.P. Bancroft. 1992. Ospreys use local enhancement and flock foraging to locate prey. *Auk* 109: 649–654.

Foronda, P., M.A. Santana-Morales, C. Feliu, and B. Valladares. 2009. New record of *Scaphanocephalus expansus* from the Canary Islands (Spain). *Helminthologia* 46: 198–200.

Forsman, D. 1999. *The Raptors of Europe and the Middle East*. London: T & D Poyser.

Francour, P., and J.C. Thibault. 1996. The diet of breeding Osprey *Pandion haliaetus* on Corsica: exploitation of a coastal marine environment. *Bird Study* 43: 129–133.

Forys, E.A., P.R. Hindsley, and S. Bryan. 2021. Predictors of Osprey nest success in a highly urbanized environment. *Journal of Raptor Research* 55: 485–495.

Fozzi, A., R. Fozzi, I. Fozzi, et al. 2020. First successful breeding of Osprey *Pandion haliaetus* in Sardinia since 1968. *Rivista Italiana di Ornitologia* 90: 85–90.

Fricke, R., W.N. Eschmeyer, and R. van der Laan (eds) 2020. *Eschmeyer's Catalog of Fishes*: Genera, Species, References. (http://researcharchive.calacademy.org/research/ichthyology/catalog/fishcatmain.asp). Accessed 21.05.2020.

Furness, R.W., S.J. Muirhead, and M. Woodburn. 1986. Using bird feathers to measure mercury in the environment: relationships between mercury content and moult. *Marine Pollution Bulletin* 17: 27–30.

Galarza, A. 2017. Osprey Restoration Project in the Urdaibai Biosphere Reserve (Basque Country) Annual Report 2017. Aranzadi Society of Sciences/County Council of Biscay. http/www.birdcenter.org

Galarza, A., and R.H. Dennis. 2009. A spring stopover of a migratory osprey (*Pandion haliaetus*) in northern Spain as revealed by satellite tracking: implications for conservation. *Animal Biodiversity and Conservation* 32: 117–122.

Galarza, A., and I. Zuberogoitia. 2012. Osprey reintroduction project in the Biosphere Reserve of Urdaibai (Basque Country). Aranzadi Society of Sciences/County Council of Biscay. http/www.birdcenter.org

Galarza, A., M. del Arco, J. Elorriaga, E. Unamuno, J. Arizaga, and I. Zuberogoitia. 2017. First evidence provided by satellite telemetry of nocturnal flight overland by an osprey (*Pandion haliaetus*). *Journal of Raptor Research* 51: 184–186.

Galarza, A., V. Alba, M. del Arco, et al. 2021. First record of albinism in an Osprey. *British Birds* 114: 485

Gallistel, C.R. 1990. *The Organization of Learning*. Cambridge, MA: Bradford Books/MIT Press.

Gandon S., and Y. Michalakis. 2001. Multiple causes of the evolution of dispersal. In: Clobert, J., J.D. Nichols, E. Danchin, and A. Dhondt (eds) *Causes, Consequences and Mechanisms of Dispersal at the Individual, Population and Community Level* (pp. 155–167). Oxford, UK: Oxford University Press.

Garrido, O.H., and F. García Montaña. 1975. *Catálogo de las Aves de Cuba*. La Habana, Cuba: Academia de Ciencias de Cuba.

Generalitat Valenciana 2019. Crónicas de la Pescadora – Reintroducción del Águila Pescadora en la Comunidad Valenciana – No 1 Oct 19.

Generalitat Valenciana 2020. Crónicas de la Pescadora – Reintroducción del Águila Pescadora en la Comunidad Valenciana – No 13 May 20.

Gerrard, J.M., D.W.A. Whitfield, and W.J. Maher. 1976. Osprey–Bald Eagle relationships in Saskatchewan. *Blue Jay* 34: 240–246.

Gilroy, J.J., J.A Gill, S.H.M. Butchart, V.R. Jones, and A.M.A. Franco. 2016. Migratory diversity predicts population declines in birds. *Ecology Letters* 19: 308–317.

Glaslyn Wildlife. 2022a. *Glaslyn Osprey Profiles*. https://www.glaslynwildlife.co.uk/glaslyn-osprey-profiles/

Glaslyn Wildlife. 2022b. *Glaslyn Osprey Egg Statistics*. https://www.glaslynwildlife.co.uk/glaslyn-osprey-egg-statistics/

Glaslyn Wildlife. 2022c. *2014–15*. https://www.glaslynwildlife.co.uk/2014-2015/

Glaslyn Wildlife. 2022d. *Glaslyn Osprey Chick Statistics*. https://www.glaslynwildlife.co.uk/.

Glass, K.A., and B.D. Watts. 2009. Osprey diet composition and quality in high- and low-salinity areas of Chesapeake Bay. *Journal of Raptor Research* 43: 27–36.

Gorez, B., and D. Foday Saine. 2021. The Gambia: Looming clouds. *Yemaya: ICSF'S Newsletter on Gender and Fisheries* 64: 4–6.

Goudie, A.S., and N.J. Middleton. 2006. *Desert Dust in the Global System*. Berlin: Springer.

Green, R. 1976. Breeding Biology of Ospreys *Pandion haliaetus* in Scotland. *Ibis* 118: 475–490.

Green, D.J., and R.C. Ydenberg. 1994. Energetic expenditure of male Ospreys provisioning natural and manipulated broods. *Ardea* 82: 249–262.

Green, D.J., and E.A. Krebs. 1995. Courtship feeding in ospreys *Pandion haliaetus*: a criterion for mate assessment? *Ibis* 137: 35–43.

Greene, E. 1987. Individuals in an osprey colony discriminate between high and low quality information. *Nature* 329: 239–241.

Greene, E.P., A.E. Geene, and B. Freedman. 1983. Foraging behaviour and prey selection by Ospreys in coastal habitats of Nova Scotia, Canada. In: Bird, D.M. (ed.) *Biology and Management of Bald Eagles and Ospreys* (pp. 243–56). Ste Anne de Bellevue, Quebec: Harpell Press.

Greenwood, P.J., and P.H. Harvey. 1982 The natal and breeding dispersal of birds. *Annual Review of Ecology and Systematics* 13: 1–21.

Grove, R.A., C.J. Henny, and J.L. Kaiser. 2009. Osprey: worldwide sentinel species for assessing and monitoring environmental contamination in rivers, lakes, reservoirs, and estuaries. *Journal of Toxicology and Environmental Health, Part B* 12: 25–44.

Grover, K.E. 1984. Nesting distribution and reproductive status of Ospreys along the upper Missouri River, Montana. *The Wilson Bulletin* 96: 496–498.

Grubb, Jr., T.G. 1977. Weather-Dependent Foraging in Ospreys. *The Auk* 94: 146–149.

Guilford, T., S. Åkesson, A. Gagliardo, et al. 2011. Migratory navigation in birds: new opportunities in an era of fast-developing tracking technology. *Journal of Experimental Biology* 214: 3705–3712.

Guigueno, M.F., K.H. Elliott, J. Levac, M. Wayland, and J.E. Elliott. 2012. Differential exposure of alpine Ospreys to mercury: Melting glaciers, hydrology or deposition patterns? *Environment International* 40: 24–32.

Guillemain, M., A.D. Fox, H. Poysa, et al. 2013. Autumn survival inferred from wing age ratios: Wigeon juvenile survival half that of adults at best? *Journal of Ornithology* 154: 351–358.

Gwinner, E. 1996. Circadian and circannual programmes in avian migration. *Journal of Experimental Biology* 199: 39–48.

Habib, M.I. 2019. Breeding Status of Ospreys in Egypt (Red Sea) from 2012 to 2018. *Raptors Conservation* 38: 43–58.

Hagan, J.M. 1986. Temporal patterns in pre-fledgling survival and brood reduction in an Osprey colony. *Condor* 88: 200–205.

Hagan, J.M., and J.R. Walters. 1990. Foraging behavior, reproductive success, and colonial nesting in Ospreys. *Auk* 107: 506–521.

Hake, M., N. Kjellén, and T. Alerstam. 2001. Satellite tracking of Swedish Ospreys *Pandion haliaetus*: autumn migration routes and orientation. *Journal of Avian Biology* 32: 47–56.

Häkkinen, I. 1978. Diet of the Osprey in Finland. *Ornis Scandinavica* 9: 11–116.

Harness, R.E., and K.R. Wilson. 2001. Utility structures associated with raptor electrocutions in rural areas. *Wildlife Society Bulletin* 29: 612–623

Harris, M.L., and J.E. Elliott. 2011. Effects of polychlorinated biphenyls, dibenzo-p-dioxins and dibenzofurans, and polybrominated diphenyl ethers in wild birds. In: Beyer, W.N., and J.P. Meador (eds) *Environmental Contaminants in Biota* (pp. 477–530). Boca Raton, Florida: CRC Press.

Harrison, I.J. 2002. Mugilidae. In: Carpenter, K. (ed.) *FAO Species Identification Guide for Fisheries Purposes. The Living Marine Resources of the Western Central Atlantic, Vol. 2   Bony Fishes part 1 (Acipenseridae to Grammatidae)* (pp. 1071–1085). Rome: FAO.

Harrison, C.J.O., and P. Castell. 2002. *Bird Nests, Eggs and Nestlings of Britain and Europe with North Africa and the Middle East* (second revised edition). London: HarperCollins.

References

Harrison, W. 1586. *A Description of England*. London: W. Scott Publishing Co.

Hedenström, A. 2002. Aerodynamics, evolution and ecology of avian flight. *Trends in Ecology & Evolution* 17: 415–422.

Heinz, G.H., D.J. Hoffman, J.D. Klimstra, K.R. Stebbins, S.L. Kondrad, and C.A. Erwin. 2009. Species differences in the sensitivity of avian embryos to methylmercury. *Archives of Environmental Contamination and Toxicology* 56: 129–138.

Henny, C.J. 1977. Research, management, and status of the osprey in North America. In: Chancellor, R.D. (ed.) Proc. World Conf. on Birds of Prey, Internal Council Bird Preservation, Vienna, Austria (pp. 199–222). Ste Anne de Bellevue, Quebec: Harpell Press.

Henny, C.J., and H.M. Wight. 1969. An endangered Osprey population: estimates of mortality and production. *The Auk* 86: 188–198.

Henny, C.J., M.M. Smith, and V.D. Stotts. 1974. The 1973 distribution and abundance of breeding Ospreys in the Chesapeake Bay. *Chesapeake Science* 15: 125–133.

Henny, C.J. 1983. Distribution and abundance of nesting Ospreys in the United States. In: Bird, D. (ed.) *Biology and Management of Bald Eagles and Ospreys* (pp. 175–186). Ste. Anne de Bellevue, Quebec: Harpell Press.

Henny, C.J., and H.M. Wight. 1969. An endangered Osprey population: estimates of mortality and production. *The Auk* 86: 188–198.

Henny, C.J., D.J. Dunawayr, D. Mulette, and J.R. Koplin. 1978. Osprey distribution, abundance, and status in western North America: I. The northern California population. *Northwest Science* 52: 261–271.

Henny, C.J., and D. Anderson. 1979. Osprey distribution, abundance, and status in western North America: III. The Baja California and Gulf of California population. *Bulletin of the Southern California Academy of Sciences* 78: 89–106.

Henny, C.J., and J.L. Kaiser. 1986. Osprey population increase along the Willamette River, Oregon, and the role of utility structures, 1976–1993. In: Bird, D., D. Varland, and J. Negro (eds) *Raptors in Human Landscapes*. London: Academic (1996): 97–108.

Henny, C.J., and D. Anderson. 2004. Status of nesting ospreys in coastal Baja California, Sonora and Sinaloa, Mexico, 1977 and 1992–1993. *Bulletin of Southern California Academy of Sciences* 103: 95–114.

Henny, C.J., R.A. Grove, J.L. Kaiser, and V.R. Bentley. 2004. An evaluation of osprey eggs to determine spatial residue patterns and effects of contaminants along the lower Columbia River, USA. In: Chancellor, R.D., and B.-U. Meyburg (eds) *Raptors Worldwide, Sixth World Conference Birds of Prey and Owls* (pp. 369–338). Budapest, Hungary: WWGBP/MME.

Henny, C.J., D.W. Anderson, A.C. Vera, and J-L. E. Carton. 2007. Population Size and Trends for Nesting Ospreys in Northwestern Mexico: Region-wide Surveys, 1977, 1992/1993 and 2006: *U.S. Geological Survey, Open-File Report 2007-1298*.

Henny, C.J., R.A. Grove, and J.L. Kaiser. 2008a. Osprey distribution, abundance, reproductive success and contaminant burdens along the lower Columbia River, 1997/1998 versus 2004. *Archives of Environmental Contamination and Toxicology* 54: 525–534.

Henny, C.J., D.W. Anderson, A.C. Vera, and J-L. E. Cartron. 2008b. Region-wide trends of nesting Ospreys in northwestern Mexico: a three-decade perspective. *Journal of Raptor Research* 42: 229–242.

Henny, C.J., J.L. Kaiser, R.A. Grove, B.L. Johnson, and R.J. Letcher. 2009. Polybrominated diphenyl ether flame retardants in eggs may reduce reproductive success of ospreys in Oregon and Washington, USA. *Ecotoxicology* 18: 802–813.

Henny, C.J., R.A. Grove, J.L. Kaiser, and B.L. Johnson. 2010. North American Osprey populations and contaminants: historic and contemporary perspectives. *Journal of Toxicology and Environmental Health, Part B* 13(7-8): 579–603.

Henny, C.J., R.A. Grove, J.L. Kaiser, B.L. Johnson, C.V. Furl, and R.J. Letcher. 2011. Wastewater dilution index partially explains observed polybrominated diphenyl ether flame retardant concentrations in osprey eggs from Columbia River Basin, 2008–2009. *Ecotoxicology* 20: 682–697.

Henny, C.J., and M.S. Martell. 2017. Satellite-tagged Osprey nearly sets longevity record and productivity response to initial captures. *Journal of Raptor Research* 51: 180–183.

Hindman, L.J., W.F. Harvey IV, G.R. Costanzo, K.A. Converse, and G. Stein Jr. 1997. Avian Cholera in Ospreys: First Occurrence and Possible Mode of Transmission (Colera Avicola en *Pandion haliaetus*: Primera Incidencia y Posible Metodo de Transmision). *Journal of Field Ornithology* 68: 503–508.

Horton, T.W., R.O. Bierregaard, P. Zawar-Reza, R.N. Holdaway, and P. Sagar. 2014. Juvenile osprey navigation during trans-oceanic migration. *PloS ONE* 9: e114557.

Houghton, L.M., and L.M. Rymon. 1997. Nesting distribution and population status of U.S. ospreys, 1994. *Journal of Raptor Research* 31: 44–53.

Hudson, G.E. 1948. Studies on the muscles of the pelvic appendage in birds II: the heterogeneous order falconiformes. *American Midland Naturalist* 39: 102–127.

Hughes, K.D., P.J. Ewins, and K.E. Clark. 1997. A comparison of mercury levels in feathers and eggs of Osprey *Pandion haliaetus* in the North American Great Lakes. *Archives of Environmental Contamination and Toxicology* 33: 441–452.

Hunt, W.G. 1998. Raptor floaters at Moffat's equilibrium. *Oikos* 82: 191–197.

Iankov, P. 2007. *Atlas of Breeding Birds in Bulgaria.* Sofia, Bulgaria: Bulgarian Society for the Protection of Birds.

Ivanovski, V.V. 2012. *Predator Birds of Belarusian Lake District.* Vitjepsk. [In Russian]

Jamieson, I., N. Seymour, and R.P. Bancroft. 1982. Time and activity budgets of ospreys nesting in northeastern Nova Scotia. *Condor* 84: 439–441.

Jennings, M.C. 2010. Atlas of the breeding birds in the Arabia Peninsula. *Fauna of Arabia, Vol. 25*, p. 751. Basel: Karger Libri.

Jiménez, B., R. Merino, E. Abad, E., J. Rivera, and K. Olie. 2007. Evaluation of organochlorine compounds (PCDDs, PCDFs, PCBs and DDTs) in two raptor species inhabiting a Mediterranean Island in Spain. *Environmental Science and Pollution Research* 14: 61–68.

Johnson, D.R., W.E. Melquist, and G.J. Schroeder. 1975. DDT and PCB levels in Lake Coeur d'Alene, Idaho, Osprey eggs. *Bulletin of Environmental Contamination and Toxicology* 13: 401–405.

Johnson, M.L., and M.S. Gaines. 1990. Evolution of dispersal: theoretical models and empirical tests using birds and mammals. *Annual Review of Ecology and Systematics* 21: 449–480.

Johnston, N.N., J.E. Bradley, A.C. Pomeroy, and K.A. Otter. 2013. Flight paths of migrating golden eagles and the risk associated with wind energy development in the Rocky Mountains. *Avian Conservation and Ecology* 8: 12.

Jollie, M.T. 1976. A contribution to the morphology and phylogeny of the Falconiformes, part 1. *Evolutionary Theory* 1: 285–298.

Judge, D.S. 1983. Productivity of Ospreys in the Gulf of California. *Wilson Bulletin* 95: 243–255.

Kalvans, A., and J. Bajinskis. 2016. The diet composition of breeding Ospreys (*Pandion haliaetus*) in Latvia. *Environmental and Experimental Biology* 14: 107–111.

Karasov, W.H. 1990. Digestion in birds: chemical and physiological determinants, and ecological implications. In: Morrison, M.L., C.J. Ralph, J. Verner, and J.R. Jehl (eds) *Studies in Avian Biology 13: Avian Foraging, Theory, Methodology, and Application* (pp. 391–415). Kansas: Cooper Ornithological Society

Kerlinger, P. 1989. *Flight Strategies of Migrating Hawks.* Chicago: University of Chicago Press.

Khan, S.B., S. Javed, and J.N. Shah. 2008. Ospreys in the Abu Dhabi Emirate: current breeding status and role of platforms as an aid to nesting. *Falco* 32: 14–16.

Kinkead, T.P. 1985. Age Structure and Dispersal of Chesapeake Bay Ospreys. MA thesis, The College of William and Mary, Virginia.

Kinsella, J.M., R.A. Cole, D.J. Forrester, and C.L. Roderick. 1996. Helminth parasites of the osprey, *Pandion haliaetus*, in North America. *Journal of Helminthological Society Washington* 63: 262–265.

Kjellén, N., M. Hake, and T. Alerstam. 2001. Timing and speed of migration in male, female and juvenile Ospreys *Pandion haliaetus* between Sweden and Africa as revealed by field observations, radar and satellite tracking. *Journal of Avian Biology* 32: 57–67.

Klaassen, R.H.G., R. Strandberg, M. Hake, and T. Alerstam. 2008. Flexibility in daily travel routines causes regional variation in bird migration speed. *Behavioural Ecology and Sociobiology* 62: 1427–1432.

Klaassen, R.H.G., M. Hake, R. Strandberg, and T. Alerstam. 2011. Geographical and temporal flexibility in the response to crosswinds by migrating raptors. *Proceedings of Royal Society B: Biological Sciences* 278: 1339–1346.

Klaassen, R.H.G., M. Hake, R. Strandberg, *et al.* 2014. When and where does mortality occur in migratory birds? Direct evidence from long-term satellite tracking of raptors. *Journal of Animal Ecology* 83: 176–184.

Kokko, H. 1999. Competition for early arrival in migratory birds. *Journal of Animal Ecology* 68: 940–950.

Kokko, H., and W.J. Sutherland. 1998. Optimal floating and queuing strategies: consequences for density dependence and habitat loss. *The American Naturalist* 152: 354–366.

Krone, O. 2000. Endoparasites in free-ranging birds of prey in Germany. *Raptor Biomedicine III*: 101–116.

Kuchin A.P. 2004. Pticy Altaya. Gorno-Altajsk, 1–777. [in Russian]

Kuznetsov, A.V., and M.V. Babushkin. 2005. The White-tailed Sea Eagle *Haliaeetus albicilla* and the Osprey *Pandion haliaetus* in the Vologda lake district and southeastern Onego area. In: *Status of raptor populations in Eastern Fennoscandia. Proceedings of the Workshop, Kostomuksha, Karelia, Russia* (pp. 80–90).

Laffaille, P., E. Feunteun, and J.C. Lefebvre. 2000. Composition of fish communities in a European macrotidal saltmarsh (the Mont Saint-Michel Bay, France). *Estuarine, Coastal and Shelf Science* 51: 429–438.

Lambert, R.A. 2001. The osprey on Speyside: an environmental history. In: *Contested Mountains 1880–1980*. Cambridge: White Horse Press.

Langner, H.W., E. Greene, R. Domenech, and M.F. Staats. 2012. Mercury and other mining-related contaminants in Ospreys along the Upper Clark Fork River, Montana, USA. *Archives of Environmental Contamination and Toxicology* 62: 681–695.

Lebreton, B., P. Richard, E. Parlier, G. Guillou, and G. Blanchard. 2011. Trophic ecology of mullets during their spring migration in a European salt marsh: a stable isotope study. *Estuarine, Coastal and Shelf Science* 91: 502–510.

Lemarchand, C., R. Rosoux, C. Talon, and P. Berny. 2014. Flagship species conservation and introduced species invasion: toxic aspects along Loire River (France). In: Larramendy, M.L. and S. Soloneski (eds) *Pesticides – toxic aspects* (pp. 53–79). London: IntechOpen.

Lenz, T.L., A. Jacob, and C. Wedekind. 2007 Manipulating sex ratio to increase population growth: the example of the Lesser Kestrel. *Animal Conservation* 10: 236–244.

Lesclaux, P. 2019. Programme en faveur du Balbuzard pêcheur en Aquitaine 2018–2021. Bilan d'Activite 2019.

Levenson, H. 1979. Time and activity budgets of Ospreys nesting in Northern California. *Condor* 81: 364–369.

Levenson, H., and J.R. Koplin. 1984. Effects of human activity on productivity of nesting ospreys. *The Journal of Wildlife Management* 48: 1374–1377.

Lewin, W. 1795. *The Birds of Great Britain*. London: J. Johnson.

Liechti, F. 1995. Modelling optimum heading and airspeed of migrating birds in relation to energy expenditure and wind influence. *Journal of Avian Biology* 26(4): 330–336.

Liechti, F., and B. Bruderer. 1995. Direction, speed, and composition of nocturnal bird migration in the south of Israel. *Israel Journal of Zoology* 41: 501–515.

Liminana, R., M. Romero, U. Mellone, and V. Urios, V. 2013. Is there a different response to winds during migration between soaring and flapping raptors? An example with the Montagu's harrier and the lesser kestrel. *Behavioural Ecology and Sociobiology* 67: 823–835.

Liston, T. 1997. Bald Eagle attacks Osprey nestlings. *Loon* 68: 238–239.

Lorente, L. 2005. Águila pescadora. Noticiario Ornitológico. *Ardeola* 52: 427.

MacDonald, J., and N.R. Seymour. 1994. Bald eagle attacks adult osprey. *Journal of Raptor Research* 28: 122.

Machmer, M.M., and R.C. Ydenberg. 1990. Weather and osprey foraging energetics. *Canadian Journal of Zoology* 68: 40–43.

Machmer, M.M., and R.C. Ydenberg. 1998. The relative roles of hunger and size asymmetry in sibling aggression between nestling ospreys, *Pandion haliaetus*. *Canadian Journal of Zoology* 76: 181–186.

Mackrill, T.R. 2017. Migratory behaviour and ecology of a trans-Saharan migrant raptor, the Osprey *Pandion haliaetus*. Doctoral dissertation, University of Leicester.

Mackrill, T.R. 2019. *RSPB Spotlight Ospreys*. London: Bloomsbury.

Mackrill, T.R., T. Appleton, and H. McIntyre. 2013. *The Rutland Water Ospreys*. London: Bloomsbury.

Männik, R. 2006. The Osprey and its conservation in Estonia. *Hirundo Supplementum* 10: 1–58.

Marchant, S., and P.J. Higgins (eds). 1993. *Handbook of Australian, New Zealand and Antarctic Birds. Vol. 2: Raptors to Lapwings*. Melbourne, Australia: Oxford University Press.

Marquiss, M., L. Robinson, and E. Tindal. 2007. Marine foraging by Ospreys in southwest Scotland. *British Birds* 100: 456–465.

Martell, M.S., C.J. Henny, P.E. Nye, and M.J. Solensky. 2001. Fall migration routes, timing, and wintering sites of North American ospreys as determined by satellite telemetry. *The Condor* 103: 715–724.

Martell, M.S., J.V. Englund, and H.B. Tordoff. 2002. An urban Osprey population established by translocation. *Journal of Raptor Research* 36: 91–6.

Martell, M.S., M.A. McMillian, M.J. Solensky, and B.K. Mealey. 2004. Partial migration and wintering use of Florida by Ospreys. *Journal of Raptor Research* 38: 55–61.

Martell, M.S., R.O. Bierregaard, B.E. Washburn, J.E. Elliott, C.J. Henny, R.S. Kennedy, and I. MacLeod. 2014. The spring migration of adult North American Ospreys. *Journal of Raptor Research* 48: 309–324.

Martín, B., C.A. Torralvo, G. Elias, J. Tomás, A. Onrubia, and M. Ferrer. 2019. Are Western European ospreys (*Pandion haliaetus*) shortening their migration distances? Evidence from trends of the wintering population in the Iberian Peninsula. *European Journal of Wildlife Research* 65: 1–13.

Martins, S., R. Freitas, L. Palma, and P. Beja. 2011. Diet of breeding ospreys in the Cape Verde archipelago, northwestern Africa. *Journal of Raptor Research* 45: 244–251.

Martínez-Abraín, A. 2018. Satellite factors influencing the impact of recreational activities on wildlife. *Animal Conservation* 21: 461–462.

McMillian, M.A. 2013. Long-term trends in Osprey (*Pandion haliaetus*) nesting populations on Lake Istokpoga, Florida. *Florida Field Naturalist* 41: 71–79.

Mees, G.F., 2006. The avifauna of Flores (Lesser Sunda Islands). *Zoologische Mededelingen* 80: 1–261.

Mellone, U., R.H.G. Klaassen, C. García-Ripollés, et al. 2012. Interspecific comparison of the performance of soaring migrants in relation to morphology, meteorological conditions and migration strategies. *PloS ONE* 7: e39833.

Mellone, U., P. López-López, R. Limiñana, G. Piasevoli, and V. Urios. 2013. The trans- equatorial loop migration system of Eleonora's falcon: differences in migration patterns between age classes, regions and seasons. *Journal of Avian Biology* 44: 417–426.

Mellone, U., J. De La Puente, P. Lopez-Lopez, R. Liminana, A. Bermejo, and V. Urios. 2015. Seasonal differences in migration patterns of a soaring bird in relation to environmental conditions: a multi-scale approach. *Behavioural Ecology and Sociobiology* 69: 75–82.

Meyburg, B.U., O. Manowsky, and C. Meyburg. 1996. The Osprey in Germany: its adaptation to environments altered by man. In: Bird, D.M., D.E. Varland, and J.J. Negro (eds) *Raptors in Human Landscapes: Adaptations to Built and Cultivated Environments* (pp. 125–135). London: Academic Press.

Miller, M.J.R., P.J. Ewins, and T.D. Galloway. 1997. Records of ectoparasites collected on Ospreys from Ontario. *Journal of Wildlife Diseases* 33: 373–376.

Miller, T.A., R.P. Brooks, M.J. Lanzone, et al. 2016. Limitations and mechanisms influencing the migratory performance of soaring birds. *Ibis* 158: 116–134.

Mischenko, A. (ed.). 2004. *Estimation of numbers and trends for birds of the European part of Russia.* Moscow, Russia: Russian Bird Conservation Union.

Mitkus, M., S. Potier, G. Martin, O. Duriez, and A. Kelber. 2018. Raptor Vision. *Oxford Research Encyclopedia of Neuroscience.* Retrieved 27 Mar. 2020, from https://oxfordre.com/neuroscience/.

Mitrofanov, O.B. 2016. Monitoringovye nablyudeniya za gnezdovaniem skopy na Teletckom ozere. Monitoring sostoyaniya prirodnykh kompleksov I mnogoletnie issledovaniya na osobo okhranyaemykh prirodnykh territoriyakh. 1. *Shushenskoe*: 79–81. [In Russian]

Mittermeier, J.C., H.E.W Cottee-Jones, E.C. Purba, N.M. Ashuri, E. Hesdianti, and J. Supriatna. 2013. A survey of the avifauna of Obi Island, North Moluccas, Indonesia. *Forktail* 29: 128–137.

Moffatt, R. 2009. Ospreys and artificial nest structures. *Wingspan* 19: 24–27.

Moll, K.-H. 1967. Der Fischadler. *Falke* 14: 134–135. [In German]

Molter, T., D. Schindler, A.T. Albrecht, and U. Kohnle. 2016. Review on the projections of future storminess over the North Atlantic European Region. *Atmosphere* 7: UNSP 60.

Monti, F. 2012. The Osprey, *Pandion haliaetus*: State of knowledge and conservation of the breeding population of the Mediterranean basin. Initiative PIM.

Monti, F., H. Nibani, J.-M. Dominici, et al. (2013) The vulnerable Osprey breeding population of the Al Hoceima National Park, Morocco: present status and threats. *Ostrich* 84: 199–204.

Monti, F., J.-M. Dominici, R. Choquet, O. Duriez, G. Sammuri, and A. Sforzi. 2014. The Osprey reintroduction in Central Italy: dispersal, survival and first breeding data. *Bird Study* 61: 465–473.

Monti, F., O. Duriez, A. Véronique et al. 2015. Being cosmopolitan: evolutionary history and phylogeography of a specialized raptor, the Osprey *Pandion haliaetus*. *BMC Evolutionary Biology* 15: 1–15.

Monti, F., D. Grémillet, A. Sforzi, et al. 2018a. Migration and wintering strategies in vulnerable Mediterranean Osprey populations. *Ibis* 160: 554–567.

Monti, F., D. Grémillet, A. Sforzi, et al. 2018b. Migration distance affects stopover use but not travel speed: contrasting patterns between long- and short-distance migrating ospreys. *Journal of Avian Biology* 49: e01839.

Monti, F., O. Duriez, J.-M.Dominici, et al. 2018c. The price of success: integrative long-term study reveals ecotourism impacts on a flagship species at a UNESCO site. *Animal Conservation* 21: 448–458.

Monti, F., F. Delfour, V. Arnal, S. Zenboudji, O. Duriez, and C. Montgelard. 2018d. Genetic connectivity among Osprey populations and consequences for conservation: philopatry versus dispersal as key factors. *Conservation Genetics* 19: 839–851.

Monti, F., P.L. Cascio, and A. Sforzi. 2019. Nocturnal activity of insect fauna in Osprey nests: insights from video-cameras. *Journal of Raptor Research* 53: 212–214.

Morandini, V., R. Muriel, I. Newton, and M. Ferrer. 2019. Skewed sex ratios in a newly established osprey population. *Journal of Ornithology* 160: 1025–1033.

Moreau, R.E. 1972. *The Palearctic-African Bird Migration Systems.* New York: Academic Press.

Mougeot, F., J.C. Thibault, and V. Bretagnolle. 2002. Effects of territorial intrusions, courtship feedings and mate fidelity on the copulation behaviour of the osprey. *Animal Behaviour* 64: 59–769.

Mouritsen, H. 2003. Spatiotemporal orientation strategies of long-distance migrants. In: Berthold, P. Gwinner, E., and Sonnenschein, E. (eds), *Avian migration* (pp. 493–513). Berlin: Springer-Verlag.

Mouritsen, H., D. Heyers, and O. Güntürkün. 2016. The neural basis of long-distance navigation in birds. *Annual Review of Physiology* 78: 610.

Mueller, T., R.B. O'Hara, S.L. Converse, R.P. Urbanek, and W.P. Fagan. 2013. Social learning of migratory performance. *Science* 341: 999–1002.

Munday, M., C. Jones, and N. Roch. 2015. *The Economic Impact of the Communities and Nature Environment for Growth (E4G) Strategic Project*. Report for Natural Resources Wales.

Muriel, R., M. Ferrer, E. Casado, E., and D. Schmidt. 2006. First breeding success of Osprey (*Pandion haliaetus*) in mainland Spain since 1981 using cross-fostering. *Journal of Raptor Research* 40: 303–304.

Muriel, R., E. Casado, D. Schmidt, C.P. Calabuig, and M. Ferrer. 2010a. Morphometric sex determination of young Ospreys *Pandion haliaetus* using discriminant analysis. *Bird Study* 57: 336–343.

Muriel, R., M. Ferrer, E. Casado, and C.P. Calabuig. 2010b. First successful breeding of reintroduced Ospreys *Pandion haliaetus* in mainland Spain. *Ardeola* 57: 175–180.

Nagai, K., T. Sakakibara, and A. Azuma. 2021. Genetic diversity and structure in Japanese populations of the osprey (*Pandion haliaetus*), Based on mtDNA. *Open Journal of Genetics* 11: 43–56.

Nasser, M., M. Ansari, E. Adly, M. Shobrak, and A. Alahmed. 2019. An analysis of osprey/chewing lice interaction, with a new record for Saudi Arabia. *African Entomology* 27: 178–184.

Naurois, R. de. 1969. Notes brèves sur l'avifaune de l'archipel du Cap Vert. Faunistique, endémisme, écologie. *Bulletin de l'Institut Fondamental d'Afrique Noire*. Tome XXXI, série A, 1: 8–218.

Naurois, R. de. 1987. Le balbuzard (*Pandion haliaetus*) aux îles du Cap Vert. *Annali del Museo Civico di Storia Naturale* 86: 657–682.

Negro, J.J., F. Hiraldo, and J.A. Donázar. 1997. Causes of natal dispersal in the lesser kestrel: inbreeding avoidance or resource competition? *Journal of Animal Ecology* 66: 640–648.

Nelson, J.S. 2006. *Fishes of the World*. 4th edition. Hoboken, New Jersey, USA: John Wiley and Sons.

Newton, I. 1988. Population regulation in peregrines: an overview. In: Cade, T.J., J.H. Enderson, C.G. Thelander, C.M. White (eds) *Peregrine Falcon Populations: their management and recovery* (pp. 761–770). Boise: The Peregrine Fund Inc.

Newton, I. 1998. *Population Limitation in Birds*. London: Academic Press.

Newton, I. 2008. *The Migration Ecology of Birds*. London: Academic Press.

Newton, I. 2010. *Population Ecology of Raptors*. London: A&C Black.

Niang, A.J., A. Ozer, and P. Ozer. 2008. Fifty years of landscape evolution in southwestern Mauritania by means of aerial photos. *Journal of Arid Environments* 72: 97–107.

Nilsson, C., R.H.G. Klaassen, and T. Alerstam. 2013. Differences in speed and duration of bird migration between spring and autumn. *American Naturalist* 181: 837–845.

Nordlöf, U, B. Helander, A. Bignert, and L. Asplund. 2010. Levels of brominated flame retardants and methoxylated polybrominated diphenylethers in eggs of white-tailed sea eagles breeding in different regions of Sweden. *Science of the Total Environment* 409: 238–246.

Odsjö, T. 1982. Eggshell thickness and levels of DDT, PCB and mercury in eggs of Osprey and Marsh Harrier in relation to their breeding success and population status in Sweden. PhD dissertation, University of Stockholm.

Odsjö, T., and J. Sondell. 2001. Population status and breeding success of Osprey *Pandion haliaetus* in Sweden, 1971–1998. *Vogelwelt* 122: 155–166

Odsjö, T., and J. Sondell. 2014. Eggshell thinning of osprey (*Pandion haliaetus*) breeding in Sweden and its significance for egg breakage and breeding outcome. *Science of the total environment* 470: 1023–1029.

Ogden, J.C. 1975. Effects of Bald Eagle territoriality on nesting Ospreys. *Wilson Bulletin* 87: 496–505.

Ogden, J.C. 1977. Preliminary report on a study of Florida Bay ospreys. In: Ogden, J.C. (ed.) *Transactions of the North American Osprey Research Conference* (pp. 143–152). Washington DC: US National Park Service.

Olendorff, R.R., and R.N. Lehman. 1986. *Raptor collisions with utility lines: an analysis using subjective field observations. Final report*. Pacific Gas and Electric Co., Research and Development, San Ramon.

Olsen, P., P. Fuller, and T.G. Marples. 1993. Pesticide-related eggshell thinning in Australian raptors. *Emu* 93: 1–11.

Österlöf, S. 1973. Fiskgjusen *Pandion haliaetus* i Sverige 1971. *Vår Fågelvärld* 32: 100–106. [In Swedish]

Østnes, J.E., R.T. Kroglund, O. Kleven, and T. Nygård. 2019. Migratory patterns of Ospreys (*Pandion haliaetus*) from central Norway. *Ornis Fennica* 96: 101–111.

Ottosson, U., R. Ottvall, J. Elmberg, *et al.* 2012. *Birds in Sweden – numbers and distribution*. Halmstad: Swedish Ornithological Society.

Palma, L., J. Ferreira, R. Cangarato, and P.V. Pinto. 2004. Current status of the Osprey in the Cape Verde Islands. *Journal of Raptor Research* 38: 141–147.

Palma, L., J. Safara, A. Dias, J. Ferreira, M. Mirinha, and P. Beja. 2019. The Portuguese Osprey Reintroduction Project: Achievements, Lessons and Perspectives. *Raptors Conservation* 38: 23–42.

Palma, L., S. Martins, R. Fortes, I. Rodrigues, M. Hernández-Montero, and R. Freitas. 2020. Twenty years later: updating the status of the osprey *Pandion haliaetus* in the Cabo Verde Islands, West Africa. *Zoologia Caboverdiana* 8: 3–10

Panuccio, M., N. Agostini, U. Mellone, and G. Bogliani. 2016. Circannual variation in movement patterns of the Black Kite *Milvus migrans*: a review. *Ethology Ecology and Evolution* 26: 1–18.

Paradis, E., S.R. Baillie, W.J. Sutherland, and R.D. Gregory. 1998: Patterns of natal and breeding dispersal in birds. *Journal of Animal Ecology* 67: 518–536.

Pearsall, W.H. 1950. *Mountains and Moorlands*. London: Collins.

Peele, G. 1594. *The Battle of Alcazar*.

Pennycuick, C.J. 1969. The mechanics of bird migration. *Ibis* 111: 525–556.

Pennycuick, C.J. 1998. Field observations of thermals and thermal streets, and the theory of cross-country soaring flight. *Journal of Avian Biology* 29: 33–43.

Pennycuick, C.J. 2008. *Modelling the Flying Bird*. London: Elsevier.

Penteriani, V., F. Otalora, and M. Ferrer. 2005. Floater survival affects population persistence. The role of prey availability and environmental stochasticity. *Oikos* 108: 523–534.

Penteriani, V., M. Ferrer, and M.M. Delgado. 2011. Floater strategies and dynamics in birds, and their importance in conservation biology: towards an understanding of nonbreeders in avian populations. *Animal Conservation* 14: 233–241.

Pitcher, T.J. 1998. Shoaling and Schooling in Fishes In: Greenberg, G. and M.M. Hararway (eds) *Comparative Psychology: a Handbook* (pp. 748–760). New York: Garland.

Philips, J.R. 2000. A review and checklist of the parasitic mites. *Journal of Raptor Research* 34: 210–231.

Philips, J.R., and D.L. Dindal. 1977. Raptor nests as a habitat for invertebrates: a review. *Raptor Research* 11: 87–96.

Pohja-Mykrä, M., T. Vuorisalo, and S. Mykrä. 2012. Organized persecution of birds of prey in Finland: historical and population biological perspectives. *Ornis Fennica* 89: 1–19.

Poole, A.F. 1982. Breeding Ospreys feed fledglings that are not their own. *The Auk* 99: 415–417.

Poole, A.F. 1984. Reproductive limitation in coastal Ospreys: an ecological and evolutionary perspective. PhD thesis, Boston University.

Poole, A.F. 1985. Courtship feeding and osprey reproduction. *The Auk* 102: 479–492.

Poole, A.P. 1989. *Ospreys: A Natural and Unnatural History*. Cambridge, UK: Cambridge University Press.

Poole, A.P. 2019. *Ospreys: The Revival of a Global Raptor*. Baltimore, MD: Johns Hopkins University Press.

Poole, A.F., Bierregaard, R.O., and Martell, M.S. 2002. Osprey. In: Poole, A.F. (ed.) *Birds of North America*. Ithaca, NY: Cornell Laboratory of Ornithology.

Postupalsky, S. 1977. Status of the osprey in Michigan. In: Ogden, J.C. *Transactions of the North American Osprey research conference* (pp. 153–165). Washington, DC: U.S. National Park Service.

Prevost, Y.A. 1977. Feeding ecology of Ospreys in Antigonish County, Nova Scotia. M.Sc. thesis, McGill Univ., Montreal, Quebec.

Prevost, Y.A. 1982. The wintering ecology of Ospreys in Senegambia. PhD thesis, University of Edinburgh.

Prevost, Y.A. 1983. The moult of the Osprey. *Ardea* 71: 199–209.

Puillat, I., P. Lazure, A.M. Jégou, L. Lampert, and P. Miller. 2006. Mesoscale hydrological variability induced by northwesterly wind on the French continental shelf of the Bay of Biscay. *Scientia Marina* 70: 15–26.

Puleston, D. 1975. Return of the Osprey. *Natural History* 84: 52–59.

Rabøl, J. 1978. One-direction orientation versus goal area navigation in migratory birds. *Oikos* 30: 216–223.

Raikow, R.J. 1985. The locomotor system. In King, A.S, and J. McLelland J(eds) *Form and Function in Birds* Vol. 3 (pp. 57–147). London: Academic Press.

Ratcliffe, D.A. 1967. Decrease in eggshell weight in certain birds of prey. *Nature* 215: 208–210.

Reese, J.G. 1977. Reproductive success of Ospreys in central Chesapeake Bay. *The Auk* 94: 202–221.

Restani, M. 2021. *Ecology and conservation of ospreys along the Yellowstone River, Montana. 2021 Annual Report*. Yellowstone Valley Audubon Society.

Richardson, P.L. 2011. How do albatrosses fly around the world without flapping their wings? *Progress in Oceanography* 88: 46–58.

Ridgway, K., and K. Hill. 2009. The East Australian Current. In: Poloczanska, E.S., A.J. Hobday and A.J. Richardson (eds) *A Marine Climate Change Impacts and Adaptation Report Card for Australia 2009*. NCCARF Publication 05/09.

River, M.S.M.C.R., M. C. nr Krema, and W. Kasiwa. 2007. New Guinea Island Systematic Bird Check List.

Roberts, T.S. 1932. *The Birds of Minnesota. Volume 1*. Minneapolis, MN, USA: University of Minnesota Press.

Robinson, R.A., Leech, D.I., and Clark, J.A. 2021. The Online Demography Report: bird ringing and nest recording in Britain & Ireland in 2010. BTO, Thetford (http://www.bto.org/ringing-report, created on 27 June 2020)

Rogers, L.L., and E.L. Lindquist. 1993. Supercanopy white pine and wildlife. In: *Proceedings of the white pine symposium: history, ecology, policy and management* (pp. 39–43). Duluth, MN, USA.

Rodgers Jr, J.A. and S.T Schwikert. 2002. Buffer-zone distances to protect foraging and loafing waterbirds from disturbance by personal watercraft and outboard-powered boats. *Conservation Biology* 16: 216–224.

Rotics, S., M. Kaatz, Y.S. Resheff, et al. 2016. The challenges of the first migration: movement and behaviour of juvenile vs. adult white storks with insights regarding juvenile mortality. *Journal of Animal Ecology* 85: 938–947.

RSPB. 2022. *Birdcrime 2021*. https://www.rspb.org.uk/globalassets/downloads/documents/birds-and-wildlife/crime/2021/bc2021_report.pdf

Ruddock, M., and Whitfield, D.P. 2007. A review of disturbance distances in selected bird species. *Report from Natural Research (Projects) Ltd to Scottish Natural Heritage 181*.

Rymon, L.M. 1989. The restoration of Ospreys (*Pandion haliaetus*) to breeding status in Pennsylvania by hacking (1980–1986). In: *Raptors in the modern world: proceedings of the III World Conference on Birds of Prey and Owls* (pp. 359–362). Berlin: World Working Group on Birds of Prey.

Ryslavy, T. 2011. Zur Bestandssituation ausgewählter Vogelarten in Brandenburg – Jahresbericht 2008 [Report 2008 on the status of selected bird species in Brandenburg]. *Naturschutz und Landschaftspflege Brandenburg* 20: 49–62.

Sachs, G. 2005. Minimum shear wind strength required for dynamic soaring of albatrosses. *Ibis* 147: 1–10.

Sakakibara, T., M. Noguchi, C. Yoshii, and A. Azuma. 2020. Diet of the Osprey *Pandion haliaetus* in Inland Japan. *Ornithological Science* 19: 81–86.

Safriel, U.N., Y. Ben-Hur, and A. Ben-Tuvia. 1985. The diet of the Osprey on Tiran Island: management implications for populations on the northern Red Sea islands. In: Moors, P.J. (ed.) *Conservation of Island Birds* (pp. 181–193). Cambridge, UK: ICBP Technical Publication No. 3.

Saggese, M.D., I. Roesler, and C.F. Marano. 2014. Wintering of Ospreys in Argentina: insights from new records between 1993–2008. *Journal of Raptor Research* 48: 345–360.

Saurola, P. 1997. The Osprey (*Pandion haliaetus*) and modern forestry: a review of population trends and their causes in Europe. *Journal of Raptor Research* 31: 129–137.

Saurola, P. 2005. Monitoring and conservation of Finnish Ospreys *Pandion haliaetus* in 1971–2005. In: *Status of Raptor Populations in Eastern Fennoscandia* (pp. 8–10), proceedings of the workshop, Kostomuksha, Karelia, Russia.

Saurola, P. 2008. Monitoring birds of prey in Finland: a summary of methods, trends and statistical power. *Ambio* 37: 413–419.

Saurola, P. 2011. Finnish Ospreys (*Pandion haliaetus*) 2010. Linnut vuosikirja – Yearbook BirdLife Finland 2010: 29–35.

Saurola, P. 2014. Finnish satellite Ospreys. LUOMUS Finnish Museum of Natural History. https://www.luomus.fi/en/finnish-satellite-ospreys [accessed 3/5/21]

Saurola, P. 2015. Suomen sääkset 2014 (Finnish Ospreys (*Pandion haliaetus*) 2014). Linnut-vuosikirja. 2014: 18–31 [In Finnish with an English summary].

Saurola, P. 2021. Viisi vuosikymmentä Suomen sääksikannan seurantaa: historiaa ja alustavia tu- loksia. Linnut-vuosikirja 2020: 86–93. [In Finnish with an English summary]

Saurola, P. and J. Koivu. 1987. Sääksi. Kanta-Hämeen Lintumiehet, Forssa, Finland. [In Finnish]

Schaadt, C.P., and D.M. Bird. 1993. Sex-specific growth in ospreys: the role of sexual size dimorphism. *The Auk* 110: 900–910.

Schmidt, D. 2010. Der Brutbestand des Fischadlers *Pandion haliaetus* in Deutschland im fru hen 21. Jahrhundert [Breeding pair numbers of the Osprey *Pandion haliaetus* in Germany in the early 21st century]. *Charadrius* 46: 10–17.

Schmidt, D., and R. Wahl. 2001. Nest-site and mate tenacity of Ospreys *Pandion haliaetus* ringed in eastern Germany and central France. *Vogelwelt* 122: 129–140.

Schmidt-Rothmund, D., R. Dennis, and P. Saurola. 2014. The Osprey in the Western Palearctic: Breeding population size and trends in the early 21st century. *Journal of Raptor Research* 48: 375–386.

Scott, T.G., Y.L. Willis, and J.A. Ellis. 1959. Some effects of a field application of dieldrin on wildlife. *The Journal of Wildlife Management* 23: 409–427.

Seacor, R., K. Ostovar, and M. Restani, M. 2014. Distribution and abundance of baling twine in the landscape near Osprey (*Pandion haliaetus*) nests: implications for nestling entanglement. *The Canadian Field Naturalist* 128: 173–178.

Semere, D., T. Hagos, G. Seleba, Y. Gebrezgabhier, Z. Haile, G. Chiozzi, and G. De Marchi. 2008. The status of breeding seabirds and waterbirds on the Eritrean Red Sea islands. *Bulletin of the African Bird Club* 15: 228–237.

Sergio, F., G. Tavecchia, A. Tanferna, et al. 2015. No effect of satellite tagging on survival, recruitment, longevity, productivity and social dominance of a raptor, and the provisioning and condition of its offspring. *Journal of Applied Ecology* 52: 1665–1675.

Serrano, D. and J.L. Tella. 2012. Lifetime fitness correlates of natal dispersal distance in a colonial bird. *Journal of Animal Ecology* 81: 97–107.

Sforzi, A., G. Sammuri, and F. Monti. 2019. From a regional reintroduction project to a country-wide conservation approach: scaling up results to promote osprey conservation in Italy. *Avocetta: Journal of Ornithology* 43: 81–85.

Shamoun-Baranes, J., O. Liechti, Y. Yom-Tov, and Y. Leshem. 2003. Using a convection model to predict altitudes of white stork migration over central Israel. *Boundary Layer Meteorology* 107: 673–681.

Sharma, A., D. Singh, S. Malik, N.J. Gupta, S. Rani, and V. Kumar. 2018. Difference in control between spring and autumn migration in birds: insight from seasonal changes in hypothalamic gene expression in captive buntings. *Proceedings of the Royal Society B* 285: 20181531.

Shepard, E.L.C., R.P. Wilson, W.G. Rees, E. Grundy, S.A. Lambertucci, and S.B. Vosper. 2013. Energy landscapes shape animal movement ecology. *American Naturalist* 182: 298–312.

Shikalova E.A., and V.V. Vinogradov. 2021. The current state of the Osprey (*Pandion haliaetus*) grouping on the territory of the Sayano-Shushensky biosphere reserve and its protected area. *Vestnik IrGSHA*. 3: 107–119.

Shobrak, M., A. Alsuhaibany, and O. Al-Segheir. 2003. *Regional status of breeding seabirds in the Red Sea and the Gulf of Aden*. The Regional Organization for the Conservation of the Environment of the Red Sea and Gulf of Aden (PERSGA).

Shobrak, M.Y., and A.A. Aloufi. 2014. Status of breeding seabirds on the Northern Islands of the Red Sea, Saudi Arabia. *Saudi Journal of Biological Sciences* 21: 238–249.

Shoji, A., Sugiyama, A., and Brazil, M.A. 2011. The status and breeding biology of Ospreys in Hokkaido, Japan. *The Condor* 113: 762–767.

Shukov, P.M. 2019. New data on distribution and population number of large birds of prey in Nizhny Novgorod Region, Russia. *Raptors Conservation* 38: 127–136.

Skorka, P. and J.D. Wojcik. 2008. Habitat utilisation, feeding tactics and age-related feeding efficiency in the Caspian Gull (*Larus cachinnans*). *Journal of Ornithology* 149: 31–39.

Simnor, A. 2015. A nationwide evaluation of the hacking programs for Osprey *Pandion haliaetus* in the United States, Environmental Studies, University of Illinois at Springfield, Springfield, IL.

Siverio, M. 2006. Population status and breeding biology of Osprey *Pandion haliaetus* in Tenerife, Canary Islands (1997–2004). *Alauda* 74: 413–419.

Siverio, M., B. Rodríguez, A. Rodríguez, and F. Siverio. 2011. Inter-insular variation of the diet of Osprey *Pandion haliaetus* in the Canarian archipelago. *Wildlife Biology* 17: 240–247.

Siverio, M., P. López-Suárez, F. Siverio, B. Rodríguez, N. Varo-Cruz, and L.F. López-Jurado. 2014. Density, nest-site characteristics and breeding rates of the Osprey (*Pandion haliaetus*) in the southern limit of its range in the Western Palearctic (Boa Vista, Cape Verde Islands). *African Journal of Ecology* 52: 50–58.

Siverio, M., F. Siverio, B. Rodríguez, and J.C. Del Moral (eds) 2018. El águila pescadora en España y Portugal: población invernante 2016–2017, reproductora en 2018 y método de censo. SEO/BirdLife. Madrid. [In Spanish]

Sivonen, T. 2014. The effects of supplementary feeding and weather factors on the breeding success of Osprey, *Pandion haliaetus*, in Finland. MSc thesis, University of Helsinki.

Snow, D.W., and C.M. Perrins. 1998. *The Birds of the Western Palearctic* (Concise Edition). Oxford: Oxford University Press.

Snyder, N.R.F., J.C. Ogden, J.D. Bittner, and G.A. Grau. 1984. Larval dermestid beetles feeding on nestling Snail Kites, Wood Storks and Great Blue Herons. *The Condor* 86: 170–174.

Solonen, T. 1993. Spacing of birds of prey in southern Finland. *Ornis Fennica* 70: 129–143.

Somveille, M., A.S.L. Rodrigues, and A. Manica, A. 2015. Why do birds migrate? A macroecological perspective. *Global Ecology and Biogeography* 24: 664–674.

Spaar, R., and B. Bruderer. 1996. Soaring migration of Steppe Eagles *Aquila nipalensis* in southern Israel: Flight behaviour under various wind and thermal conditions. *Journal of Avian Biology* 27: 289–301.

Spitzer, P.R. 1980. Dynamics of a discrete coastal breeding population of Ospreys in the northeastern USA, 1969–1979. PhD thesis, Cornell University, Ithaca, NY.

Spitzer, P.R., R.W. Risebrough, W. Walker, et al. 1978. Productivity of Ospreys in Connecticut-Long Island increases as DDE residues decline. *Science* 202: 333–335.

Spitzer, P.R., A.F. Poole and M. Scheibel. 1983. Initial population recovery of breeding Ospreys between New York City and Boston. In: Bird, D.M. (ed.) *Biology and Management of Bald Eagles and Ospreys* (pp. 231–241). Ste. Anne de Bellevue, Quebec: Harpell Press.

Sprunt IV, S. 1977. Report on Osprey sightings and nest locations in coastal Mexico and British Honduras. In: *Transactions of the North American Osprey Research Conference*. Williamsburg, VA, USA: US National Park Service.

Steeger, C., H. Esselink, and R.C. Ydenberg. 1992. Comparative feeding ecology and reproductive performance of Ospreys in different habitats of southeastern British Columbia. *Canadian Journal of Zoology* 70: 470–475.

Steeger, C., and R.C. Ydenberg. 1993. Clutch size and initiation date of Ospreys: natural patterns and the effect of a natural delay. *Canadian Journal of Zoology* 71: 2141–2146.

Steen, O.F., and G. Hansen. 2001. Osprey *Pandion haliaetus* management in Vestfold county, SE Norway 1984–98. *Vogelwelt* 122: 223–226.

Steidl, R.J., and C.R. Griffin. 1991. Growth and brood reduction of mid-Atlantic coast Ospreys. *The Auk* 108: 363–370.

Steidl, R.J., C.R. Griffin, and L.J. Niles. 1991. Contaminant levels of Osprey eggs and prey reflect regional differences in reproductive success. *Journal of Wildlife Management* 55: 601–608.

Strandberg, R., T. Alerstam and M. Hake. 2006. Wind-dependent foraging flight in the Osprey *Pandion haliaetus*. *Ornis Svecica* 16: 150–163.

Strandberg, R., and T. Alerstam. 2007. The strategy of fly-and-forage migration, illustrated for the Osprey. *Behavioural Ecology Sociobiology* 61: 1865–1875.

Strandberg, R., R.H.G. Klaassen, M. Hake, and T. Alerstam. 2010. How hazardous is the Sahara Desert crossing for migratory birds? Indications from satellite tracking of raptors. *Biology Letters* 6: 297–300.

Strandberg, R. 2013. Ageing, sexing and subspecific identification of Osprey, and two Western Palearctic records of American Osprey. *Dutch Birding* 35: 69–87.

Stone, W. 1937. *Bird Studies at Old Cape May: an Ornithology of Coastal New Jersey*. Delaware Valley Ornithology Club, Philadelphia, PA, USA.

Strahm, W., and D. Landenbergue. 2013. La réintroduction du Balbuzard pêcheur *Pandion haliaetus* en Suisse romande – un projet pour le centenaire de Nos Oiseaux. *Nos Oiseaux* 60: 123–142.

Strahm, W., and D. Landenbergue. 2021. Projet Balbuzard *Pandion haliaetus*: retours et lâchers en Suisse en 2020. *Nos Oiseaux* 68: 69–79.

Sutherland, W.J., and K. Norris. 2002. Behavioural models of population growth rates: implications for conservation and prediction. *Philosophical Transactions of the Royal Society of London. Series B: Biological Sciences* 357: 1273–1284.

Swenson, J. 1978. Prey and Foraging Behavior of Ospreys on Yellowstone Lake, Wyoming. *The Journal of Wildlife Management* 42: 87–90.

Swenson, J. 1979. The relationship between prey species ecology and dive success in ospreys. *The Auk* 96: 408–412.

Swenson, J. 1981. Osprey nest-site characteristics in Yellowstone National Park. *Journal of Field Ornithology* 52: 67–69.

Sustaita D, Y. Gloumakov, L.R. Tsang, and A.M. Dollar. 2019. Behavioral correlates of semi-zygodactyly in Ospreys (*Pandion haliaetus*) based on analysis of internet images. *PeerJ* 7: e6243 DOI 10.7717/peerj.6243

Tapper, S. 1992. *Game Heritage*. Fordingbridge, UK: Game Conservancy.

Tennessee Wildlife Resources Agency (TWRA). 2022. https://www.tn.gov/twra/wildlife/birds/waterbirds/osprey.html. Accessed 14.02.2022.

J.-C. Thibault and O. Patrimonio. 1991. Some aspects of breeding success of the Osprey *Pandion haliaetus* in Corsica, West Mediterranean. *Bird Study* 38: 98–102.

Thiao, D., and S.W. Bunting. 2022. *Socio-economic and biological impacts of the fish-based feed industry for sub-Saharan Africa*. FAO Fisheries and Aquaculture Circular No. 1236. Food & Agriculture Org. Rome: FAO, Worldfish and University of Greenwich, Natural Resources Institute.

Thibault, J.-C., V. Bretagnolle, and J.-M. Dominici. 2001. *Le Balbuzard Pêcheur en Corse. Du Martyre au Symbole de la Protection de la Nature*. Ajaccio, France: Editions Alain Piazzola

Thomas, A.L.R. 1996. The flight of birds that have wings and tail: variable geometry expands the envelope of flight performance. *Journal of Theoretical Biology* 183: 237–245.

Thomson, V.K., D. Jones, and T. Stevens. 2019. Nest-site selection by Eastern Osprey '*Pandion haliaetus cristatus*' in coastal south-eastern Queensland. *Australian Field Ornithology* 36: 142–147.

Thorup, K., T. Alerstam, M. Hake, and N. Kjellén. 2003. Can vector summation describe the orientation system of juvenile ospreys and honey buzzards? An analysis of ring recoveries and satellite tracking. *Oikos* 103: 350–359.

Treinys, R., D. Dementavičius, G. Mozgeris, S. Skuja, S. Rumbutis, and D. Stončius. 2011. Coexistence of protected avian predators: does a recovering population of White-tailed Eagle threaten to exclude other avian predators? *European Journal of Wildlife Research* 57: 1165–1174.

Tsang, L.R. 2012. Facultative zygodactyly in the Black-shouldered Kite *Elanus axillaris*. *Australian Field Ornithology* 29: 89–92.

Tucker, V.A. 1987. Gliding birds: the effect of variable wing span. *Journal of Experimental Biology* 133: 33–58.

Tucker, V.A. 1998. Gliding flight: speed and acceleration of ideal falcons during diving and pull out. *Journal of Experimental Biology* 201: 403–414.

Tuvi, J. and Ü. Väli. The impact of the White-tailed Eagle *Haliaeetus albicilla* and the Osprey *Pandion haliaetus* on Estonian Common Carp *Cyprinus carpio* production: How large is the economic loss? *Proceedings of the Estonian Academy of Sciences: Biology, Ecology* 56: 209–223.

Tyler, A.V. 1971. Surges of Winter Flounder into the intertidal zone. *Journal of the Fisheries Research Board of Canada* 28: 1727–1732.

Ueoka, L. and J.R. Koplin. 1973. Foraging behavior of Ospreys in northwestern California. *Raptor Research* 7: 32–38.

UNEP. 2009. Stockholm Convention on Persistent Organic Pollutants, as amended in 2009 – text and annexes. http://chm.pops.int/TheConvention/Overview/TextoftheConvention/tabid/2232/Default.aspx

Van Daele, L.J. and H.A. Van Daele. 1982. Factors affecting the productivity of Ospreys nesting in west-central Idaho. *The Condor* 84: 292–299.

Vansteelant, W.M.G., W. Bouten, R.H.G. Klaassen, et al. 2015. Regional and flight speeds of soaring migrants and the role of weather conditions at hourly and daily scales. *Journal of Avian Biology* 46: 25–39.

Väli, Ü., and U. Sellis. 2016. Migration patterns of the Osprey *Pandion haliaetus* on the Eastern European–East African flyway. *Ostrich* 87: 23–28.

Väli, Ü., A. Kalvāns, and J. Tuvi. 2021. Apparent survival and dispersal in a recovered Osprey population: effects of age, sex and social status. *Journal of Ornithology* 162: 1025–1034.

Vardanis, Y., J.A. Nilsson, R.H.G. Klaassen, R. Strandberg, and T. Alerstam. 2016. Consistency in long-distance bird migration: contrasting patterns in time and space for two raptors. *Animal Behaviour* 113: 177–187.

Vergara, Pablo, and J.A. Fargallo. 2008. Sex, melanic coloration, and sibling competition during the postfledging dependence period. *Behavioral Ecology* 19: 847–853.

Wahl, R. and C. Barbraud. 2014. The demography of a newly established Osprey *Pandion haliaetus* population in France. *Ibis* 156: 84–96.

Walcott, J. 1789. *Synopsis of British Birds*. Vol 1. London: W. Justins.

Washburn, B.E. 2014. Human–Osprey conflicts: industry, utilities, communication, and transportation. *Journal of Raptor Research* 48: 387–395.

Washburn, B.E., M.S. Martell, R. Bierregaard, C.J. Henny, B.S. Dorr, and T.J. Olexa. 2014. Wintering ecology of North American Ospreys. *Journal of Raptor Research* 48: 325–333.

Waterston, G. 1971. *Ospreys in Speyside* (third edition). Edinburgh: RSPB.

Watts, B.D., and B.J. Paxton. 2007. Ospreys of the Chesapeake Bay: population recovery, ecological requirements, and current threats. *Waterbirds* 30: 39–49.

Weatherhead, P.J., and M.R.L. Forbes. 1994. Natal philopatry in passerine birds: genetic or ecological influences? *Behavioral Ecology* 5: 426–433.

Wetmore, S.T., and D.I. Gillespie. 1976. Osprey and Bald Eagle populations in Labrador and north-eastern Quebec, 1969–1973. *Canadian Field Naturalist* 90: 330–337.

Whalley, J.I. 1982. *Pliny the Elder, Historia Naturalis*. London: Sidgwick and Jackson.

Whishaw, I.Q., and D.G. Wallace. 2003. On the origins of autobiographical memory. *Behavioural Brain Research* 138: 113–119.

Widén, P., and M. Richardson. 2000. Copulation behavior in the Osprey in relation to breeding density. *The Condor* 102: 349–354.

Wiemeyer, S.N., P.R. Spitzer, W.C. Krantz, T.G. Lamont, and E. Cromartie. 1975. Effects of environmental pollutants on Connecticut and Maryland Ospreys. *Journal of Wildlife Management* 39: 124–139.

Wiemeyer, S.N., T.G. Lamont, and L.N. Locke. 1980. Residues of environmental pollutants and necropsy data for eastern United States Ospreys, 1964–1973. *Estuaries* 3: 155–167.

Wiemeyer, S.N., C.M. Bunck, and A.J. Krynitsky. 1988. Organochlorine pesticides, polychlorinated biphenyls, and mercury in Osprey eggs 1970–79 and their relationships to shell thinning and productivity. *Archives of Environmental Contamination and Toxicology* 17: 767–787.

Wiley, J.W., A.F. Poole, and N.J. Clum. 2014. Distribution and natural history of the Caribbean Osprey (*Pandion haliaetus ridgwayi*). *Journal of Raptor Research* 48: 396–407.

Wink, M., H. Sauer-Gürth, and H.H. Witt. 2004. Phylogenetic differentiation of the Osprey (*Pandion haliaetus*) inferred from nucleotide sequences of the mitochondrial cytochrome b gene. In: Chancelor, R.D. and Meyburg, B.U. (eds) *Raptors Worldwide*. Berlin: WWGBP.

Witherby, H.F. 1912. A swallow ringed in Staffordshire and recovered in Natal. *British Birds* 6: 277–278.

Woodford, J.E., W.H. Karasov, M.W. Meyer, and L. Chambers. 1998. Impact of 2, 3, 7, 8 TCDD exposure on survival, growth, and behavior of Ospreys breeding in Wisconsin, USA. *Environmental Toxicology and Chemistry* 17: 1323–1331.

Wotzkow, C. 1985. Status and distribution of Falconiformes in Cuba. *Bulletin of the World Working Group on Birds of Prey* 2: 1–10.

Woźniak, B., M. Zygmunt, Ł. Porębski, P. Woźniak, and D. Anderwald. 2022. Red Spot on the European Green Map: Will the extra catastrophic phenomenon take the Polish poaching-pressured Ospreys to the brink of extinction? *Animals* 12: 69.

Yalden, D.W., and Albarella, U. 2009. *The History of British Birds*. Oxford: Oxford University Press.

Yosef, R., M. Markovets, L. Mitchell, and P. Tryjanowski. 2006. Body condition as a determinant for stopover in bee-eaters *Merops apiaster* on spring migration in the Arava Valley, southern Israel. *Journal of the Arid Environment* 64: 401–411.

Zachos, F.E., and U. Schmölcke. 2006. Archeozoological records and distribution history of the Osprey (*Pandion haliaetus*) in central Europe. *Journal of Ornithology* 147: 565–568.

Zubcov, N., A. Munteanu, V. Crudu, L. Bogdea, and N. Vasilashcu. 2014. On the state of rare and endangered diurnal birds of prey in Moldova. *Buletinul Academiei de Ştiinţe a Moldovei. Ştiinţele vieţii* 322: 78–85.

Zwarts, L., R.G. Bijlsma, J. Van der Kamp, and E. Wymenga. 2012. *Living on the edge: wetlands and birds in a changing Sahel*. Zeist, The Netherlands: KNNV Publishing.

Acknowledgements

I wish to express my sincere thanks to everyone who contributed to this book, and to my wider work with Ospreys. Roy Dennis has been a friend, mentor and source of great inspiration since I first met him over 25 years ago. Much of my work with Ospreys has been at Rutland Water where I have been fortunate to work with people who have become lifelong friends. John Wright, Paul Stammers and Lloyd Park deserve special mention, as do Helen McIntyre and Barrie and Tricia Galpin who encouraged me to start my journey with Ospreys. So many other staff and volunteers have contributed to the Osprey story, as have the farmers and other land managers who have helped to protect their local Ospreys. More recently, it has been a pleasure to work with the staff and volunteers at Birds of Poole Harbour. Particular thanks to Mark and Mo Constantine, Paul Morton, Brittany Maxted, Liv Cooper, Lucy Allen and also Jason Fathers. Ian Perks and Fraser Cormack have also become great friends and colleagues on different projects. Ken Davies provided great help and assistance in researching references to Ospreys in historical literature, while Tim Appleton, who set up the Rutland Osprey Project with Roy, has been another great influence on my career. Pete Davies and Joanna Dailey who monitor Ospreys in Cumbria and Kielder Forest respectively, generously allowed me to include data they have collected, and were a great help with various queries. Thanks also to Charlie Hewitt for his valuable assitance with GIS analysis.

The global nature of the Osprey's range means that I have consulted Osprey experts in many countries while researching and writing this book. Particular thanks are due to those who read and commented on sections of text, including Mark Martell, Terry Dennis, Rolf Wahl, Paul Lesclaux, Daniel Schmidt-Rothmund, Pertti Saurola, Luis Palma, Urmas Sellis, Miguel Ferrer, Roberto Muriel, Aitor Galarza, Itziar Colodro, Frank Hailer, Flavio Monti, Wendy Strahm and Denis Landenbergue. I am also very grateful to other Osprey friends and colleagues at the various Osprey projects around the UK, as well as those further afield including Alan Poole, Rob Bierregaard, Ian MacLeod, Andrea Sforzi, Yossi Leshem and anyone else I may have inadvertently omitted from this list. Thanks also to Robert Steidl and Thomas Alerstam for allowing me to reproduce figures from their research on Ospreys. Jean-Marie Dupart provided valuable data on Ospreys in Senegal and The Gambia, Frederic Bacuez welcomed myself, John Wright and Paul Stammers to his home in St Louis, northern Senegal, and showed us the local area and the wintering Ospreys, while Junkung Jadama, Gambian bird guide, has led numerous trips to The Gambia and Senegal for myself and friends and colleagues from Rutland Water and the Osprey Leadership Foundation; thanks also to Alhagie for driving the 'Osprey bus' on those visits. My sincere thanks, as well, to Barry Dore for his support and encouragement.

Very many thanks to the team at Bloomsbury, particularly Jim Martin for commissioning the book, and tolerating my delays; Katy Roper, David Campbell and particularly Amy Hodkin for her brilliant work in bringing the book together; to David Hawkins, Alex Berryman and Cathryn Pritchard for their respective work copy-editing, proofreading and indexing this book; and to Mark Heslington for the excellent layout and design. I am also very grateful to Alison Copland for proof reading and copyediting an initial draft of the manuscript.

This book is brought alive by wonderful photos and I am extremely grateful to all the credited photographers. Particular thanks go to John Wright for contributing a huge number of his images and brilliant artwork. Sincere thanks to John Davis for his stunning cover artwork.

Finally, and very importantly, I would like to thank my wife Louise, and parents Bob and Bev Mackrill for all their support and encouragement, and my sons Harry and Laurie, both already budding Osprey fieldworkers.

Index

Page numbers in *italic* indicate figures and in **bold** indicate tables.

age
  at first breeding 59
  hunting success and 33
  see also juveniles
alarm calls 24–5
albatrosses 151
aldrin 198
Algerian breeding birds 11, 75, 167–8, 262
Alqueva reservoir, Portugal 229–31, **230**, *230*, **235**, 249
Andalucía translocation project 17, 50–1, 106, 226–7, **235**
annual survival rates 241–4, **242**, **243**
anthropogenic threats
  climate change 258–9
  dieldrin and aldrin pollution 198
  dioxin pollution 198
  egg-collecting 191, 192–4, 247
  electrocution 245–6, *246*
  forestry and agriculture 250
  human disturbance 248–50
  illegal killing 246–8
  mercury pollution 199
  on migration 124
  netting and entanglement 251–3, *251*, *252*
  organochlorine pesticides 194–8, 253
  overfishing 254–5
  persecution 189–92, 246–8
  poisoning/contamination 194–9, 253–4
  polybrominated diphenyl ethers (PBDEs) 253–4
  polychlorinated biphenyls (PCBs) 198–9, 253
  powerline collisions 244–5
  tourism development 250
  water-based disturbance 249–50
  wind turbines 124, 245
Antigonish Harbour, Nova Scotia, Canada 34–5, 36, 46–7
aquaculture 38–9, 192, 247, 248
Aquitaine translocation project 233–4, *233*, **235**, 261
Aristotle 187, *187*
arrival dates
  at breeding sites 60–2, 87–90, *88*, *89*, **90**, *90*
  at wintering grounds 164
artificial nests 200–11
  Australia 211
  breeding density 81–2
  Canada 211
  construction 201–4, *201*, *202*, *203*

England 201, *202*, 203, 204–5, *205*, **206**, *207*, *208*, *209*
Finland 83, 208–10
Germany 76, 207–8, *210*
height of 81, *82*
lining materials 201–2, *203*
Scotland 200, *201*, 204
United States 50, 77, *78*, 81–2, *82*, 200, 210–11, *211*, *246*, *265*, 266
Wales 55, 66, 205–7, **206**, *209*
see also translocation projects
artificial structures, nesting on 76–8, *77*, 81, *82*, 207–8, *210*, 215, 266
aspergillosis 256
Auburn Lake, Minnesota 95
Australian breeding birds 12
  artificial nest provision 211
  breeding season 87
  clutch size 98
  DDT exposure 197–8
  human disturbance 250
  individual variation 21
  laying dates 97
  nest-building 83
  nest-sites 75, *77*, *268*
  polyandry 67
  population estimates 267–8
  prey species 41
  site fidelity 69
  wintering 171
avian cholera 256

Bahamian breeding birds 12, 266–7
Bahrain breeding birds 11, 264
Balearic Islands breeding birds 11
  arrival at breeding sites 87
  cliff nesting 75
  laying dates 97
  migration of juveniles 137
  migratory distances 134, **135**, 136
  migratory routes 129–31
  natal dispersal 50–1
  population estimates 262
  sea crossings 148–9
  stop-overs 142
  translocation of 229
  wintering 164, 167–8, 171
Barbate Reservoir, Spain 50–1, 226–7, **235**

293

Basque Country translocation project 158, 227–8, 228, 233–4, **235**
Beaver 73
Belarus breeding birds **40**, 71, 247, 260, *260*
Belizean breeding birds 12, 267
BirdLife Australia 12–13
bodyweight 13, 17, 108
breeding 71–123
    age at first breeding 59
    arrival dates 60–2, 87–90, *88*, *89*, **90**, *90*
    breeding density 50, 65–7, 81–3, 95
    breeding dispersal 48–9, 67–70
    brood reduction 111–12, *112*
    cliff nesting 65–6, 75–6, 83, 249
    clutch size 98–100, *99*, 105, *112*
    colonial nesting 46–7, 50, 59, 79, 81–2
    conspecific attraction 50, 65, 81–2, 200, 214, 225, 226–7, 231, 234–6
    copulation 93–5, *93*
    density dependence 65–7
    departure dates 121–2, 132–3
    disputes over nest-sites 23–4, 59–65, 95, 103–5, *104*
    dominance of older siblings 110–12
    extra-pair copulations 95
    feeding fledged juveniles 25, 117–18, 120, 121, *121*, 132
    feeding nestlings 107–8, 113
    fledging 114–17, *115*, **116**, *116*
    ground nesting 74–6, *74*, *75*, 81, 82, *262*
    growth of nestlings 109–14, *110*, *111*, *113*
    hatching 105–8, *106*, *107*
    height of nests 73, 81, *82*
    human disturbance and 248–50
    incubation 101–5, *102*, *103*, *104*
    interspecific competition and 80–1, *80*, 100–1, *100*, *101*, 257–8
    intrusions by non-breeding birds 23–4, 51, 59–65, 95, 103–5, *104*
    laying dates 97–8
    moulting and 20
    nest-building 55, 83–6, *84*, *85*, *86*, *87*
    nesting on artificial structures 76–8, *77*, 81, 82, 207–8, *210*, 215, 266
    polygyny and polyandry 66–7
    post-fledging period 117–22, *118–19*, *121*, *122*, 132, 241, 257
    predation of eggs and nestlings 71, 74, *74*, 79, 80, 105, 114, 256–7
    provisioning rates 108–9, *108*
    proximity to water 73, 78–9, *78*
    replacement clutches 105
    site fidelity 67–70, *67*, *69*
    tree nesting 71–4, *71*, *72*, *73*, 81, 200
    vocalisations 23–5, *24*, 60, 103
    see also artificial nests; courtship behaviour
breeding phenology 87–90, *88*, *89*, **90**, *90*
brood reduction 111–12, *112*

Buzzard, Common 13, 114, 191, 257

cacti 74
calls 23–5
    alarm calls 24–5
    display calls 23, *24*, 48
    excited calls 23, 24, 25, 103
    food solicitation call 24, 25, 92, 102
    guard calls 23–4, 60, 103
    screaming calls 23–4, 103
camouflage 113–14, *114*
Canadian breeding birds 11–12
    arrival at breeding sites 89
    artificial nest provision 211
    breeding density 83
    clutch size 100
    colonial nesting 46–7
    courtship feeding 92
    departure from breeding sites 132
    hunting 34–5, 36, 46–7
    incubation 105
    interspecific competition 80, 100
    mercury exposure 199
    migration 12, 131, *131*, 132, 152
    nest-sites 73, 78, 79
    nestlings 108–9, 110–11
    persecution of 192
    population estimates 266
    prey species 44
    sea crossings 147, 152
    site fidelity 168
    wintering 163, 164, 167, 168, 170
Canary Islands breeding birds 11, 97, 262, *263*
Cape Verde breeding birds 11, 146
    breeding season 87
    egg-collecting 194
    laying dates 97
    nest-sites 75–6, *75*, 83, *262*
    population estimates 194, 262
    prey species 42–3
Caribbean breeding birds 12, 266–7
carpal patch 13, 14, 15, 16, 22–3, *22*, *23*
Casamance region, Senegal 140, **166**, 167, 173, 177
central fovea 26
channel markers 76, *77*, 81
Chesapeake Bay 11, 44, 59, 69–70, 76, 79, 80, 86, 89, 97, 195, 196, 212, 213, 244, 251, 256, 258, 265–6
cholera, avian 256
'chord and clock' navigation 147–8
cliff nesting 65–6, 75–6, 83, 249
climate change 87, 168, 258–9
clock-and-compass concept 129
clutch size 98–100, *99*, 105, *112*
colonial nesting 46–7, 50, 59, 79, 81–2
competition, interspecific 80–1, *80*, 100–1, *100*, *101*
conservation
    legal protection 192, 194

nest guarding 192–4, *193*, *194*
  see also artificial nests; translocation projects
conspecific attraction 50, 65, 81–2, 200, 214, 225, 226–7, 231, 234–6
copulation 93–5, *93*
  extra-pair copulations 95
Cormorant, Great 32
Corn Creek Marsh, British Colombia 34, 36
Cors Dyfi, Wales 53–5, *54*, 66–7, 68, 101, 105, 106, *106*, **116**, 117, 225, 236, 238, *239*, 259, 274
Corsican breeding birds 11
  arrival at breeding sites 87
  cliff nesting 65–6, 75, 83, 249
  copulation 94
  density dependence 65–6
  display flights 91
  extra-pair copulations 95
  food solicitation by females 24, 92
  human disturbance 249
  individual variation 21
  laying dates 97
  males guarding females 94–5
  migration of juveniles 137
  migratory distances 134, **135**, 136
  migratory routes 129–31
  movements during second calendar year 184–5
  population estimates 65, 262
  prey species **40**, 41
  sea crossings 148–9
  stop-overs 142
  translocation of 59, 231, **235**, 241
  wintering 164, 167–8, 171, 184–5
courtship behaviour 91–3
  copulation 93–5, *93*
  courtship feeding 92, 93–4, 98
  display flights 23, *24*, 48, *48*, *90*, 91, 92–3
  extra-pair copulations 95
  food solicitation 24, 92, 93–4
  males guarding females 94–5
  mantling display 91, *91*
crocodiles 28, 258
crosswinds 127–8, *128*, 144–8, *145*
Cuban breeding birds 12, 267

Danish breeding birds 190, 192, 259
DDE 195
DDT 194–8, 253
De Biesbosch National Park, Netherlands 55, *56*, 261, *261*
density
  breeding 50, 65–7, 81–3, 95
  wintering 164–5, 167, *167*, 168, 169, 173–6, *173*, *174*, 180
density dependence 65–7
departure dates
  from breeding sites 121–2, 132–3
  from wintering grounds 185, *185*
Deschutes National Forest, Oregon 73, 81

dichlorodiphenyldichloroethylene (DDE) 195
dichlorodiphenyltrichloroethane (DDT) 194–8, 253
dieldrin 198
diet *see* prey
dioxin 198
disease 216, 256
dispersal and settlement patterns 48–70
  age at first breeding 59
  breeding density 50, 65–7, 81–3, 95
  breeding dispersal 48–9, 67–70
  colonial nesting 46–7, 50, 59, 79, 81–2
  conspecific attraction and 50, 65, 81–2, 200, 214, 225, 226–7, 231, 234–6
  density dependence 65–7
  impact of floaters in populations 59–65
  natal dispersal 48, 49–58
  polygyny and polyandry 66–7
  site fidelity 67–70, *67*, *69*
display calls 23, *24*, 48
display flights 23, *24*, 48, *48*, *90*, 91, 92–3
disputes over nest-sites 23–4, 59–65, 95, 103–5, *104*
distribution 9–13, *10*
Djoudj National Bird Sanctuary, Senegal 153, 155, 157, 165, **166**, 176–7, *178*, 184
drag 125

Eagle
  Bald 32, 80, *80*, 114, 257–8
  Booted 138–9
  Golden 144, 191, 195, 245
  Greater Spotted 12
  Imperial 12
  Lesser Spotted 12
  Spanish Imperial 12
  Steppe 127
  White-tailed 32, 80–1, 114, 216, 257, 258
ecotourism 193, 236–8, *238*, 273–4
education programmes 238–9, *239*
egg-collecting 191, 192–4, 247
eggs
  clutch size 98–100, *99*, 105, *112*
  colour and markings 98
  eggshell thinning due to DDT 195, *196*
  hatching 105–8, *106*, *107*
  incubation 101–5, *102*, *103*, *104*
  laying dates 97–8
  predation of 74, *74*, 79, 105
  replacement clutches 105
  size and shape 98
Egret
  Cattle 9
  Great 9
Egyptian breeding birds 11, 74, 250, 263
electrocution 245–6, *246*
English breeding birds
  age at first breeding 59
  arrival at breeding sites 87, *88*, 89–90, *89*, **90**

295

artificial nest provision 201, *202*, 203, 204–5, *205*, **206**, *207*, *208*, *209*
clutch size 98–9, *99*, 105
collisions 245
colour rings *165*
courtship behaviour *24*, *91*, 92–3
departure from breeding sites 132
departure from wintering grounds 185, **185**
disease 216, 256
display flights *24*, *48*
disputes over nest-sites 60–5, *61*, *62*, *64*, 105
egg markings *99*
electrocution 246
entanglement 251, 253
fledging 115, 116, **116**, 117
food solicitation by females 94, 102
hatching 108
human disturbance 250
hunting 28, *28*, *31*, 34, 35–6, *37*, 45–6
incubation 102, *102*, 103, 105, 106
individual variation 21, *21*, 22–3, *22*, *23*
interspecific competition 100–1, *101*
kleptoparasitism 32, *32*
laying dates 97–8
migration *161*
migration of juveniles 136, *137*
migratory distances 134
mortality of juveniles 181–2, **183**, 241–3, **242**, **243**
movements during second calendar year 184
natal dispersal 51–6, *52*, *54*, *56*
nest-sites 72, *72*, 79
nestlings 113
other bird species using nests 86
persecution of 189–90, 247–8
population estimates *209*, 261–2
post-fledging period 118, 120–1, *122*
predation of 257, *257*
prey species 37, 38, *38*, 39, **40**, 41
Sahara crossing 160, *161*
sea crossings 148
site fidelity *67*, 68–9, *69*, 168
stop-overs 140
viewing sites 236, 247, *247*, 273–4
wintering 168, 169–70, *169*, *170*, 171–2, 181–2, **183**, 184
English translocation projects
 Poole Harbour, Dorset 224–6, *226*
 see also Rutland Water translocation project
Estonian breeding birds 11
 migration 129, *130*, 133–4, **135**, 158
 payments to protect 247
 population estimates 260
 predation of 257
 prey species 39
 wintering 163, 164
excited calls 23, 24, 25, 103
extra-pair copulations 95

eyes
 central fovea 26
 iris markings 21–2
 retina 26
 size 26
 temporal fovea 26
 visual acuity 26, 29

feather mites 256
feeding
 courtship 92, 93–4, 98
 fledged juveniles 25, 117–18, 120, 121, *121*, 132
 nestlings 107–8, 113
feet and toes 29, 30–1
fights *see* disputes over nest-sites
Findhorn Bay, Scotland 45, 46
Finnish breeding birds 11
 annual survival rates 241
 arrival at breeding sites 87–9
 artificial nest provision 83, 208–10
 DDT exposure 196, 197
 entanglement 251
 human disturbance 249, 250
 laying dates 97
 legal protection of 192
 longevity 244
 migration 129, *130*, 133, 158
 natal dispersal 49
 nest-sites 72, 78–9, 83
 payments to protect 247
 persecution of 190, 247, 248
 population estimates 208, 259
 predation of 257
 prey species **40**
 translocation of 17, 50, 226, 229, 231, **235**
 wintering 163, 165
fish *see* prey
fish farms 38–9, 192, 247, 248
flapping flight 125–6, *126*, 136, 138, 149, 151
fledging 114–17, *115*, **116**, *116*
flight
 display flights 23, *24*, *48*, *48*, *90*, *91*, 92–3
 flapping 125–6, *126*, 136, 138, 149, 151
 helicoptering 115, *115*
 hovering 26–8, *27*, 33–4, 46
 maiden flights 114–17, *115*, *116*
 migration flight modes 125–7, *126*, *127*, 136, 138, 149–51, *150*, 158–9, *159*
 nocturnal 127, 138, 151–2, 158
 slope soaring 125
 soaring-gliding 125, 126–7, *126*, 136, 138, 149–51, *150*
floater-to-breeder ratio 60
floaters
 interference at nests 23–4, 51, 59–65, 95, 103–5, *104*
 polygyny and polyandry 66–7
fly-and-forage strategy 132, 141–2

food solicitation
  by females 24, 92, 93–4, 102
  by juveniles 25, 117–18, 121, *121*, 132
food solicitation call 24, 25, 92, 102
forearm length 17, *18*
Foulshaw Moss, Cumbria 273
French breeding birds 11
  annual survival rates 241
  arrival at breeding sites 87
  colour rings *165*
  DDT exposure 197
  density dependence 66
  laying dates 97
  legal protection of 192
  migration 129, 158
  natal dispersal 50
  nest-sites 76
  population estimates 261
  wintering 165
French translocation project 233–4, *233*, **235**, 261

Gambia, The
  densities *173*
  education programmes 239
  hunting 33, 35
  juveniles 177, *179*
  overfishing threats 254
  predation by crocodiles 258
  prey species 41–2, *43*, **44**
Gardiner's Island, New York 75, 82, 195–6
Garganey 176, *178*
genetic analysis 10, 12–13
German breeding birds 11
  artificial nest provision 76, 207–8, *210*
  colour rings *165*
  DDT exposure 196
  interspecific competition 101
  legal protection of 192
  migration 129, 158
  natal dispersal 49–50
  nest-sites 72, 76, 78, 81, 82
  persecution of 192
  population estimates 207, 260
  prey species 39, **40**
  pylon nesting 76, 81, 82, 207–8, *210*
  translocation of 17, 50, 106, 226, 227, 232, 233, **235**
  wintering 163, 165, *179*
Glaslyn, Wales 52–3, 56, 66–7, 69, 105, 106, **116**, 207, 224, 225, 274
gliding flight 125, 126–7, *126*, 136, 138, 149–51, *150*
goal areas 152–7, *154*
Goose
  Canada 100, *100*, 203
  Egyptian 53–5, 100–1, *101*, 116, 203
  Greylag 100, 203
Goshawk, Northern 71, 114, 247, 256–7, *257*

Gould, John *188*, 189
Grand Côte, Senegal 165, 166, **166**, 176
Grebe, Little 37
ground nesting 74–6, *74*, *75*, 81, 82, *262*
guard calls 23–4, 60, 103
Gull
  Slender-billed 176, *182*
  Yellow-legged 249

Harrier
  Marsh 124, 127, 128, 152
  Montagu's 124, 127
Harrison, William 189
hatching 105–8, *106*, *107*
Hawk
  Ferruginous 245
  Red-tailed 245
head markings 14, *15*, 20–1, *21*
helicoptering 115, *115*
helminths 256
Heron, Great Blue 87
historical writings 187–9
Horn Mill Trout Farm, Rutland 28, *28*, 63, 64, 247, *247*, 274
hovering 26–8, *27*, 33–4, 46
human relationship with Ospreys
  ecotourism 193, 236–8, *238*, 273–4
  education programmes 238–9, *239*
  historical writings 187–9
  live streaming and tracking 237–8, *239*
  tolerance of human activity *175*, 176, 248, *265*
  see also anthropogenic threats; conservation
hunting 26–47
  age and 33
  angle of dive 28–9
  distances flown 29, 45–6, 79
  dive success *32–5*
  duration of foraging trips 35–6
  in estuaries 34–5, 39–40
  grasping prey 29–32, *31*
  hovering 26–8, *27*, 33–4, 46
  juveniles 33, *33*, 120–1, *122*
  kleptoparasitism and 32, *32*, 257–8
  out at sea 29
  from perches 28, *28*
  prey species 35, 37–44, *38*, 39, **40**, *42*, *43*, **44**
  pulling out of dives 29
  refraction and 29
  resting on water 30, *30*
  sea surface temperatures and 39–40
  seasonal changes in foraging behaviour 45–6
  size of prey 30, 34
  social interactions at foraging sites 46–7
  taking off with prey 30, *31*
  tides and 34–5
  timing of foraging trips 36, *37*
  visual acuity 26, 29
  wind conditions and 26–8, 33–4

Iberian Peninsula wintering population 129, 163, 168, 258
Ibis, Glossy 9
Île d'Oléron, France *140*, 141
incubation 101–5, *102*, *103*, *104*
individual Ospreys *see* tracked Ospreys
induced drag 125
interspecific competition 80–1, *80*, 100–1, *100*, *101*, 257–8
invertebrates in nests 86–7
iris markings 21–2
Irish breeding birds 190
Italian breeding birds
   age at first breeding 59
   invertebrates in nests 86–7
   legal protection of 192
   migration 134, **135**, 136, 150–1
   population estimates 262
   post-fledging period 120
   sea crossings 150–1
   wintering 164, 171
Italian translocation project 59, 120, 231–2, *232*, **235**, 241

Jackal, Side-striped 183
Japanese breeding birds 11, 74, 75, 80, 81, 258, 264
juveniles
   calls 25
   fledging 114–17, *115*, **116**, *116*
   food solicitation 25, 117–18, 121, *121*, 132
   helicoptering 115, *115*
   hunting 33, *33*, 120–1, *122*
   maiden flights 114–17, *115*, *116*
   migration 124–5, 129, 132–3, 136–7, *137*, 142–4, *143*, 144–8, *145*, 160, 161, **161**, 183
   mortality 144, 147, 160, **161**, 181–4, **183**, 241–3, **242**, **243**
   movements during second calendar year 184–5
   natal dispersal 48, 49–58
   navigation 129, 147–8
   plumage 15–16, *16*, *17*
   post-fledging period 117–22, *118–19*, 121, *122*, 132, 241, 257
   post-juvenile moult 18–19, *19*
   predation of 183, 257, *257*
   Sahara crossing 160, 161
   sea crossings 144–8, *145*
   sexing 16–17, 114
   sexual dimorphism 16–17, *18*
   stop-overs 136, 142–4, *143*
   survival rates 241–3, **242**, **243**
   wintering 164, *172*, 173, *173*, 176–85, *177*, *179*, *181*, *182*, **183**
   see also nestlings

Kangaroo Island, Australia 21, 67, 69, 171, 268
Kestrel
   American 245

Lesser 49
Kielder Forest, Northumberland 55–6, 68–9, *69*, 99, *99*, **116**, 204, *205*, **206**, *209*, 273
Kite
   Black 138–9, 227
   Red 32, *32*, 191, 216
   Snail 87
kleptoparasitism 32, *32*, 257–8
Kootenay, British Colombia 100, 108–9, 110–11

Lake Asnen, Sweden 83
Lake Ellis Simon, North Carolina 79, 81
Lake George, Florida 34, 36
Lake Hammarsjön, Sweden 27
Lake Helgasjön, Sweden 83
Lake Istokpoga, Florida 81
Langue de Barbarie National Park, Senegal 165, **166**, 173
Latvian breeding birds 11, 38–9, **40**, 165, *165*, 257, 260
laying dates 97–8
legal protection 192, 194
Lewin, William *188*, 189
Ligula intestinalis 38
Lithuanian breeding birds 258, 260
live streaming and tracking 237–8, *239*
Loch Garten, Scotland 37, 38, 70, 94, 95, 115, **116**, 117, 192–3, *193*, 204, 219, 236, 273
Loch of the Lowes, Scotland 21–2, 194, 204, 273
longevity 244

maiden flights 114–17, *115*, *116*
mammalian predators 74, 79, 183, 204, 258
mantling display 91, *91*
Marten, Pine 204, 258
Mediterranean breeding birds 11, 13
   annual survival rates 241
   arrival at breeding sites 87
   cliff nesting 65–6, 75, 83, 249
   copulation 94
   density dependence 65–6
   display flights 91
   egg-collecting 194
   extra-pair copulations 95
   food solicitation by females 24, 92
   human disturbance 249
   individual variation 21
   laying dates 97
   males guarding females 94–5
   migration of juveniles 137
   migratory distances 134, **135**, 136
   migratory routes 129–31
   movements during second calendar year 184–5
   natal dispersal 50–1
   population estimates 65, 262
   prey species **40**, 41
   sea crossings 148–9
   stop-overs 142

translocation of 59, 231, **235**
  wintering 164, 167–8, 171, 184–5
mercury pollution 199
Mexican breeding birds 12
  fledging 115
  migration 132
  nest-sites 74, 75, 77–8, 79, 81, 82
  persecution of 192
  population estimates 266, 267
  prey species 41
Middle East breeding birds 11, 197, 250, 259, 263–4
migration 9–10, 124–62
  daily distances flown 134–7, **135**, *137*
  departure dates 121–2, 132–3
  duration of 133–4, **135**, 136, 137–44
  flight modes 125–7, *126*, *127*, 136, 138, 149–51, *150*, 158–9, *159*
  fly-and-forage strategy 132, 141–2
  goal areas 152–7, *154*
  juveniles 124–5, 129, 132–3, 136–7, *137*, 142–4, *143*, 144–8, *145*, 160, 161, **161**, *183*
  migratory routes and destinations *10*, 11, 12, 129–32, *130*, *131*, 163–4
  morphology and 125–7, *127*, 144
  mortality on 124, 141, 144, 147, 159–61, **161**, *161*
  navigation 129, 147–8, 152
  nocturnal flight 127, 138, 151–2, 158
  Sahara crossing 129–31, 148, 157–61, *159*, *160*, *161*
  sea crossings 127, 144–52, *145*, *149*
  stop-overs 136, 138, 140–4, *140*, *142*, *143*
  studies of repeated journeys 152–7, *154*
  tailwind support 128, 132, 138, 139, 144, 147, 148, 149, 151–2
  time-minimisation strategies 137–9, *139*
  timing of departure 121–2, 132–3
  total distances flown 133–4, **135**
  weather conditions and 127–8, *128*, 132, 138, 139, 144–8, *145*, 149, 151–2
Minnesota translocation project 50, 213–15, *214*
mites 256
mitochondria DNA (mtDNA) studies 12–13
Mono Lake, California 79
Moroccan breeding birds 11, 50–1, 75, 167–8, 255, 262
morphology
  bodyweight 13, 17, 108
  feet and toes 29, 30–1, 189
  forearm length 17, *18*
  iris markings 21–2
  migration and 125–7, *127*, 144
  sexual dimorphism 13–15, 16–17, *16*, *18*
  tarsus length 17, *18*
  wing chord length 17
  wing loading 126–7
  wing shape 127, *127*, 144
  wingspan 13
  see also plumage

mortality
  juveniles 144, 147, 160, **161**, 181–4, **183**, 241–3, **242**, **243**
  on migration 124, 141, 144, 147, 159–61, **161**, *161*, **183**
  on wintering grounds 181–4, **183**
  see also survival
moult 18–20, *19*, *20*
mullet 41–2, *42*, *43*, **44**, 46

natal dispersal 48, 49–58
navigation 129, 147–8, 152
navigational markers 76, *77*, 81
nestlings
  bodyweight 17, 108
  brood reduction 111–12, *112*
  camouflage 113–14, *114*
  dioxin exposure 198
  dominance of older siblings 110–12
  feeding 107–8, 113
  fledging 114–17, *115*, **116**, *116*
  growth 109–14, *110*, *111*, *113*
  hatching 105–8, *106*, *107*
  mercury exposure 199
  parasites 86–7
  plumage 109, 112, 113–14, *113*, *114*
  predation of 71, 74, *74*, 79, 80, 114, 256–7
  provisioning rates 108–9, *108*
  sexing 114
  weather conditions and 109
nests 71–90
  arrival dates 60–2, 87–90, *88*, *89*, **90**, *90*
  on artificial structures 76–8, *77*, 81, 82, 207–8, *210*, 215, 266
  breeding density 50, 65–7, 81–3, 95
  on cliffs 65–6, 75–6, 83, 249
  colonial nesting 46–7, 50, 59, 79, 81–2
  disputes over nest-sites 23–4, 59–65, 95, 103–5, *104*
  egg-collecting 191, 192–4, 247
  on ground 74–6, *74*, *75*, 81, 82, *262*
  guard calls 23–4, 60, 103
  guarding from egg-collectors 192–4, *193*, *194*
  height of *73*, 81, *82*
  interspecific competition and 80–1, *80*, 100–1, *100*, *101*, 257–8
  intrusions by non-breeding birds 23–4, 51, 59–65, 95, 103–5, *104*
  invertebrates in 86–7
  lining materials 85, 86, *87*, 201–2, *203*
  mantling display 91, *91*
  nest-building 55, 83–6, *84*, *85*, *86*, *87*
  other bird species using 86
  predation of 71, 74, *74*, 79, 80, 105, 114, 256–7
  proximity to water *73*, 78–9, *78*
  site fidelity 67–70, *67*, *69*
  in trees 71–4, *71*, *72*, *73*, 81, 200
  see also artificial nests; eggs

Netherlands breeding birds 55, 56, 73, 261, *261*
netting and entanglement 251–3, *251*, *252*
New Caledonian breeding birds 12, 268–9
Newnan's Lake, Florida 33, 121
nocturnal flight 127, 138, 151–2, 158
non-breeding birds
    interference at nests 23–4, 51, 59–65, 95, 103–5, *104*
    polygyny and polyandry 66–7
Norwegian breeding birds 11
    legal protection of 192
    migration 129
    natal dispersal 49
    persecution of 190
    population estimates 259
    translocation of 232–3, **235**
    wintering 171

ocean crossings 127, 144–52, *145*, *149*, *150*
Odiel Marshes, Spain 50–1, 226, 227, **235**
Oman breeding birds 11, 264
organochlorine pesticides 194–8, 253
overfishing threats 254–5
Owl
    Barn 9, 245
    Eagle 114
    Eurasian Eagle 257
    Great Horned 258

Pandion haliaetus carolinensis 11–12, 13, 14–15, *15*, 16
Pandion haliaetus cristatus 12–13, 14–15, *15*, 16
Pandion haliaetus haliaetus 10–11, 13, 14–15, *15*, 16, *16*, 22
Pandion haliaetus ridgwayi 12, 13, 14–15, *15*, 16
parasites 86–7, 256
    see also kleptoparasitism
parasitic drag 125
PBDEs 253–4
PCBs 198–9, 253
Peele, George 188
Pelican, Brown 253
Pennsylvania translocation project 213
perches, hunting from 28, *28*
Peregrine 9, 60, 195, 197–8, 253
persecution 189–92, 246–8
Pliny the Elder 187–8
plumage 13–16
    camouflage 113–14, *114*
    carpal patch 13, 14, 15, 16, 22–3, *22*, *23*
    differences between subspecies 14–15, *15*
    head markings 14, *15*, 20–1, *21*
    individual variation 20–3, *21*, *22*, *23*
    juveniles 15–16, *16*, *17*
    moult 18–20, *19*, *20*
    nestlings 109, 112, 113–14, *113*, *114*
    sexual dimorphism 13–15, 16, *16*
    underwing-coverts 13–16, 22–3, *22*, *23*

Pointe Sarène lagoon, Senegal 165, **166**
poisoning/contamination 194–9, 253–4
Polish breeding birds 39, 192, 247, 256–7, 258, 260
pollutants
    aldrin 198
    dieldrin 198
    dioxin 198
    mercury 199
    organochlorine pesticides 194–8, 253
    polybrominated diphenyl ethers (PBDEs) 253–4
    polychlorinated biphenyls (PCBs) 198–9, 253
polyandry 67
polybrominated diphenyl ethers (PBDEs) 253–4
polychlorinated biphenyls (PCBs) 198–9, 253
polygyny 66–7
Poole Harbour, Dorset 55, 98, 224–6, **225**, *226*, **235**, 236, *237*, 256, 257, *257*, 258, 274
population declines, persecution 189–92
population density
    breeding 50, 65–7, 81–3, 95
    wintering 164–5, 167, *167*, 168, 169, 173–6, *173*, *174*, 180
population estimates 259–69
    Australia and Pacific Islands 267–8
    Caribbean 266–7
    Europe 65, 189, 194, 197, 204, 207, 208, *209*, 259–63
    Japan and Eastern Eurasia 264
    Middle East and Northeast Africa 197, 263–4
    North America and Mexico 76, 197, 265–6
    see also current status of breeding populations
Portugese translocation project 229–31, **230**, *230*, **235**, 249
post-fledging period 117–22, *118–19*, *121*, *122*, 132, 241, 257
powerline collisions 244–5
predation 256–8
    by crocodiles 28, 258
    of eggs and nestlings 71, 74, *74*, 79, 80, 105, 114, 256–7
    of juveniles 183, 257, *257*
    by mammalian predators 74, 79, 183, 204, 258
    on Sahara crossing 160, *161*
prey
    overfishing threats 254–5
    size 30, 34
    species 35, 37–44, *38*, *39*, **40**, *42*, *43*, **44**, 254
    see also hunting
prey species foraging index 35
profile drag 125
pylon nesting 76–7, 81, 82, 207–8, *210*, 266

Raccoon 79, 258
Raven 98
Red Sea breeding birds 11, 14
    breeding season 87
    laying dates 97
    nest-sites 74, 81

population estimates 263–4
threats to 250
refraction 29
reintroduction projects *see* translocation projects
replacement clutches 105
retina 26
Rothiemurchus Fishery, Scotland 45, 46, 247, 274
RSPB 193–4
Russian breeding birds 10–11
   interspecific competition 81, 257, 258
   migration 129, 163
   nest-sites 72
   population estimates 259, 264
   prey species **40**
Rutland Water breeding birds
   age at first breeding 59
   annual survival rates 241–4, **242**, **243**
   arrival at breeding sites 88, 89–90, *89*, **90**
   artificial nest provision *202*, 203, 204–5, **206**, *207*, **208**, *209*
   clutch size 98–9, 105
   collisions 245
   courtship behaviour *24*, *91*, 92–3
   departure from wintering grounds 185, **185**
   disease 216, 256
   display flights *24*, 48
   disputes over nest-sites 60–5, *61*, *62*, *64*, 105
   egg markings 99
   electrocution 246
   entanglement 251, 253
   fledging 116, **116**, 117
   food solicitation by females 94, 102
   hatching 108
   human disturbance 250
   hunting 28, *28*, *31*, 34, 35–6, *37*, 45–6
   incubation 102, *102*, 103, 105, 106
   individual variation 21, *21*, 22–3, **22**, 23
   interspecific competition 100–1, *101*
   kleptoparasitism 32, *32*
   laying dates 97–8
   migration 140, 160, *161*
   mortality of juveniles 181–2, **183**, 241–3, **242**, **243**
   movements during second calendar year 184
   natal dispersal 51–6, *52*, *54*, 56
   nest-sites 79
   nestlings 113
   other bird species using nests 86
   persecution of 247–8
   post-fledging period 118, 120–1, *122*
   prey species 37, 38, *38*, 39
   Sahara crossing 160, *161*
   site fidelity *67*, 68, 168
   viewing sites 247, *247*, 273
   wintering 168, 169–70, *169*, *170*, 171–2, 181–2, **183**
Rutland Water translocation project 46, 51–2, 59, 106, 215–24, *217*, **218**, *218*, *219*, **235**

dispersal to Wales 52–5, *52*, *54*, 69, 223–4, **225**
highly productive individuals *221*, 222–3, **223**, *223*, *224*
population establishment 219–22, **220**, **221**, **222**, **271–2**

Sahara 129–31, 148, 157–61, *159*, *160*, *161*
Salmonella infection 216, 256
San Ignacio Lagoon, Mexico 77–8, 82
Sandpiper, Spoon-billed 264
Scandola Marine Protected Area, Corsica 249
Scottish breeding birds
   age at first breeding 59
   annual survival rates 243, **243**
   arrival at breeding sites 87
   artificial nest provision 200, *201*, 204
   clutch size 98, *112*
   colour rings *165*
   copulation 94, 95
   density dependence 66
   departure from breeding sites 132
   departure from wintering grounds **185**
   disease 256
   display flights 91
   disputes over nest-sites 60
   egg-collecting 192–4, 247
   entanglement 251, 253
   extra-pair copulations 95
   fledging 115, **116**
   guarding nests of 192–4, *193*, *194*
   human disturbance 249
   hunting 29, 45, 46
   individual variation 21–2
   interspecific competition 258
   longevity 244
   migration of juveniles 136, *137*, 142–3, 144–7, *145*, 153–7, *154*, 160
   migratory distances 134
   mortality of juveniles 181–4, **183**
   movements during second calendar year 184
   natal dispersal 49, 53, 56–8, *57*, **58**
   nest-sites 72, 74, *74*, 76, 79, 81, 83, 200
   nestlings 112, *112*
   nocturnal flight 138
   persecution of 190–2, 247
   population estimates 204, 261–2
   post-fledging period 117, *118–19*, 120
   predation of 257, 258
   prey species 37, 38, 40, **40**
   Sahara crossing 158–9, *159*
   sea crossings 144–7, *145*, 148, 149–52, *149*, *150*
   site fidelity 68, 70, 168
   stop-overs 140–1, 142–3, *142*
   studies of repeated migration journeys 153–7, *154*
   time-minimisation strategies 138
   translocation of 17, 50, 51, 169, 216, 225, 226, 227, 228, 232, **235**

viewing sites 193, 236, 247, 273, 274
   wintering 168, 176–84, *177*, *181*, **183**, 184
screaming calls 23–4, 103
sea crossings 127, 144–52, *145*, *149*, *150*
sea surface temperatures 39–40
semi-zygodactyl feet 29, 30–1
Senegal wintering birds
   densities 165, 167, *167*, 169, 173, *174*
   hunting 29, 33, 35, 36, 46
   juveniles 173, 176–81, *177*, *181*, 182–4, *182*
   migration of tracked birds 55, 57, 134, 140, 141, 143, 153, 155, 157, 158, 160–1, **161**, 229
   migratory routes 129, *130*
   mortality of juveniles 182–4
   moulting 19–20, *20*
   prey species 41–2, *43*, **44**
   site fidelity 168
   social behaviour 173–6, *174*
   surveys 165–7, **166**
   tolerance of human activity *175*, 176
   winter range of tracked birds 163, 169–70, *169*
sexing juveniles 16–17, 114
sexual dimorphism 13–15, 16–17, *16*, *18*
Shakespeare, William 188
Shoveler 176
Sine-Saloum delta, Senegal 42, 55, 153, 155, 163, **166**, 167, 168, 173, 174–6, *174*, 177, 180, 181, 182, *182*
site fidelity
   breeding 67–70, *67*, *69*
   wintering 139, 168
slope soaring 125
soaring-gliding flight 125, 126–7, *126*, 136, 138, 149–51, *150*
social behaviour
   at foraging sites 46–7
   in winter 172–6, *172*, *173*, *174*, *175*
   see also conspecific attraction
Somone Marine Protected Area (MPA), Senegal 165, **166**, 176
Spanish breeding birds
   collisions 245
   electrocution 246
   incubation 106
   interspecific competition 101
   legal protection of 192
   longevity 244
   natal dispersal 50–1
   nest-sites 76
   population estimates 262
   wintering 165
Spanish translocation projects 226–9
   Andalucía 17, 50–1, 106, 226–7, **235**
   Basque Country 158, 227–8, *228*, 233–4, **235**
   Valencia 229, **235**
Sparrow
   House 86
   Tree 86
Sparrowhawk 195
St Louis Marine Protected Area (MPA), Senegal 165, **166**
Starling, European 86
stop-overs, on migration 136, 138, 140–4, *140*, *142*, *143*
Stork
   Black 127
   White 101, 127, 232
   Wood 87
survival
   annual survival rates 241–4, **242**, **243**
   longevity 244
   see also mortality
Swallow, Tree 86
Swedish breeding birds 11
   annual survival rates 241, 243
   arrival at breeding sites 87
   copulation 94, 95
   DDT exposure 196, 197
   departure from breeding sites 132
   extra-pair copulations 95
   goal areas 152
   hunting 27, 36
   legal protection of 192
   migration of juveniles 136–7
   migratory distances 134, **135**
   migratory routes 129
   natal dispersal 49
   nest-sites 83
   PCB exposure 198–9
   persecution of 190
   population estimates 259
   predation of 258
   Sahara crossing 157–8
   sea crossings 149, 150
   stop-overs 142, 144
   studies of repeated migration journeys 152
   time-minimisation strategies 138
   translocation of 229, 231, **235**
   wintering 164, 183
Swiss translocation project 232–3, **235**, 260–1

tailwinds 128, 132, 138, 139, 144, 147, 148, 149, 151–2
tapeworms 38
tarsus length 17, *18*
taxonomy 9–13
temporal fovea 26
Tennessee Valley translocation project 212
Tern, Arctic 126
territorial intrusions 23–4, 51, 59–65, 95, 103–5, *104*
thermals 125, 126–7, 150, 151, 152, 158–9, *159*
threats to Ospreys 244–59
   climate change 258–9
   dieldrin and aldrin pollution 198
   dioxin pollution 198

disease 216, 256
egg-collecting 191, 192–4, 247
electrocution 245–6, *246*
forestry and agriculture 250
human disturbance 248–50
illegal killing 246–8
interspecific competition 80–1, *80*, 100–1, *100*, *101*, 257–8
kleptoparasitism 32, *32*, 257–8
mercury pollution 199
on migration 124
netting and entanglement 251–3, *251*, *252*
organochlorine pesticides 194–8, 253
overfishing 254–5
parasites 86–7, 256
persecution 189–92, 246–8
poisoning/contamination 194–9, 253–4
polybrominated diphenyl ethers (PBDEs) 253–4
polychlorinated biphenyls (PCBs) 198–9, 253
powerline collisions 244–5
tourism development 250
water-based disturbance 249–50
wind turbines 124, 245
see also predation
tides 34–5
tolerance of human activity *175*, 176, 248, **265**
tourism development 250
tracked Ospreys
  2AF(16) 92, 94
  03(97) *23*, 35–6, 46, *67*, 68, *88*, *89*, 217, 219, 220, 221, 222, **223**, *223*
  05(00) 92–3, 102, 222, **223**
  5N(04) *32*, 68, *89*, 102, 220, *221*, 222, **223**
  06(01) 184, 220, 222
  09(98) 160, *161*, 169–70, *169*, 220
  30(05) *89*, 140, *169*, 170, *170*, 185, **185**
  32(11) *89*, 163, 168
  33(11) *89*, *91*, 98, **223**
  AW(06) 45–6, 171–2
  Beatrice (Green 5B) 141, *142*, 185, **185**
  Blue XD 138, 140, *150*, *159*, **185**
  CJ7(15) 55, 221–2, 225
  Fiddich 143, 183–4
  Green J 168, **185**, 244
  Joe 182–3
  Maya *21*, 68, *89*, *91*, 98, 222, **223**
  Nimrod (Red 7J) 45, 140–1, **185**
  Red 8T 45, **185**
  Rothiemurchus (Blue AE) 57–8, *57*, **58**, 59, *145*, 146–7, 153–7, *154*, 176–81, *177*, *181*, **185**
  S1(15) 62–5, *62*, *64*, 92
  S2(15) 55, *56*, 221
  Stan *137*, 142–3, 144–6, *145*, 147, 160
  T3(16) 184
  White PE (Morven) 68, **185**
translocation projects 11, 117, 212–36
  Andalucía, Spain 17, 50–1, 106, 226–7, **235**
  Aquitaine, France 233–4, *233*, **235**, 261

Basque Country, Spain 158, 227–8, *228*, 233–4, **235**
future of 234–6
Italy 59, 120, 231–2, *232*, **235**, 241
Minnesota, United States 50, 213–15, *214*
Pennsylvania, United States 213
Poole Harbour, England 224–6, *226*
Portugal 229–31, **230**, *230*, **235**, 249
Switzerland 232–3, **235**, 260–1
Tennessee Valley, United States 212
Valencia, Spain 229, **235**
see also Rutland Water translocation project
tree nesting 71–4, *71*, *72*, *73*, 81, 200
triangle of velocities theory 127–8, *128*, 147, 148
Turks and Caicos Islands breeding birds 12, 267
Turner, William 188, 189
Turnstone, Ruddy *174*, 176

Ukrainian breeding birds 260
underwing-coverts 13–16, 22–3, *22*, *23*
United Arab Emirates breeding birds 11, 250, 264
United Kingdom breeding birds 11
  arrival at breeding sites 60
  artificial nest provision 55, 66, 200, 201, *201*, *202*, 203, 204–7, *205*, **206**, *207*, *208*, *209*
  departure from breeding sites 132
  departure from wintering grounds 185, **185**
  laying dates 97–8
  legal protection of 192, 194
  migration of juveniles 136, *137*, 142–3, 144–7, *145*, 160, **161**
  migratory distances 134, **135**
  migratory routes 129, *130*
  nocturnal flight 138, 151–2
  persecution of 189–92
  population estimates 189, 204, *209*, 261–2
  Sahara crossing 158–9, *159*, 160, *161*
  sea crossings 144–7, *145*, 148, 149–52, *149*, *150*
  site fidelity 67–9, *67*, *69*, 139
  stop-overs 140–1, *140*, 142–3, *142*
  time-minimisation strategies 138
  viewing sites 193, 236, 237, 247, *247*, 273–4
  wintering 163, 164, 165, 168, 169–70, *169*, *170*, 171–2, 176–84, *177*, 181–2, *181*, **183**, 184
  see also English breeding birds; Scottish breeding birds; Welsh breeding birds
United Kingdom translocation projects
  Poole Harbour, Dorset 224–6, *226*
  see also Rutland Water translocation project
United States breeding birds 11–12
  age at first breeding 59
  annual survival rates 241, 244
  arrival at breeding sites 89
  artificial nest provision 50, 77, *78*, 81–2, *82*, 200, 210–11, *211*, 246, **265**, 266
  clutch size 98–100
  collisions 245
  copulation 92, 95

courtship feeding 92
DDT exposure 195–7
departure from breeding sites 132
dioxin exposure 198
disease 256
electrocution 245, *246*
entanglement 251, *252*, 253
extra-pair copulations 95
fledging 115
human disturbance 249–50, *250*
hunting 28, 33, 34, 35, 36, 45, 46
incubation 102
interspecific competition 80, *80*, 257–8
laying dates 97, 98
longevity 244
mercury exposure 199
migration 12, 131–2, *131*, 134–6, **135**
natal dispersal 50
nest-building 85, 86
nest-sites 73–4, *73*, 75, 76–7, *77*, 78, *78*, 79, 81–2, *82*, 215
nestlings 109, 110, 111–12, *111*
parasites 256
PBDE exposure 253–4
PCB exposure 198
persecution of 192, 248
polygyny 66
population estimates 76, 197, 265–6, *265*
post-fledging period 117, 121
predation of 257, 258
prey species 39, 44
sea crossings 147, 148, 152
site fidelity 69–70, 168
time-minimisation strategies 138
wintering 163, 164, 167, 168, 170, 172
United States translocation projects 212–15
  Minnesota 50, 213–15, *214*
  Pennsylvania 213
  Tennessee Valley 212
Urdaibai Biosphere Reserve, Spain 94, 228, *228*, 233–4, **235**

Valencia translocation project 229, **235**
vector summation 129, 144
visual acuity 26, 29
vocalisations 23–5
  alarm calls 24–5
  display calls 23, *24*, 48
  excited calls 23, 24, 25, 103
  food solicitation call 24, 25, 92, 102
  guard calls 23–4, 60, 103
  screaming calls 23–4, 103
Vulture, Bearded 13

Walcott, John 189
water-based disturbance 249–50
weather conditions
  hunting and 26–8, 33–4
  migration and 127–8, *128*, 132, 138, 139, 144–8, *145*, 149, 151–2
  nestlings and 109
Welsh breeding birds
  arrival at breeding sites 87
  artificial nest provision 55, 66, 205–7, **206**, *209*
  fledging 115, **116**
  hatching 105, *106*
  incubation 101
  natal dispersal 51–6, *52*
  persecution of 190
  polygyny 66–7
  population estimates *209*, 261–2
  post-fledging period 117
  prey species 40–1, **40**
  replacement clutches 105
  site fidelity 68, 69
  viewing sites 236, 274
Whistling Duck, White-faced *178*
wind conditions
  hunting and 26–8, 33–4
  migration and 127–8, *128*, 132, 138, 139, 144–8, *145*, 149, 151–2
wind turbines 124, 245
wing chord length 17
wing loading 126–7
wing shape 127, *127*, 144
wingspan 13
wintering 163–86
  arrival dates 164
  densities 164–5, 167, *167*, 168, 169, 173–6, *173*, *174*, 180
  departure dates 185, **185**
  destinations 129–32, *130*, *131*, 163–4
  habitat requirements 164–8
  juveniles 164, *172*, 173, *173*, 176–85, *177*, *179*, *181*, *182*, **183**
  site fidelity 139, 168
  social behaviour 172–6, *172*, *173*, *174*, *175*
  tolerance of human activity *175*, 176
  winter range 163, 169–72, *169*

Yellowstone Lake, Wyoming 39
Yellowstone National Park, Wyoming 73, 196
Yellowstone River, Montana *78*, 245, *246*, *252*, 253, *265*

zygodactyl toe arrangement 29, 30–1